Geohazard in Rocky Coastal Areas

Geological Society books refereeing procedures

The Society makes every effort to ensure that the scientific and production quality of its books matches that of its journals. Since 1997, all book proposals have been refereed by specialist reviewers as well as by the Society's Books Editorial Committee. If the referees identify weaknesses in the proposal, these must be addressed before the proposal is accepted.

Once the book is accepted, the Society Book Editors ensure that the volume editors follow strict guidelines on refereeing and quality control. We insist that individual papers can only be accepted after satisfactory review by two independent referees. The questions on the review forms are similar to those for *Journal of the Geological Society*. The referees' forms and comments must be available to the Society's Book Editors on request.

Although many of the books result from meetings, the editors are expected to commission papers that were not presented at the meeting to ensure that the book provides a balanced coverage of the subject. Being accepted for presentation at the meeting does not guarantee inclusion in the book.

More information about submitting a proposal and producing a book for the Society can be found on its web site: www.geolsoc.org.uk.

It is recommended that reference to all or part of this book should be made in one of the following ways:

VIOLANTE, C. (ed.) 2009. *Geohazard in Rocky Coastal Areas*. Geological Society, London, Special Publications, **322**.

BRANDOLINI, P., FACCINI, F., ROBBIANO, A. & TERRANOVA, R. 2009. Slope instability on rocky coast: a case study of Le Grazie landslides (eastern Liguria, northern Italy). *In*: VIOLANTE, C. (ed.) *Geohazard in Rocky Coastal Areas*. Geological Society, London, Special Publications, **322**, 143–154.

GEOLOGICAL SOCIETY SPECIAL PUBLICATION NO. 322

Geohazard in Rocky Coastal Areas

EDITED BY

C. VIOLANTE
IAMC–CNR, Italy

2009
Published by
The Geological Society
London

THE GEOLOGICAL SOCIETY

The Geological Society of London (GSL) was founded in 1807. It is the oldest national geological society in the world and the largest in Europe. It was incorporated under Royal Charter in 1825 and is Registered Charity 210161.

The Society is the UK national learned and professional society for geology with a worldwide Fellowship (FGS) of over 9000. The Society has the power to confer Chartered status on suitably qualified Fellows, and about 2000 of the Fellowship carry the title (CGeol). Chartered Geologists may also obtain the equivalent European title, European Geologist (EurGeol). One fifth of the Society's fellowship resides outside the UK. To find out more about the Society, log on to www.geolsoc.org.uk.

The Geological Society Publishing House (Bath, UK) produces the Society's international journals and books, and acts as European distributor for selected publications of the American Association of Petroleum Geologists (AAPG), the Indonesian Petroleum Association (IPA), the Geological Society of America (GSA), the Society for Sedimentary Geology (SEPM) and the Geologists' Association (GA). Joint marketing agreements ensure that GSL Fellows may purchase these societies' publications at a discount. The Society's online bookshop (accessible from www.geolsoc.org.uk) offers secure book purchasing with your credit or debit card.

To find out about joining the Society and benefiting from substantial discounts on publications of GSL and other societies worldwide, consult www.geolsoc.org.uk, or contact the Fellowship Department at: The Geological Society, Burlington House, Piccadilly, London W1J 0BG: Tel. +44 (0)20 7434 9944; Fax +44 (0)20 7439 8975; E-mail: enquiries@geolsoc.org.uk.

For information about the Society's meetings, consult *Events* on www.geolsoc.org.uk. To find out more about the Society's Corporate Affiliates Scheme, write to enquiries@geolsoc.org.uk.

Published by The Geological Society from:
The Geological Society Publishing House, Unit 7, Brassmill Enterprise Centre, Brassmill Lane, Bath BA1 3JN, UK

(*Orders*: Tel. +44 (0)1225 445046, Fax +44 (0)1225 442836)
Online bookshop: www.geolsoc.org.uk/bookshop

British Library Cataloguing in Publication Data

A catalogue record for this book is available from the British Library.
ISBN 978-1-86239-282-3

Typeset by Techset Composition Ltd., Salisbury, UK
Printed by MPG Books Ltd., Bodmin, UK

Distributors

North America
For trade and institutional orders:
The Geological Society, c/o AIDC, 82 Winter Sport Lane, Williston, VT 05495, USA
Orders: Tel. +1 800-972-9892
Fax +1 802-864-7626
E-mail: gsl.orders@aidcvt.com

For individual and corporate orders:
AAPG Bookstore, PO Box 979, Tulsa, OK 74101-0979, USA
Orders: Tel. +1 918-584-2555
Fax +1 918-560-2652
E-mail: bookstore@aapg.org
Website: http://bookstore.aapg.org

India
Affiliated East-West Press Private Ltd, Marketing Division, G-1/16 Ansari Road, Darya Ganj, New Delhi 110 002, India
Orders: Tel. +91 11 2327-9113/2326-4180
Fax +91 11 2326-0538
E-mail: affiliat@vsnl.com

Contents

Preface

This book brings together contributions dealing with different aspects of hazard-related geological processes that naturally drive coastal slope evolution and deeply influence the human use of coastal resources. The editorial plan was conceived with the aim to frame the study of coastal geohazard phenomena in a context based on sea–land correlations that include marine geophysical surveys and direct field investigations. Special attention was paid to the study of documentary sources, an important source of data in the reconstruction of event chronology on a human time scale.

The collected papers focus on Italian case histories, mostly related to the Neapolitan coastal area, and cover different geological settings and morphologies which are dependent on rock type and tectonics. Contributions from key-areas concern both volcanic and non-volcanic coastal ranges and provide a significant source of information for researchers working in similar coastal environments. Particular attention received the Sorrento Peninsula, a tectonically uplifted coastal area typically exposed to stream floods, with most of the human activities located on unstable alluvial fan-deltas or along the path of floodwaters.

The introductory paper by **Violante** provides definitions and constraints for geological hazard assessment in rocky coastal areas. Processes of rapid sediment transfer by catastrophic stream-flow, sea-cliff retreat and flank collapse of volcanic and rocky slopes are described with examples based on both marine and terrestrial geological data. The introduction is followed by a set of papers dealing with the volcanic influence on coastal sedimentary system of the Naples and Salerno districts (Southern Italy). The paper by **Sacchi *et al.*** and the work by **Cinque & Robustelli**, concern the far ranging catastrophic reaction of steep coastal watersheds of the Sorrento peninsula following large explosive eruptions, noting the role of unstable pyroclastic fall-out deposits in volcanic hazard assessment. This issue was identified by the paper on historical reconstructions of stream-flow events by **Porfido *et al.***, who give a detailed flood chronology, based on the study of numerous and varied documentary sources. A further mechanism for volcaniclastic sediment delivery into the Neapolitan coastal system is provided by **de Alteriis & Violante** and **Milia *et al.***, who document catastrophic collapses of volcanic flank on Ischia island and Somma-Vesuvius respectively. The reported coastal landslides involve large volumes (>1 km^3) of volcanic debris and blocks being rapidly transferred from the coast to the sea, with a significant tsunamigenic potential.

The wide occurrence and variety of sea-cliffs and rocky slopes along Italy's coasts is reported by **Iadanza *et al.*** who illustrate the types and distribution of Italian coastal landslides based on information derived from the IFFI archive (Italian Landslide Inventory; http://www.sinanet.apat.it/progettoiffi) with examples from various coastal settings. A case history on coastal slope retreat in the Ligura district (Northern Italy) is presented by **Brandolini *et al.***, who point out the influence of both natural and anthropogenic factors on slope instability. Finally, **De Pippo *et al.*** propose a method to assess coastal hazard based on an interaction matrix.

Napoli, 27 May 2009

CRESCENZO VIOLANTE
Institute for Coastal Marine Environment – IAMC
National Research Council – CNR
Napoli – Italy

Rocky coast: geological constraints for hazard assessment

CRESCENZO VIOLANTE*

Institute for Coastal Marine Environment, Consiglio Nazionale delle Ricerche (CNR), Calata Porta di Massa, Porto di Napoli, 80133 Napoli, Italy

**Corresponding author (e-mail: crescenzo.violante@iamc.cnr.it)*

Abstract: Geological hazard along rocky coasts is basically associated with processes of rapid sediment transfers. Massive transport of rock, regolith, sedimentary cover and soil occur episodically, accounting for cliff recession, sudden increase in solid load in short coastal rivers, and flank collapse of volcanic structures and rocky slopes. In geohazard terms, rocky coasts operate as transfer zones that deliver sediment directly from slopes to the coast and open sea at intermittent time intervals. Erosion and transport of material causes major physical changes and exposes coastal communities and human activity to hazard with potential damage to property and infrastructure, and loss of life. This paper focuses on geological processes that regulate rapid sediment transfers in rocky coastal areas, with examples drawn mostly from the Italian coasts. It is stressed that proper comprehension of coastal mass wasting hazard has to include marine and historical investigations. As a main delivery area, the submerged part of rocky coasts preserves reliable sedimentary records of past geological events occurring on land, which are often only partly detectable along subaerial rocky slopes and commonly reported in historical sources.

Natural hazard on the coast is largely affected by processes of rapid sediment transfers produced by meteorological, oceanographic and geological forces. Coastal failure, mass wasting and floods are some of the processes that operate naturally in this environment and significantly influence the human use of coastal resources. It is estimated that more than 37% of the world's population live within 100 km of the coastline and that 80% of the shores are rocky (Emery & Kuhn 1982); this includes beaches that are backed by bedrock cliffs or rocky uplands. The geological processes that regulate sediment transfer in these environments also cause major physical changes both onshore and at sea, and their understanding is essential for hazard assessment and the determination of the related geological risk.

According to the coastal zone concept, the term 'rocky coast' is used here to denote a spatial zone between the landward limit of marine influence and seaward limit of terrestrial influence (Carter 1988) composed of a rocky substrate retaining at the coastline the form of a cliff with different profiles. This definition includes steep coastal watersheds, pocket beaches situated between bedrock headlands, fan-delta systems, and other non-rocky elements such as barrier spits downdrift of river mouths and estuaries. This term is also suitable for the study of physical changes and related hazard or risk as it includes coastal settlements and human activity.

Rocky coasts occur in a variety of geological settings with a wide range of morphologies depending on rock type, tectonics and climate. Rocky coastal areas can be associated with mountainous regions with active or recent tectonics or volcanic activity, or develop as low-relief cliffs along non-active margins, which limit seaward flattened areas. Steep coasts commonly occur also in glacial environments, such as fjords or lakes. In all these settings, slope instability represents the most effective hazardous process, which can erode and transfer large volumes of materials directly, or via coastal streams, into the sea, lake or fjord. Landslide activity has a significant impact on communities living on the rocky coast, commonly inducing destructive waves and massive sediment transport in short coastal rivers. Material eroded from rocky coasts is mostly delivered in the form of cliff debris, landslide accumulations, coarse-grained deltas and ultimately as fluvial turbiditic flows (hyperpycnal flows). As a result of high-gradient sea-floor topography and often a narrow or non-existent shelf along rocky coasts, the eroded deposits often go straight to the open sea and less frequently, as a wider shelf develops, can be trapped at shallow depth as sandy lobes.

Coastal evolution mainly depends on the balance between sediment availability and wave reworking processes. For rocky coasts, delivery of sediment is typically intermittent, and persistence of the

From: VIOLANTE, C. (ed.) *Geohazard in Rocky Coastal Areas*. The Geological Society, London, Special Publications, **322**, 1–31. DOI: 10.1144/SP322.1
0305-8719/09/$15.00

displaced material in the littoral environment as a natural armour for wave action is consequently low. This exposes rocky coasts to an irreversible loss of land over human-scale periods.

The main aims of this study are to discuss the processes of hazardous sediment transfer and accumulation at rocky coasts, and highlight the role of marine geophysical and sedimentological investigations in reconstructing coastal geohazard (Locat & Sanfaçon 2000; Violante *et al.* 2006). Moreover, it is acknowledged that the use of historical data combined with the above data sources is an important task in this matter, particularly for assessing damage to property and infrastructure.

Sediment transfer at rocky coast

Rocky coasts are potentially subject to mass-wasting events over a range of magnitude and period of recurrence, which are able to transfer large amounts of material into coastal and open seas (Table 1). Topographic gradients arise from volcanic and tectonic processes of deformation and uplift, which also have a primary effect on the denudation rate and coastline features. Besides coastal slope, tectonic activity shapes the sea-floor morphology of marine areas, which is characterized by high gradients, narrow and abrupt continental margins, and submarine canyons close to the shoreline.

Sediment transfer at rocky coasts is typically intermittent, involving massive transport of rock, regolith, sedimentary cover and soil. The resulting deposits have a coarse-grained texture with relatively small quantities of fine sand and mud, and are transient through the shore zone and mostly redeposited at great depth. The combination of steep continental shelves, which are unable to dissipate wave energy, and episodic coastal supply prevents the beach profile being maintained over a long period, although coastline progradation may occur as ephemeral alluvial deltas at stream mouths. Lack of extensive coastal plains on rocky coasts is further due to the capture of sand at the heads of submarine canyons, with the result that sand is carried to the deep sea out of the coastal system (Fig. 1).

Mass movement is a fundamental component of landscape evolution on rocky coasts that accounts for active cliff recession, lateral collapse of coastal volcanic structures and rocky slopes, and sudden increases in sediment load in short coastal rivers. The catastrophic delivery of materials exposes coastal communities to both mass wasting and tsunami hazards, the latter being produced by displaced waves as rock avalanches enter a lake or the sea.

Catastrophic river floods

Sediment discharge in mountainous and hilly coastal rivers occurs through episodic flood events, often with catastrophic implications. In steep coastal orogenic belts such as the Alpine and Apennine flank of the Mediterranean (Fig. 2), the open-ocean coastlines bordering the Pacific Ocean, the peninsular Gulf of California margin and the Gulf of Corinth in Greece, fluvial systems are small to medium in size, with ephemeral and torrential discharge regimes and high-elevation drainage basins. Stream paths deeply dissect the rocky substrate, resulting in high-gradient V-shaped valleys with low aggradation and most of the solid load bypassed

Table 1. *Processes, factors and forms associated with sediment transfers in rocky coastal areas*

	Cliff recession	Stream flow	Large slope failure
Geological processes	Rock fall Topple Rotational slump Mudflow	Shallow landslide Debris flow Slope-to-stream delivery	Debris avalanche Debris flow Turbidity current Creeping slump
Main promoting and triggering factors	Wave action Storm surges Weathering Unloading of cliff toe Water seepage	Localized burst of waters Volcanic watershed disturbance Small/medium, steep and high watersheds	Oversteepening Unbuttressed slope Large earthquakes Volcanic eruption Volcano-tectonic uplift Tectonic stress Dike intrusion
Associated forms and phenomena	Debris toe Shore platform	Temporary dam Translatory wave Ephemeral coastal fan Fan-delta	Tsunami Hummocky topography Amphitheatre scarp

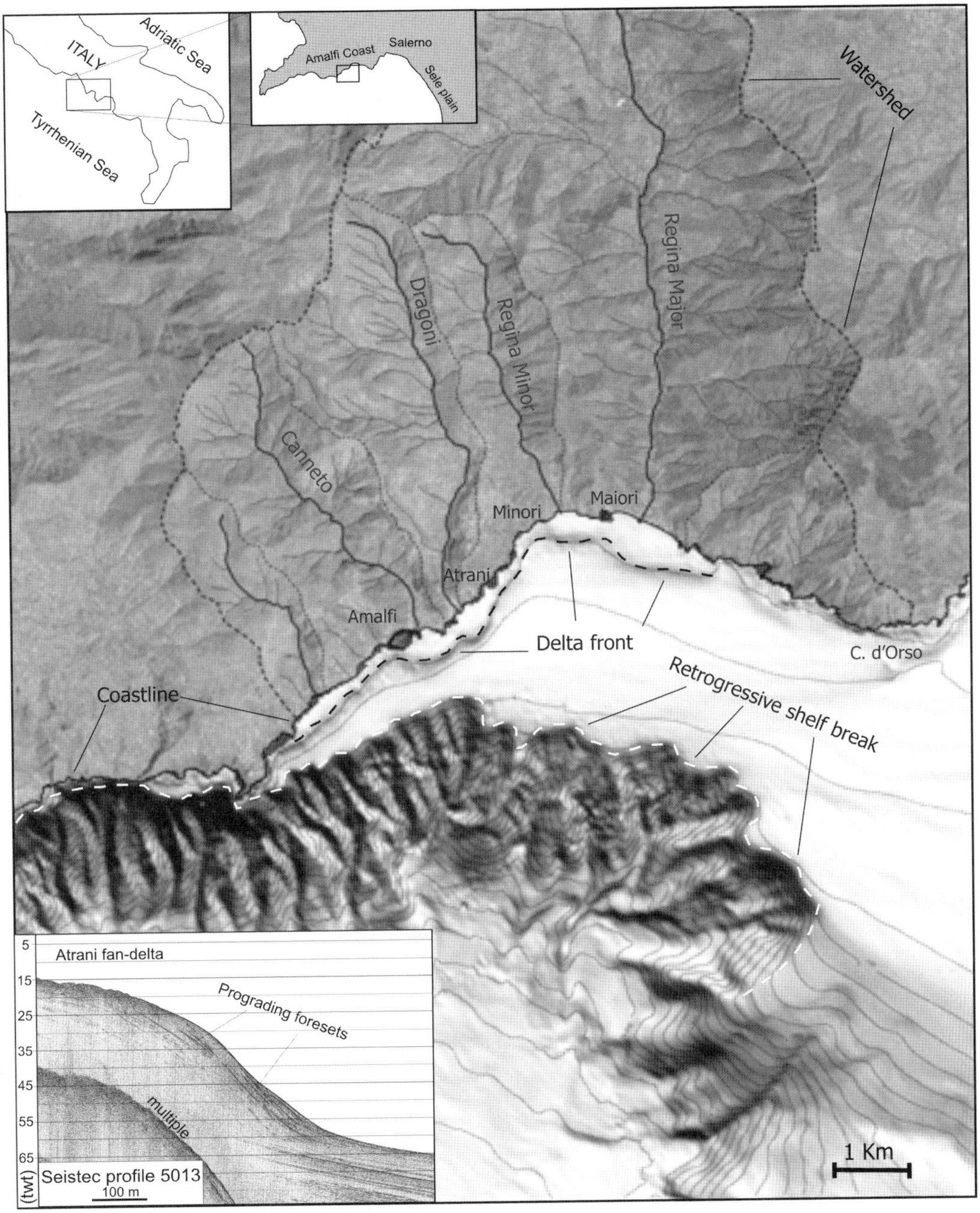

Fig. 1. The Amalfi rocky coastal system (eastern Tyrrhenian Sea), characterized by steep watersheds, fan-deltas at the mouth of coastal streams, reduced continental shelf etched by canyons, and abrupt shelf break (fault-controlled). The fan-deltas are composed of prograding clinoforms resulting from flood activity as revealed by high-resolution seismic profiles (inset map in the lower left corner). This map is based on a combination of multibeam bathymetric data and terrestrial elevation data. Inset map shows location. twt, two-way travel time.

Fig. 2. Amalfi, located at the mouth of a flood-prone stream, the Canneto, fed by an high-elevation basin (Amalfi coast, eastern Tyrrhenian Sea). (See Fig. 1 for location.)

to the coast. It is now recognized that these rivers have very high sediment yield (Milliman & Syvitski 1992; Mertes & Warrick 2001) following high-magnitude events, such as extreme rainfall and earthquakes. In such settings, flooding of the stream paths is associated with sediment supply from side slopes through a variety of mass-wasting phenomena that deliver sediment to streams (Fig. 3; Schumm 1977; Benda 1990; Anthony & Julian 1999).

Infrequent rain storms (Meade *et al.* 1990; Perez 2001; Esposito *et al.* 2004*a*, *b*; Violante *et al.* 2009), with both seasonal and longer recurrence intervals, heavy and rapid snowmelt (Julian & Anthony 1997), as well as abrupt draining of glacial lakes (Baker 1994; Clague & Evans 1994; Milliman *et al.* 1996) produce intense slope erosion, thus increasing the transported load and raising the water level in steep coastal streams. In addition,

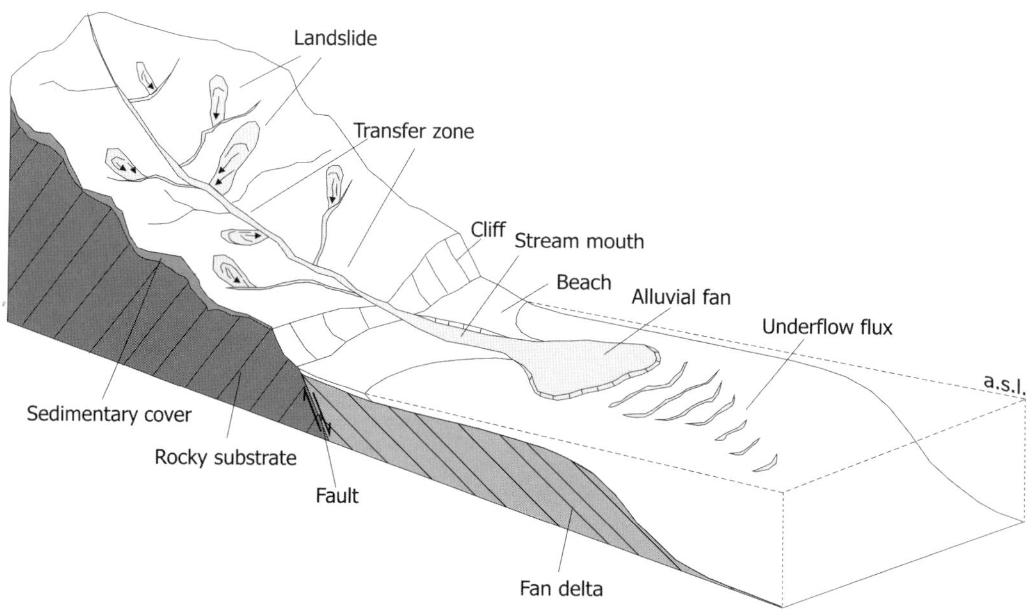

Fig. 3. Sketch depicting the relations between landslide activity, slope to stream delivery and fan-delta system on a rocky coast.

major floods can result from volcanic eruptions that can dramatically increase landslide activity and sediment load in coastal and inland river systems (Hubbel *et al.* 1983; Major *et al.* 2000; Meade & Parker 1985; Cinque & Robustelli 2009). Investigations on the Amalfi coast in southern Italy (Sacchi *et al.* 2009) have found evidence for a significant alluvial crisis lasting some decades in late Medieval times, resulting from early mobilization from steep coastal slopes of the air-fall deposits of the Vesuvius eruption of AD 79. In recent times, the same materials along with the air-fall deposits of the last Vesuvius eruption (1944) induced a catastrophic flood in the coastal stream Bonea, a few kilometres east of Amalfi, causing serious damage and more than 300 casualties. Hazardous ephemeral gravel-bed streams characterize many other rocky coastal regions of the Mediterranean and elsewhere in the world, such as the regions of Valencia and Barcelona (Spanish *ramblas*; Belmonte & Beltràn 2001; Camarasa & Segura 2001) and Calabria (Italian *fiumaras*; Sabato & Tropeano 2004), as well as South American (Montgomery *et al.* 2001; Perez 2001) and California coastal ranges. Such streams typically have periods of apparent stability with rapid transition to catastrophic events.

The critical relationship between landslide activity and sediment delivery to slope–stream systems indicates the role of slope erodibility in coastal river floods. In this context, tectonism is of primary importance, as it results in a pattern of rock fractures, oversteepened slopes, and seismic and volcanic activity. Lithology determines both the abundance and capacity of streams to mobilize more or less coarse sediments produced by landslides, as these systems commonly exhibit clast sizes ranging from large boulders to clay. However, human activity, including deforestation for agriculture and housing as well as regulation and defence works, strongly modifies slope features, often with hazardous implications for hydrological and biological equilibria.

Intense slope erosion at rocky coasts is associated with heavy rainfall in zones of limited areal extent where powerful convective cells precipitate large quantities of water in concentrated bursts (Woolhiser & Goodrich 1988; Nouh 1990; Camarasa 1994; Faurés *et al.* 1995; Anthony & Julian 1999; Belmonte & Beltràn 2001; Esposito *et al.* 2004*a*, *b*) whereas a few kilometres away, it may rain only at low intensity. Hydrological features typically include a few days of steady rains with anomalous high levels of daily totals, followed by a few hours of heavy rain commonly exceeding 200 mm but easily reaching values as high as 400–500 mm (Blair *et al.* 1985; Baldwin *et al.* 1987; Martin-Vide *et al.* 1999; Perez 2001). The slides are often shallow and very wide, extending all the way to the mountain ridge and crest with high sediment transfer to the stream paths (Fig. 4). In these conditions, landslide debris mixed with rising floodwaters can produce fast-moving debris flows of large proportion, which induce further slope instability as a result of strong bank erosion in the valley bottoms. Highly destructive peak-flows with depths as great as 8–10 m (Larsen *et al.* 2001; Perez 2001; Esposito *et al.* 2004*a*, *b*) are further related to abrupt draining of temporary debris dams formed at narrow valley gorges where the flow backs up to a critical threshold beyond which a translatory wave flood is produced (Fig. 5; see Eliason *et al.* 2007). Having occurred in a given area, the likelihood of recurrence of such events is usually very high but their confined character makes them highly erratic and does not prevent similar disasters occurring shortly after in nearby areas (Fig. 6 and Table 2).

Alluvial fans

Perhaps the most significant flood-induced geological effect at rocky coasts is the deposition of coarse terminal fans at river mouths (Fig. 7). They occur as part of fan-deltaic systems prograding seaward, with large-scale foresets (delta face) passing upwards and landwards to topset segments. Coarse alluvial fans and their subaqueous counterpart indicate that the whole fluvial system acts as a transfer zone where slide debris produced by intense erosional events is rapidly transported to the coast as concentrated flows. As the flows leave the constrictive valleys they quickly decelerate and spread laterally, dumping large quantities of alluvial sediment at sea (Nemec 1990; Nava-Sanchez *et al.* 1995, 1999; Perez 2001). The resulting shoreline progradation is largely ephemeral, as fluvial discharges are of short duration followed by long periods of non-deposition, so that waves are free to erode alluvial deposits and restore the original conditions to a varying extent.

Fan-delta systems resulting from high-energy fluvial events are composed of wedge-shaped coarse-grained deposits (Nemec 1990; Orton & Reading 1993; Soh *et al.* 1995; Nava-Sanchez *et al.* 1999) that thicken towards the sea. Well-developed alluvial fans occur where pre-existing offshore relief and bottom slope gradients are such as to allow sediment aggradation at river mouths, as in the late stage of fan-delta development or on gently sloping sea floor (e.g. Prior & Bornhold 1990). Fine-grained deposits can locally prevail as a result of Holocene sea-level rise (Dubar & Anthony 1995), but they pass upward to present-day sandy–gravel environments. Because of coarse textures, the seaward limits of subaerial fans are

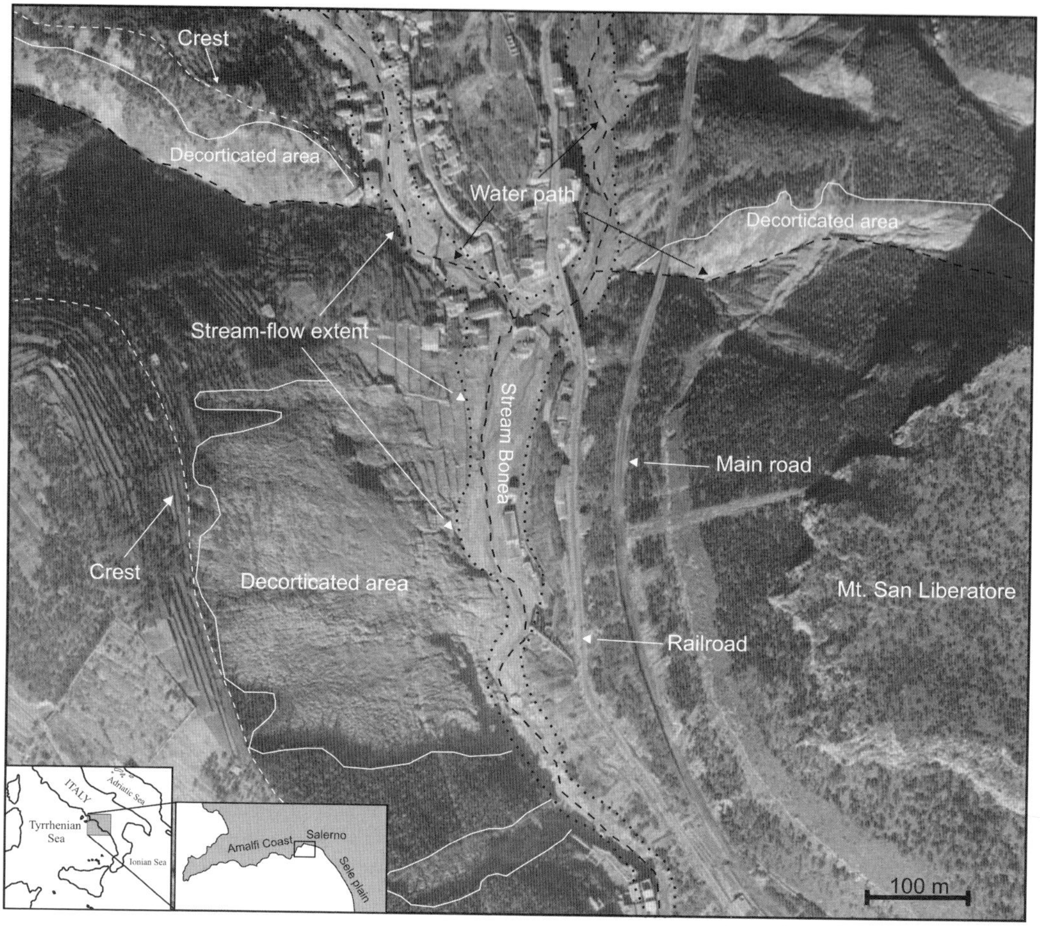

Fig. 4. Aerial photograph taken soon after the 1954 cloudburst in the basin of the Bonea stream (Vietri sul Mare, eastern Tyrrhenian Sea). The photograph clearly shows the critical relation between sediment supply from side slopes (decorticated areas extending up to the mountain crests) to stream paths with significant increase of bed load transport and consequent production of channelized hyperconcentrated flows.

composed of narrow and steep beaches, often backed by low cliffs cut by marine erosion or by narrow dune-ridges, whose characteristics depend on longshore currents and wave energy.

The alluvial fan surface can be strongly modified by human activity. Distributary streams flowing on supratidal delta areas are frequently diverted or covered for urban development. This is the case in many villages along the Amalfi coast (southern Italy), where flood-prone streams are artificially forced to flow underneath roads and squares to exploit the whole surface of narrow alluvial deltas. The consequence of this type of urbanization became evident in recent times, when very high river sediment discharges occurring in conjunction with the 25–26 October 1954 cloudburst exploded the artificial cover (Fig. 8) and causing severe damage and loss of life (Lazzari 1954; Penta *et al.* 1954; Esposito *et al.* 2004*a*, *b*).

Fan-deltas

A number of modern underwater delta slopes have been recently investigated using marine geophysical investigations aided by sea-floor sampling (Prior *et al.* 1981; Ferentinos *et al.* 1988; Prior & Bornhold 1988, 1989, 1990; Syvistki & Farrow 1989; Nemec 1990; Liu *et al.* 1995; Nava-Sanchez *et al.* 1999; Lobo *et al.* 2006; Sacchi *et al.* 2009). Interpretation of the marine data shows that sediment dispersal along the subaqueous extensions of alluvial fans is directly related to supply from rivers during floods. Flood-induced density flows have sufficient momentum and concentration to

Fig. 5. The 1954 flood of the Bonea stream (Vietri sul Mare, eastern Tyrrhenian Sea). (**a**) Aerial photograph showing stream flow extent and location of a temporary dam. Numbers indicate the same locations as in (**b**) and (**c**). (b) and (c) are photographs of the damming area before and after the 1954 event, respectively. (See Fig. 4 for location.)

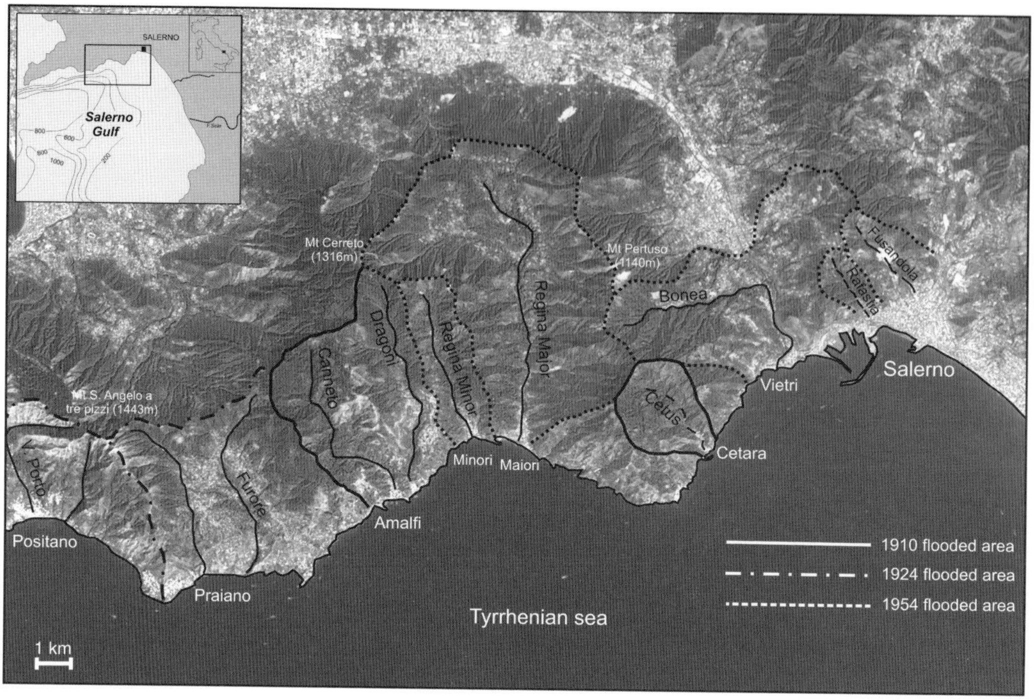

Fig. 6. Areal distribution of the three major floods that occurred in the twentieth century on the Amalfi coast (eastern Tyrrhenian Sea). Each event was associated with severe stream flows, landslide phenomena and alluvial fan deposition at stream mouths. Data from Esposito *et al.* (2004*b*).

cross the land–water boundary and continue downslope as underflows, bringing river-borne sediment to the shelf and open sea. Subaqueous sediment transfer commonly occurs through chutes and channels developed seawards of main stream mouths, and the sediment is deposited as terminal sand–debris lobes and splays in the pro-delta areas. Thus, river-derived materials entering the sea are primarily stored as prograding foreset at the delta face and then transferred further out into the low-gradient delta-toe zone. Dispersal processes range from mass flows to turbidity currents, depending on the energy of river floods and sea-floor relief or gradient, so that a single event may involve different types of movements along subaqueous slopes. In addition, fine sediments can settle from buoyant plumes, forming pelagic drapes on top of coarse-grained deposits (Fig. 9).

Table 2. *Historical floods in the Salerno province (southern Italy) since the 18th century*

Day	Month	Year	Day	Month	Year	Day	Month	Year	Day	Month	Year
	November	1738	1	April	1875	23–25	June	1905		March	1960
	November	1760	1	December	1875	1	September	1905	16	February	1963
11	November	**1773**	25	February	1879		November	1908	25	September	1963
24	December	1796	1	November	1881	24	October	**1910**	9	January	1968
13	September	1834	15–17	September	1882	23	January	1911	19	October	1970
18	July	1835	1	February	1885	13	November	1921		December	1971
27	September	1837	5	May	1885	26	March	**1924**			1980
1	January	1841	1	November	1893			1929	16–17	November	1985
17	December	1867			1896			1935		March	1986
21–22	June	1868			1898	1	October	1949			
7–12	November	1868	7–8	October	**1899**			1951			
11	December	1869			1904	25–26	October	**1954**			

Catastrophic events are indicated in bold (modified from Esposito *et al.* 2003).

Fig. 7. The mouth of the Bonea stream (Vietri sul Mare, eastern Tyrrhenian Sea) a few days after the 1954 flood event. (**a**) A coarse alluvial fan extends about 200 m from the stream mouth. Dashed box shows location of (b). (**b**) Emplacement of the fan was associated with intense erosion of the beach profile. (See Fig. 6 for location.)

Fig. 8. 'Explosion' of the main street at Maiori village (Amalfi coast, eastern Tyrrhenian Sea) after the 1954 flood event. The street was constructed by covering the path of a flood-prone stream, the Regina, at its mouth. (See Fig. 6 for location.)

Extensive shelfal sandy lobes occurring in the fossil record (Mutti *et al.* 1996, 2003) and the acknowledgment that small (basin areas $<10^4$ km^2) mountainous river systems are important sources of sediment dispersal in coastal and open seas (Milliman & Syvitski 1992; Wheacroft *et al.* 1997) have stimulated great interest in turbidity currents resulting from the continuation of concentrated fluvial flows in marine settings (hyperpycnal flows). Hyperpycnal flows have been recognized in a variety of marine environments, generated both from seasonal and extreme floods as well as from catastrophic events such as lahars, abrupt draining of glacial lakes (jökulhaups), and dam breaking and draining (see Mulder *et al.* 2003). They can locally induce high sedimentation rates (1–2 m per 100 years) and represent the ultimate means by which sediment particles eroded from the high coastal slope are transported to deep-sea environments (Brunner *et al.* 1999; Mulder *et al.* 2001). Deposits related to hyperpycnal processes (hyperpycnites; Mulder *et al.* 2003) differ from other turbidites because of a basal coarsening-up unit, deposited during the increasing or waxing period of river discharge, underlying the typical turbiditic fining upward unit, in this case deposited during the waning of river discharge (Fig. 10). It should be noted that as a river-related turbidity current travels across the continental shelf and beyond it down the slope, the resulting deposit may reflect a longitudinal gradient and reach greater depth in the form of a submarine fan.

Sea-cliffs

An important source of sediments at rocky coasts is represented by colluvial deposits resulting from cliff recession (Fig. 11). Cliff erosion is produced by both wave and weathering action, which operate with varying intensity depending on local meteorological and oceanographic features, and rock resistance, and results in an irreversible loss of land. Although basal erosion is a critical factor for cliff instability (Richards & Lorriman 1987; McGreal 1979), precipitation and infiltration of water resulting from rainfall events and groundwater may act as driving or forcing agents in the upper part of the cliff slope, significantly contributing to coastal changes (Lawrence 1994). Again, a major role is played by landslide activity of various magnitude, which can involve the rock substrate and loose superficial terrain transporting significant amount of materials to the cliff toe. The form and stability of rocky coasts is further related to factors inherited from past environmental conditions, characterized by different sea level and climate, which interact with contemporary erosive agents so that the sea may rework steep slopes initially formed by non-marine processes (Fig. 12; Sunamura 1992; Bray & Hooke 1997; Trenhaile 2002).

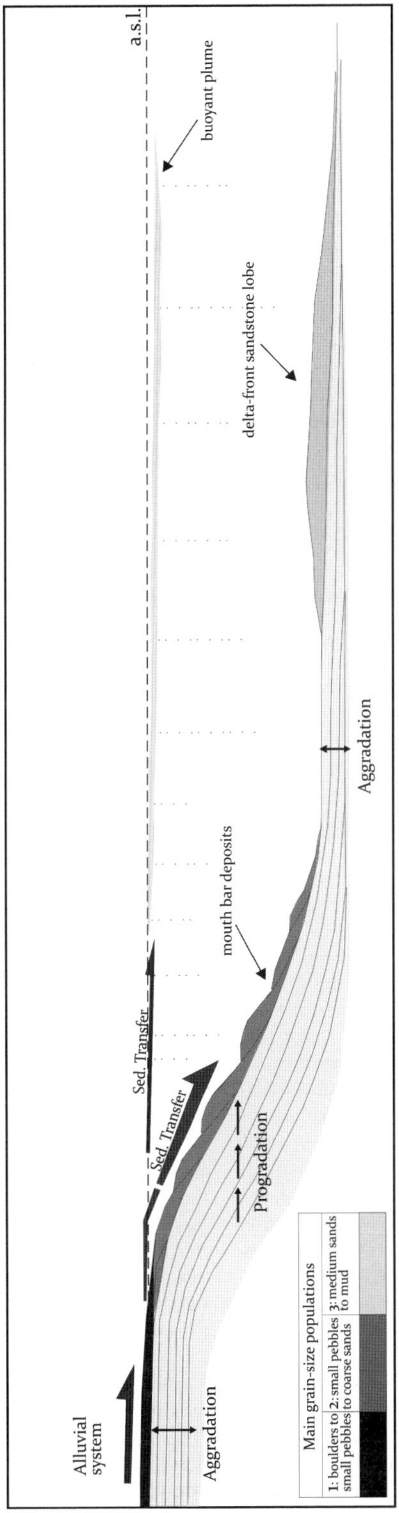

Fig. 9. Sedimentary features of a delta–pro-delta system at the mouth of a flood-prone stream (after Nemec 1990; Mutti *et al.* 2003).

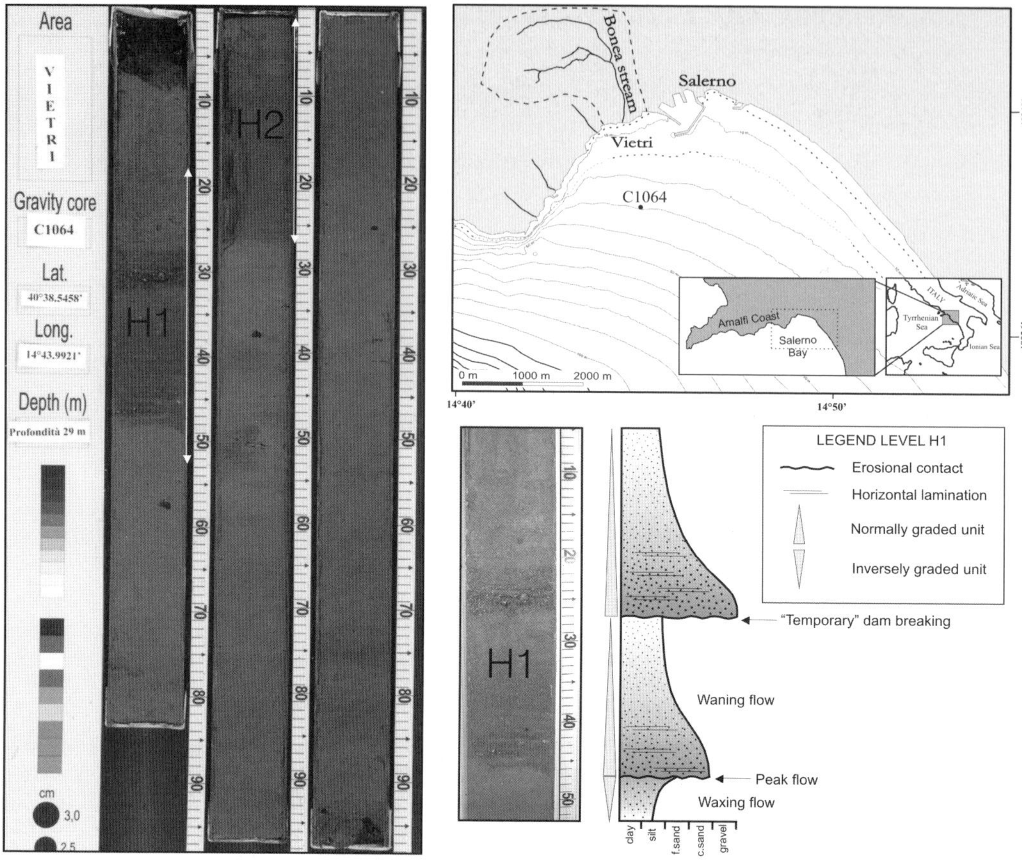

Fig. 10. Piston core 1064 collected off Vietri sul Mare (eastern Tyrrhenian Sea). Two sandy levels occur at 46 cm (H1) and 103 cm (H2) below the sea floor. H1 is composed of a basal coarsening upward unit followed by a fining upward units. The two units are separated by an erosional contact and result from a hyperpycnal flow at the mouth of the Bonea stream triggered by the 1954 cloudburst. A third, normally graded sharp-based unit develops in the upper part of H1, and is associated with the abrupt draining of a fluvial debris dam (see also Fig. 5). The level H2, composed of normally graded sand, probably results from a previous stream-flow event (data from Violante *et al.* 2004; Budillon *et al.* 2006).

Mechanical strength and wave energy are the main elements affecting the recession of cliffed coasts (Sunamura 1992). The wave factor is greatly influenced by the occurrence of loose sediments in coastal waters, which increase mechanical abrasion and wave impact. However, as the solid load increases to high values, wave energy is dissipated in moving and reworking sediments, and the coast is consequently protected. Therefore, fallen and/or fluvial-derived debris that accumulates in the form of a beach or as landslide deposits at the cliff toe significantly reduces cliff instability. The persistence of such basal protective sediments depends upon the balance between hydrodynamic forcing (waves, tides, cross- and long-shore currents), and the type and amount of materials supplied. A general model for the evolution of a rocky coast, first proposed by Sunamura (1983), involves a cyclic process with phases of cliff retreat followed by failures and mass movements and longshore transport of the accumulated material, such that the cliff is exposed again to the wave action (Fig. 13).

Another consequence of cliff retreat is the creation of shore platforms, which is mainly related to quarrying and abrasion activities with significant aid from bio-erosion and weathering (Fig. 14; Sunamura 1992; Haslett 2000; Trenhaile 2002). These structures are seldom horizontal, and often have a gentle seaward slope of up to 3°, possibly covered by a small amount of sediment (Trenhaile 2004). Although associated with the rate of cliff retreat, the occurrence of a shore platform in front of a cliff increasingly acts to dissipate wave energy as it develops landward, up to a critical platform width, beyond which waves are unable to

Fig. 11. Cliff recession on the Amalfi coastal area (eastern Tyrrhenian Sea). (**a**) View from the sea of the Amalfi sea-cliff. (**b**) Shaded relief map of the same area as in (a) obtained by merging multibeam bathymetric data and terrestrial elevation data. Inset map shows location.

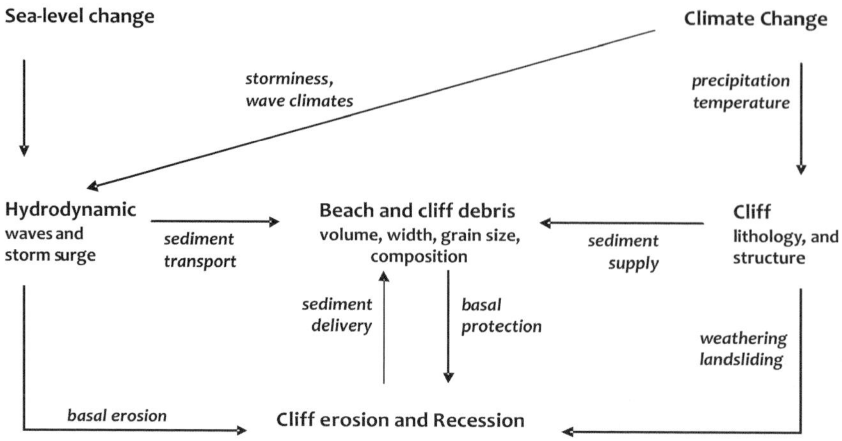

Fig. 12. Factors influencing cliff erosion and recession (modified from Bray & Hooke 1997).

erode or remove debris protecting the cliff face. Nevertheless, the dissipative effect may decrease as a result of platform downwearing, which reduces the platform height relative to sea level, thus maintaining cliff retreat. Again, this process has a finite limit, as shear stress between the platform and the waves decreases with water depth.

The retreat of a cliffed shore is the cumulative result of numerous variables acting on each other. Interaction between processes and products may result in self-regulation from negative feedback associated with cliff debris that supports, protects or loads the toe. In these cases the recession can stop, and the cliff may be degraded by subaerial processes, or evolve through a cyclic process involving debris removal and redistribution by hydrodynamic forces.

Cliff recession

The stability of a rocky coastal slope is greatly influenced by intrinsic geological features that determine material strength and rock mechanics. Lithology, patterns of fractures and faults, and strata attitude can vary significantly also at a local scale, affecting cliff response to wave energy (Allison 1989; Sunamura 1992; Bray & Hooke 1997) and the types of mass movement. Mudflows and rotational slumps regularly develop in soft and weak lithologies, whereas on firm and rocky cliffs rockfalls and topples are predominant (Fig. 15). More resistant lithologies are often characterized by the development of stress-release jointing resulting from a decrease in the confining pressure as cliff retreat proceeds. Such tension cracks can cause a high degree of freedom for block movements, often resulting in toppling failures.

The types of mass movement are especially important because they affect the nature, size and amount of sediment released by cliff erosion. These characteristics influence the proportion of cliff input retained in the littoral area, which inhibits wave impact, thus reducing further recession. Coarse and block-size durable materials are more likely be retained on the upper shoreface and act as natural armour whereas sands are more susceptible to cross-shore transport induced by seasonal storms. However, local bathymetry and oceanographic factors can allow the presence of wide sandy beaches that actively protect cliffs from marine erosion (Fig. 16). Fine deposits of silt and clay size do not contribute significantly to active shore profiles, as they are regularly removed from the littoral environment as suspended material and redeposited further offshore.

Coastal landslides are common where rocks incline or dip seawards, with the resultant cliff angle being largely determined by the dip angle. If

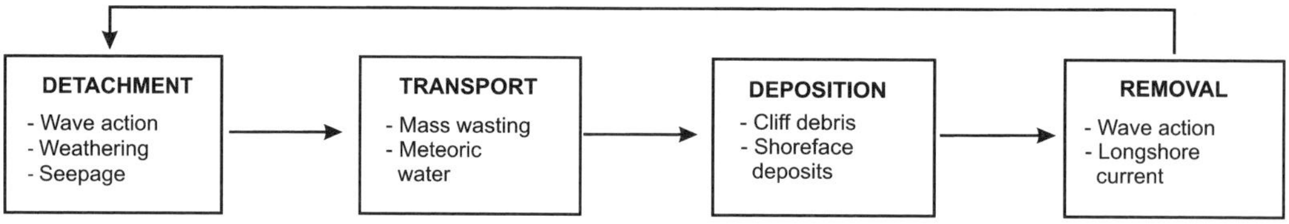

Fig. 13. Process of sea-cliff evolution.

Fig. 14. Shore platform cut into volcanic deposits at Carloforte island (southern Sardinia, Italy). Inset map shows location.

strata dip landward or have a horizontal attitude, slope instability is significantly reduced, and near-vertical cliff faces may develop. In many Mediterranean coastal environments where the rock masses are of a carbonate nature, chemical weathering may exert an important effect through the activity of karst processes. For these coasts basal erosion often produces deep notches, formed by a combination of biological and physical activities, which effectively undermine the cliff and lead to slope failure. Seepage erosion also may facilitate major mass movements at coasts characterized by groundwater circulation within permeable strata that overlie or are interbedded with impermeable units. This is the case for the coastal bluffs of New England, where ground waters remove material and reduce sediment strength, greatly enhancing slope instability (Kelley 2004).

The study of factors and eroding actions controlling the evolutionary dynamics of sea cliffs (Trehaile 1987; Sunamura 1992, 1997; Griggs 1994) suggest that retreat of coastal slopes largely depends on mechanical wave action. Besides mechanical rock strength, assessment of cliff recession

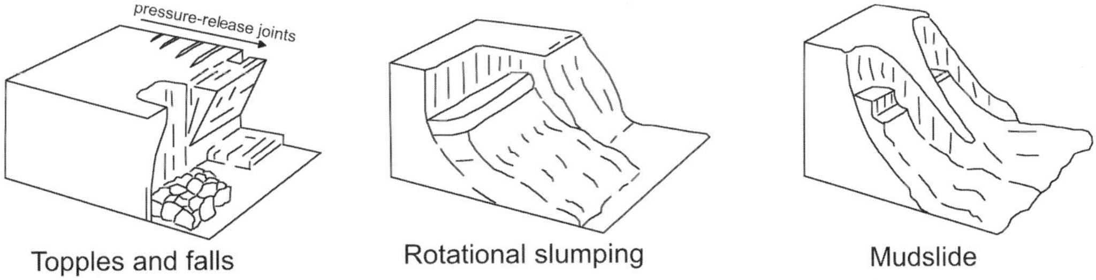

Fig. 15. Main types of mass movements affecting a sea-cliff.

Fig. 16. Cliff debris and beach deposits occurring in the littoral environment at the toe of cliffed coastal areas. (**a**) Cliff debris resting at the toe of mountainous coastal slopes (California Coastal Range, USA). (**b**) Bluff-backed shore along the Oregon coast (USA).

has to take into account all the variables that influence the persistence of cliff-derived deposits or river-borne sediments in the littoral environment in the form of beaches and/or landslide deposits. These parameters commonly vary at a local scale, and sectors with different erosional processes and types of landslide often characterize the evolution of a given rocky coastal area.

Large coastal slope failures

Coastal regions of high relief, such as young tectonically active mountain chains or volcanic edifices, are prone to catastrophic slope failure that can mobilize large volumes of shattered rocks from 10^6–10^7 m^3 to 10–100 km^3 (Moore & Moore 1984; Siebert 1984; Melosh 1987). Landslides resulting from massive rock and volcano slope failure (rock or debris avalanches) are important sources of geological hazard in many regions of the world and have been responsible for some of the most destructive natural disasters (Schuster & Highland 2001). The most widely quoted examples occurring in historical times include the volcanic sector collapses of the northern flank of Mount St. Helens on 18 May 1980 (Voight *et al.* 1983; Glicken 1998), the Bandai and Unzen volcano slope failures in 1888 and 1792, respectively (Siebert *et al.* 1987), the Bezymianny eruption in Kamchatka in 1956, and the catastrophic lateral failure at Ritter volcano (Papua New Guinea) on 13 March 1888 (Johnson 1987), as well as some rapid giant rock landslides occurring in Norway's fjords during the last 100 years (Hermanns *et al.* 2006) and the well-known Mt. Toc catastrophic collapse into Italy's Vajont reservoir on 9 October 1963 (Müller 1964).

The main consequence of coastal slope collapses is the production of tsunamis as the rock avalanches enter a lake or the sea. The displaced waves can reach several tens to hundreds of metres in height and may propagate for long distances across the sea (Ward 2001), with a destructive impact on coastal areas. Rock avalanches and related tsunamis represent one of the most serious natural hazards in Norway, where more than 170 casualties have occurred in fjord areas in the last century (Blikra *et al.* 2005). Other examples of well-documented landslide tsunamis are those from the above-mentioned Unzen debris avalanche, which flowed into Ariake Bay and caused more than 11 000 victims, from the Hokkaido coast (Japan) after the collapse of the volcano Oshima-Oshima in 1741 (Satake & Kato 2001), and from Papua New Guinea in July 1998, where the tsunami, probably generated by a submarine slump triggered by an earthquake (Tappin *et al.* 2001), devastated the nearby coasts causing total destruction and thousands of victims. In the case of Italy, destructive waves of water swept away the villages of Longarone, Pirago, Villanova, Rivalta and Fae after a mountainside collapsed into the Vajont reservoir in northern Italy, causing more than 2000 victims (Müller 1964), and the small village of Scilla in Calabria, southern Italy, where a collapse of a mountain sector along the coast triggered a major tsunami in February 1783 (Fig. 17; Bozzano *et al.* 2006; Graziani *et al.* 2006). It has been estimated that of all tsunami events nearly 5% result from volcanic landslides (Smith & Shepherd 1996).

Rock avalanches, whether volcanic or non-volcanic in origin, possess high mobility in

Fig. 17. Historical maps and drawings illustrating the 1783 slope failure and related tsunami event along the Calabrian coast (eastern Tyrrhenian Sea). (**a**) and (**b**) show the slope before and after the landslide event, respectively. (**c**) and (**d**) are representations of the waves induced by the coastal failure (Biblioteca Nava, Reggio Calabria, Italy). Inset map shows location.

terrestrial (Hsü 1975; Ui *et al.* 1986) as well as marine environments (e.g. Moore *et al.* 1989; Watts & Masson 2001). The travel distance of displaced materials is of the order of some tens of kilometres and seems to be greater for subaqueous than for subaerial events (Hampton *et al.* 1996). The mobility of large mass movements has been attributed to the collisions between grains (Hsü 1975), to layers of compressed air trapped beneath the sliding mass (Shreve 1968) or to mechanical and acoustic fluidization (Melosh 1979). For subaqueous landslides the hydroplaning effect (Mohrig *et al.* 1998) resulting from the presence of a basal layer of water offers a plausible and widely accepted explanation for the long travel distances and high velocities of many submarine flows even on very gentle slopes. Evaluation of the runout of slope failures is particularly important for subaerial rock avalanches, as it allows the potential distribution of hazard intensity from the source area to be determined (Crosta *et al.* 2006).

Tectonic deformation and uplift play a crucial role in the development of threshold conditions for rock slope instability by increasing hillslope inclination or height. Also, tectonic stress along with lithology and weathering intensity exert a major control on geometric and strength characteristics of discontinuities and intact rock, with a profound influence on the stability of rock slopes. As emphasized by numerical simulations (Bhasin & Kaynia 2004), decrease in the residual friction angle along rock discontinuities under active environmental and earthquake conditions is an important factor for sliding and rotation of blocks that may evolve into catastrophic rock avalanches.

Large failure of rock slopes may be triggered by intense rainfall or earthquakes with Richter magnitudes greater than 6.5, or by high pore-water pressures associated with rapid snowmelt (Blodgett *et al.* 1998; Geertsema *et al.* 2006). In other cases, triggers are not well known and failure may result from progressive long-term degradation of the tectonically deformed and altered rock mass (Boultbee *et al.* 2006). With the exception of those collapses occurring in conjunction with large earthquakes, catastrophic slope collapse can naturally evolve from prolonged intervals of accelerating creep phenomena (Voight 1978) as a result of slow rock cracking, normally preceding the failure. The landside event is then related to self-accelerating fractures that are readily catalysed by internal circulating waters and the resultant increase in pore pressure (Kilburn & Petley 2003).

Instability at coastal and island volcanoes

Rocky coasts dominated by volcanic landforms may be built up of lava flows, pyroclastic flows, peninsular and island volcanoes, or calderas. Besides processes of cliff retreat affecting all these coasts, the erosion of volcanic coastal structures and island volcanoes is strongly related to their growth evolution, which is characterized by rapid vertical accretion leading to slope collapse. The mass of accumulated volcanic products resulting from sustained rates of volcanic construction can fail under its own weight, producing structural failure at any scale, from small rock falls of some hundred to a few thousand cubic metres to giant ocean-island megaslides that may involve up to 5000 km^3 of material (Moore *et al.* 1989).

The importance of landslide debris in the internal structure of volcanoes became clear after the climactic eruption of Mt. St. Helens during May 1980 (Lippman & Mullineaux 1981), which focused attention on the tendency of volcanic edifices to undergo lateral collapse. Subsequently, improved submarine imaging techniques revealed large-scale mass-wasting deposits on the slope of ocean island volcanoes such as those of the Hawaiian Ridge, exposed over about 100 000 km^2 (Moore *et al.* 1989), Augustine Island (Beget & Kienle 1992), Reunion (Labazuy 1996), Stromboli (Kokelaar & Romagnoli 1995), the Lesser Antilles arc (Deplus *et al.* 2001), the Canary Islands (Masson *et al.* 2002), and Ischia (Chiocci & de Alteriis 2006). Indeed, similar behaviour characterizes subaerial volcanoes, as suggested by the large number of avalanche deposits produced by lateral collapse around volcanic structures in Japan (Inokuchi 1988) and in the Andes (Francis & Wells 1988).

Lateral collapse at coastal and island volcanoes can occur catastrophically in the form of debris avalanches or more slowly as creeping slumps (Moore *et al.* 1989). These two phenomena include the largest and more dramatic landslides recognized on island slopes, with which, however, intermediate forms of mass wasting such as debris flow and turbidity currents are associated (Weaver *et al.* 1994; Garcia 1996; Masson *et al.* 2002; Talling *et al.* 2007). Slumps can involve great thickness of volcanic material, which may affect an entire flank of a volcanic island to depths as great as 10 km (Hilina slump, Hawaiian Ridge; Moore *et al.* 1989), have steep scarps at their toes, and have no connection with amphitheatre-like failure scarps. Debris avalanches are relatively thin when compared with slumps, they affect failed sections a few hundred metres to 1 km thick, and are associated with debris deposits with blocky appearance. Whereas debris avalanches and slumps normally are phenomena that cut into volcanic and intrusive rocks on the flanks of volcanic edifices, debris flow affects only the sedimentary cover of submarine slopes and develops over much greater distances (up to 600–700 km; Gee *et al.* 2001; Masson *et al.*

2002). Debris flow may occur simultaneously with debris avalanche, as reported on the western slopes of El Hierro and La Palma in the Canary Islands (Masson *et al.* 1998), or it develops after debris avalanche emplacement, leaving ridges and scours of debris on top of the blocky deposits. The transport of volcanically derived debris into the marine environment may continue downslope in the form of turbidite currents developing far from debris avalanches, which are able to transfer a complex sedimentary assemblage of volcaniclastic and pelagic slope sediments to abyssal plains (Fig. 18; Rothwell *et al.* 1992; Weaver *et al.* 1995; Garcia 1996; Talling *et al.* 2007). Finally, slumps and debris avalanches are not mutually exclusive, debris avalanches may form from disaggregation of oversteepened scarps at the toe of large slumps.

Of all landslide phenomena, debris avalanches are probably the most hazardous, as they represent single episodes of catastrophic failure moving at high velocity down volcano flanks. Evidence of high emplacement velocity includes uphill transport as far as several hundred metres (Moore *et al.* 1994; Masson *et al.* 2002; de Alteriis & Violante 2009) and the blocky nature of the resulting landslide deposits. Debris avalanches involve masses of fragmented volcanic rock (hummocky topographies; Fig. 19), with block size and degree of fragmentation being highly variable from site to site. Block structure and distribution mostly depend on the nature of landslide material (volcanic or intrusive rocks, pyroclastic or volcaniclastic deposits), speed of emplacement and degree of lateral constraint, which forces interaction between blocks and consequent disintegration (Masson *et al.* 2002; Mitchell *et al.* 2002). Generally, the displaced material spreads over wide and elongated sea-floor areas, with a high length-to-width ratio, extending from horseshoe-shaped or amphitheatre structures.

Heads of major slope failures are typically developed along the subaerial part of volcanic structures and continue across the coastline below sea level, setting up a connection with catastrophic coastal slope failures (Fig. 20). Many of the studied examples include coastal arcuate embayments and defined headwalls up to some hundred metres high, continuing underwater as lateral scarps that gradually decrease in height downslope to disappear at greater depths. Failure areas can be entirely submarine, as in the cases reported from the Hawaiian Ridge, where most of the upper submarine slopes off island volcanoes are scalloped by amphitheatre-like indentations, or entirely subaerial, as in the cases of the Unzen volcano (Japan), Mt. St. Augustine (Alaska) and Mt. St. Helens (Washington, USA).

Various factors can affect the stability of volcano flanks simultaneously, so their relative importance is difficult to evaluate. Steep and broad regional slopes widening off the unbuttressed flanks of coastal and island volcanoes facing the sea represent a common predisposing factor for the initiation of such enormous mass movements. Volcanic edifices that undergo sector collapse are often buttressed on the landward side by older edifices (such as Kilauea, Hawaii) or by thick progradational sedimentary sequences (such as Ischia, Bay of Naples) leading to the preferential release of accumulated intra-edifice stresses seaward. This condition is revealed topographically by the pronounced morphological asymmetry that characterizes the sea–land profile of these structures. However, the preparatory mechanisms for landsliding are often associated with site-specific volcanic activity, including dyke intrusion (McGuire *et al.* 1990), volcanic spreading (Borgia *et al.* 1992), accumulation of eruptive products on steep slopes (Murray & Voight 1996) magma chamber inflation and deflation (Lo Giudice & Rasa 1992), caldera collapse (Martì *et al.* 1997), fluctuation in sea level affecting the stress regime (McGuire *et al.* 1997), and volcano-tectonic uplift (Violante *et al.* 2004*a*; de Alteriis & Violante 2009). Other predisposing factors able to reduce slope stability to critical conditions are related to regional tectonics and climate, which may influence spatial distribution of landslide amphitheatres in a given area, to the effect of weak substrates (Moore *et al.* 1994) including residual soils (Hurlimann *et al.* 2001), which may act as potential slip surfaces of large volcanic landslides, and to earthquakes associated with volcanic activity (Lippman *et al.* 1988). However, debris avalanche events can either be associated with volcanic eruptions, such as the 1980 eruption of Mt. St. Helens, or have no connection to periods of unusual volcanic activity. It is worth noting that after long periods of volcanic build-up the load on volcanic edifices reaches a critical condition such that a relatively minor trigger is sufficient to initiate a landslide.

The high hazard potential of catastrophic debris avalanches is clearly evident when the total volume of a volcanic island is compared with the volume of blocky landslide deposits surrounding its lower submarine slopes. Based on data collected off oceanic islands (Holcomb & Searle 1991), including the western margin of the Canary Islands (Masson *et al.* 2002) and the submarine slopes of the Hawaiian Islands (Moore *et al.* 1994), it has been estimated that a single landslide may have been sufficiently large to remove up to 25% of the edifice volume of an island. Some of these major events occurred in prehistoric times, such as at Ischia in the Bay of Naples (Chiocci & de Alteriis 2006; de Alteriis & Violante 2009), at Stromboli in the Aeolian Islands (Tinti *et al.* 2000) and possibly at Etna on Sicily (Pareschi *et al.* 2006). Nevertheless, the risk

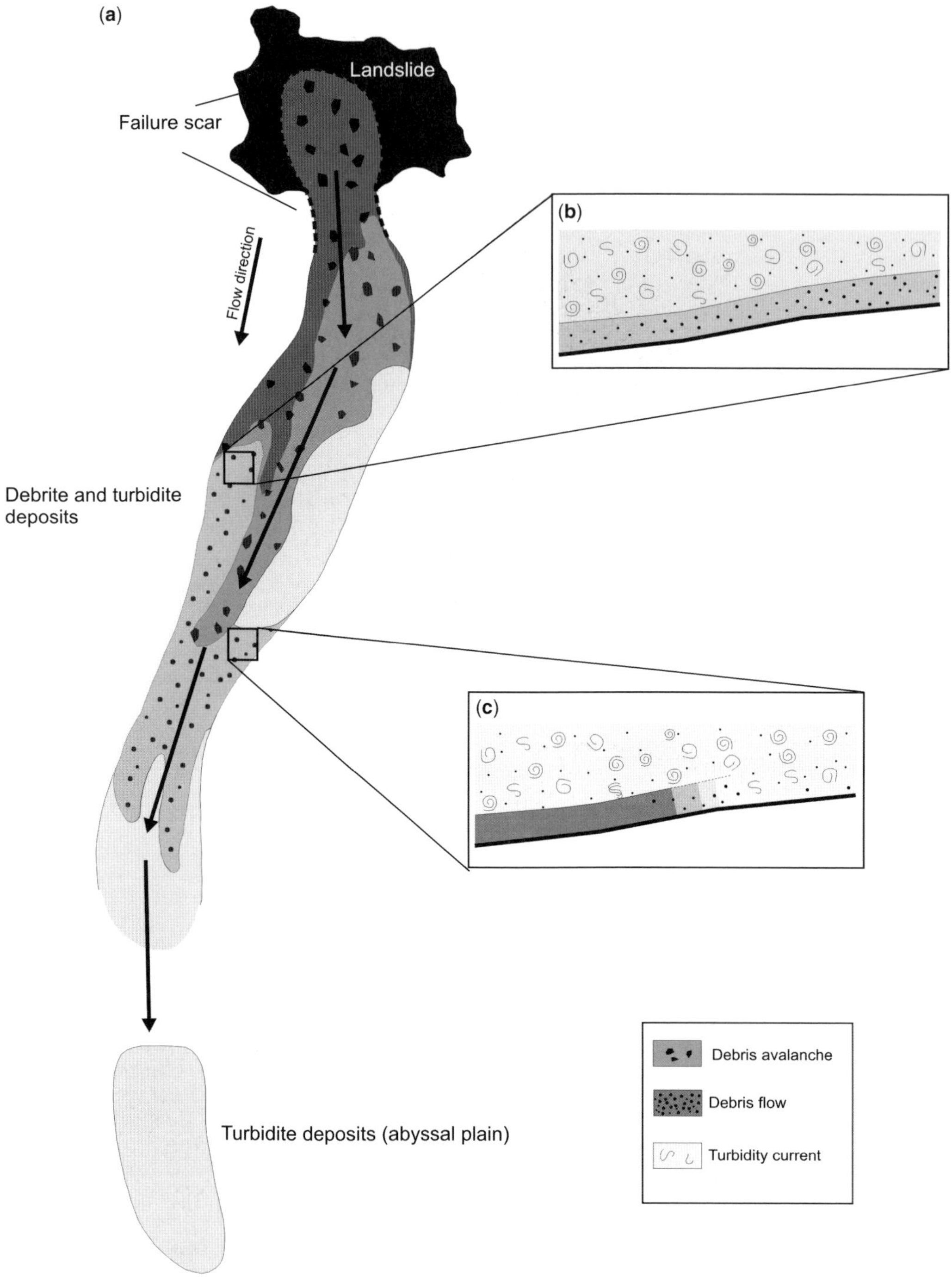

Fig. 18. Evolution of flow events from an original giant landslide. (**a**) Evolution of a submarine flow event from original volcanic flank collapse and consequent debris avalanche. (**b**) Debris flow from disintegration of initial landslide. (**c**) Debris flow generated by flow transformation from a decelerating turbidity current (after Kim *et al.* 1995; Masson *et al.* 1998; Talling *et al.* 2007).

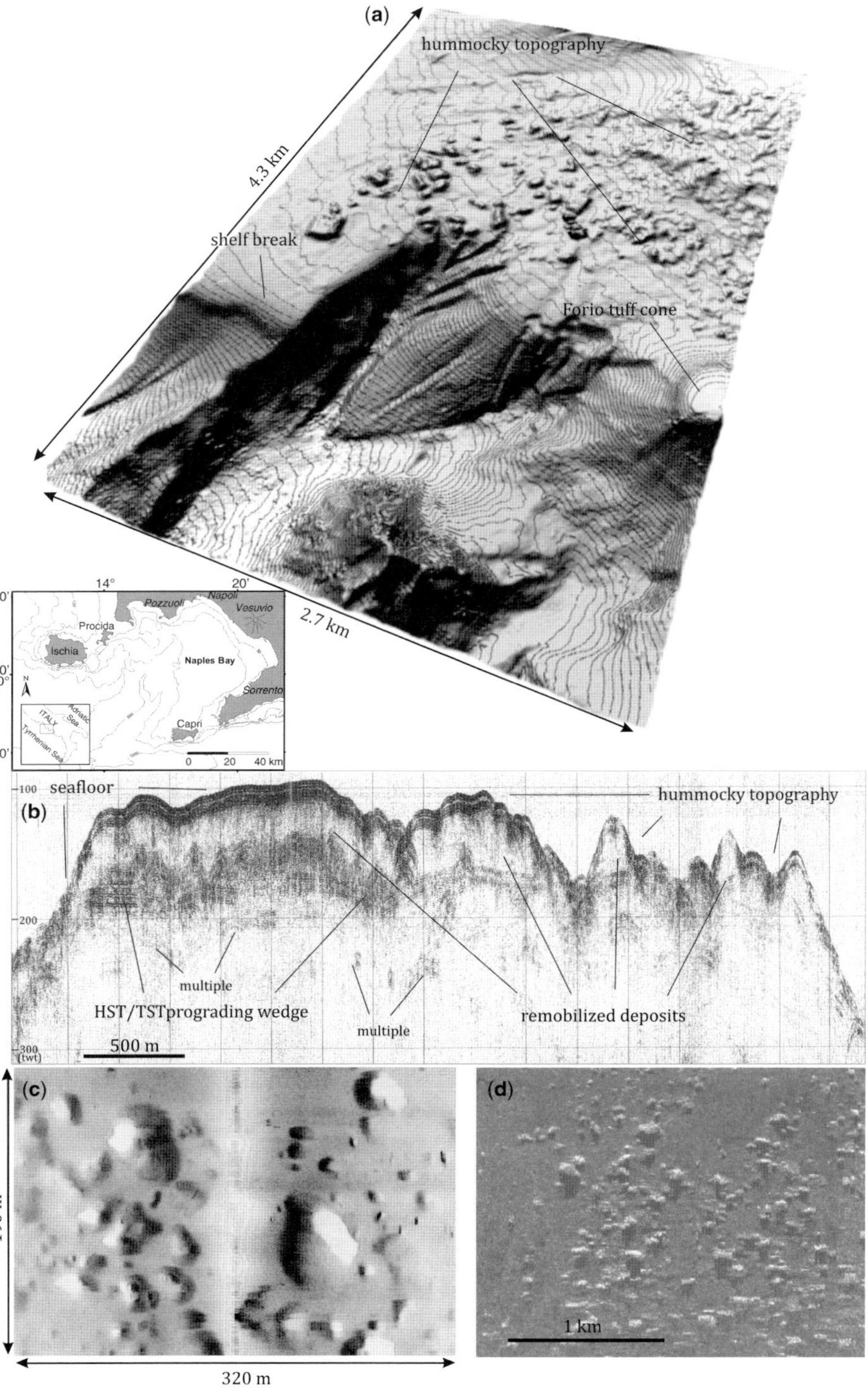

Fig. 19. Hummocky topographies off Ischia volcanic island (eastern Tyrrhenian Sea): (**a**) 3D map of a sea-floor sector off western Ischia obtained from multibeam bathymetric data; (**b**) interpreted seismic reflection profile (sparker source) shot off western Ischia; (**c**) 100 kHz side-scan sonar and (**d**) TOBI images surveyed off northern and southern Ischia, respectively. Inset map shows location. HST/TST, highstand systems tract–transgressive systems tract.

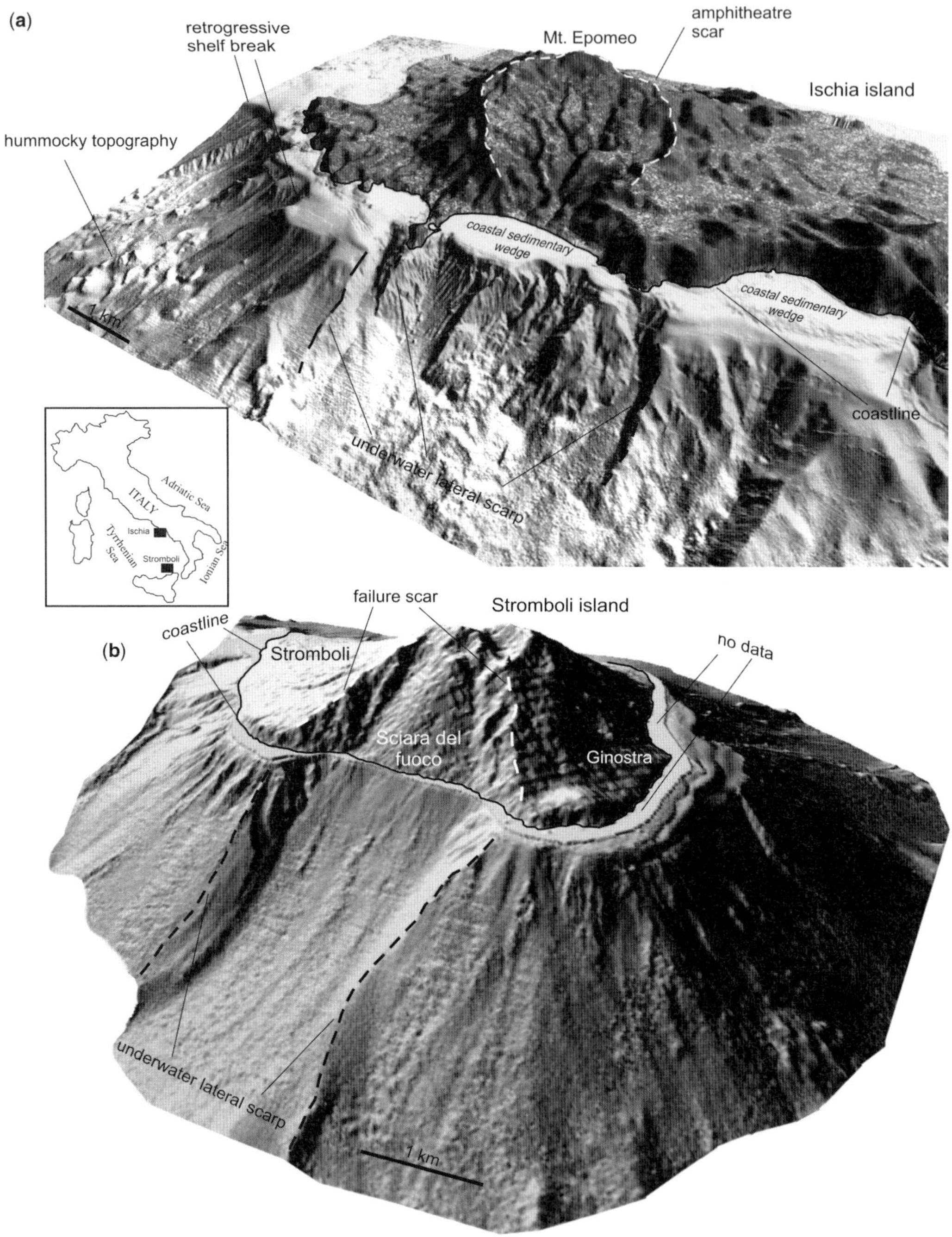

Fig. 20. Three-dimensional maps showing failure scarps and their underwater continuation, consequent to flank collapses of volcanic islands. (**a**) Southern flank of Mt. Epomeo, Ischia (eastern Tyrrhenian Sea). (**b**) Southern slope of Stromboli (southern Tyrrhenian Sea). These maps were obtained by merging multibeam bathymetric data and terrestrial elevation data. Inset map shows location.

from such landslides seems to be relatively low, as their period of recurrence is of the order of a few hundred thousand years.

Assessment of geological hazard related to catastrophic volcanic collapses must take into account the fact that the present form of a volcanic edifice is the result of complex interaction between the construction forces of volcanic processes and destructive factors causing removal by mass-wasting phenomena. Flanks of oceanic volcanic islands undergo repeated cycles of volcanic construction and failures (Beget & Kienle 1992), as shown by the large aprons of avalanche deposits with a total volume substantially greater than the volume of the relative failure scars. Probabilities of occurrence for instability events are significantly higher in the early stage of growth of a coastal or island volcano, when sector collapse is driven by higher rates of volcanic production.

Historical analysis

Records from historical documents combined with other methods can be used to enhance the reconstruction and knowledge of the occurrence of geohazards in a given coastal area. Extensive records on past stream flows, coastal landslides, cliff retreats and other natural hazards can be collected from published and unpublished documents (Fig. 21) that are widely available in countries with ancient civilizations, and allow us to characterize geologically sensitive areas with a recurrence in time and space of hazardous events. Availability of information regularly decreases as the investigations push back in time, up to a point where only the highest magnitude events are reported, mostly through oral tradition. This has led to the use of myth as a source of information to make statements about geohazard potential, particularly in regions where written history is relatively recent (Nunn & Pastorizio 2007).

The review of historical information can lead to a better understanding of the factors that influence extreme geohazards, or help to reconstruct areal distributions of hazardous phenomena and induced geological effects. Archival data may be used to estimate the recurrence interval or to build time series of extreme geological hazards, to extend the effective length of record prior to the instrumental era or systematic measurements. Identification and reconstruction of poorly monitored geohazards such as landslide phenomena or flood events in small watersheds is often possible only through the use of historical documentary sources.

Nevertheless, in statistical terms, the use of historical data is considered an informal method. Problems in the use of data from historical documents can arise from a greater availability of information in more recent times, which causes an increase in the frequency of reported geohazards. In the last few decades, enhanced accuracy in the documentation process and the greater presence of elements exposed to damage have led to increased documentation of minor events, whereas in the early 20th century and before small events were reported only occasionally. Another factor that may contribute to a selective availability of data on past geohazards concerns the conservation of historical information, which increases with the magnitude of the event. In particular, incorporation of archival data into frequency estimates has to take into account that past information usually records major phenomena, such that the historical record represents an incomplete catalogue of events, usually of unknown specific magnitude.

The format of historical documentary sources is likely to be highly variable. Information on past geohazards can be obtained from a number of sources including public record office collections, chronicles, ecclesiastical records, scientific, academic and engineering journals, newspapers, private collections and the World Wide Web. Valuable data sources refer to photographic documents (see Figs 5, 7 and 8), large-scale topographic maps (see Fig. 17), aerial photographs (see Figs 4 and 5) and repeated field measurements. Such information also tends to be of varying reliability and subject to manipulation. It is therefore important to distinguish fact from legend, as documents frequently contain erroneous and misleading information (Bell & Ogilvie 1978).

Historical aerial photography and maps have been used to study long-term sea-cliff retreat over periods of tens of years to a century (Moore *et al.* 1999). Shoreline retreat rates can be quantified using the landward migration of the top edge of the cliff by comparing a cliff edge digitized from historical maps with a recent cliff edge interpreted from Lidar (Light Detection and Ranging) topographic surveys (Hapke *et al.* 2007) or from recent aerial photography. Cliff retreat rates resulting from these methods represent conditions from a given period of time and yield amounts of cumulative retreat that are unsuitable for predicting future cliff edge positions or rates of retreats. Estimates of future retreat have to consider the effect of influencing factors such as sea-level rise (Bray & Hooke 1997), construction and existence of armouring at the cliff base, and human activities, which may strongly affect the degree of protection at the cliff toe or volume of infiltrating waters.

Historical documents from different sources combined with field data can be used to reconstruct the areal distribution of the geological effects and damage induced by catastrophic stream flows occurring in small, steep coastal watersheds

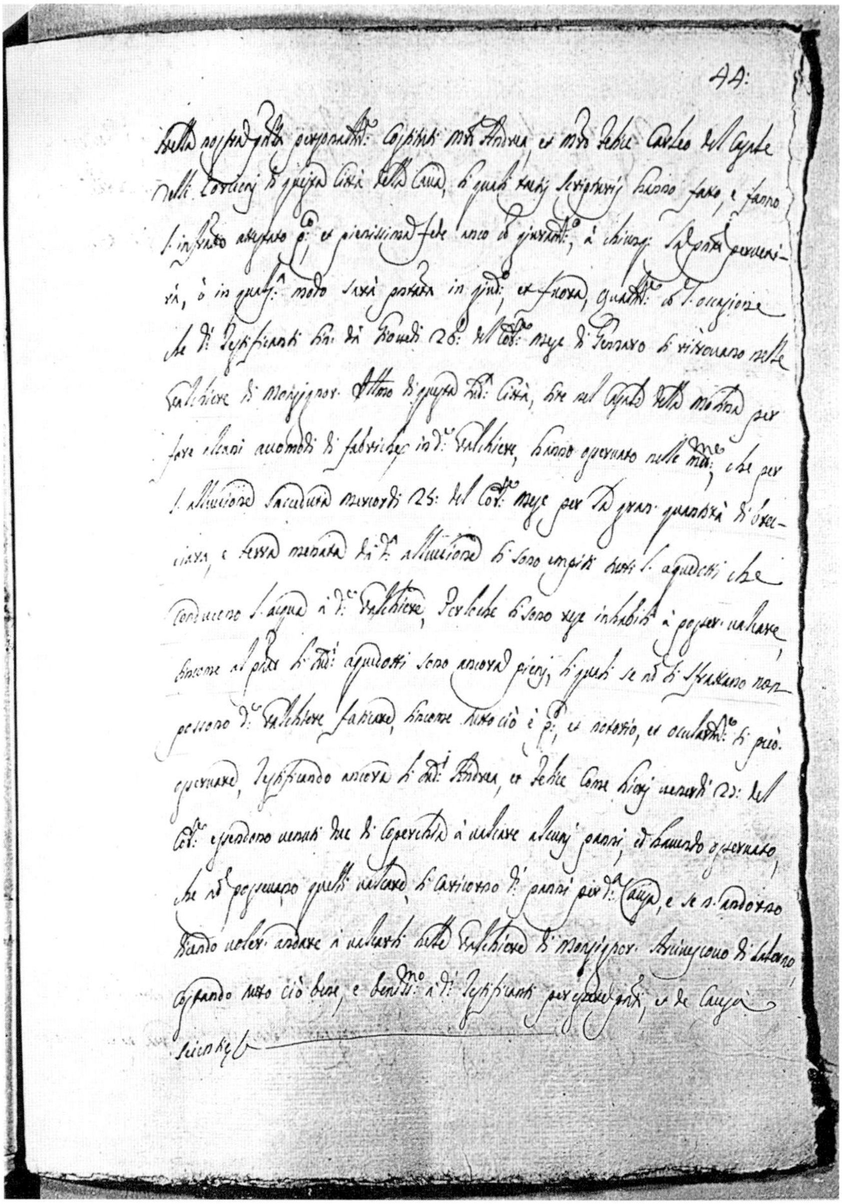

Fig. 21. Notarial document reporting reimbursement for 'valchiere' (paper factories) repairs after the 25 January 1736 flood event at Vietri sul Mare (eastern Tyrrhenian Sea). From Esposito *et al.* (2004*a*).

(Fig. 22; Esposito *et al.* 2004*b*; Porfido *et al.* 2009). Especially in these settings, the finding of slack-water deposits (Fig. 23) as palaeoflood indicators accumulated in zones of low-energy flow conditions at the valley margin can be used to reconstruct the magnitude and frequency of large floods that occurred before the establishment of instrumental stations (Baker & Kochel 1988; Benito *et al.* 2003). However, short steep streams are commonly poorly monitored or not monitored at all, as they are often dry with long periods of apparent stability, and the use of historical sources is often the only means for hazard assessment and the determination of related risk arising from very high urbanization rates mostly concentrated at a river mouth.

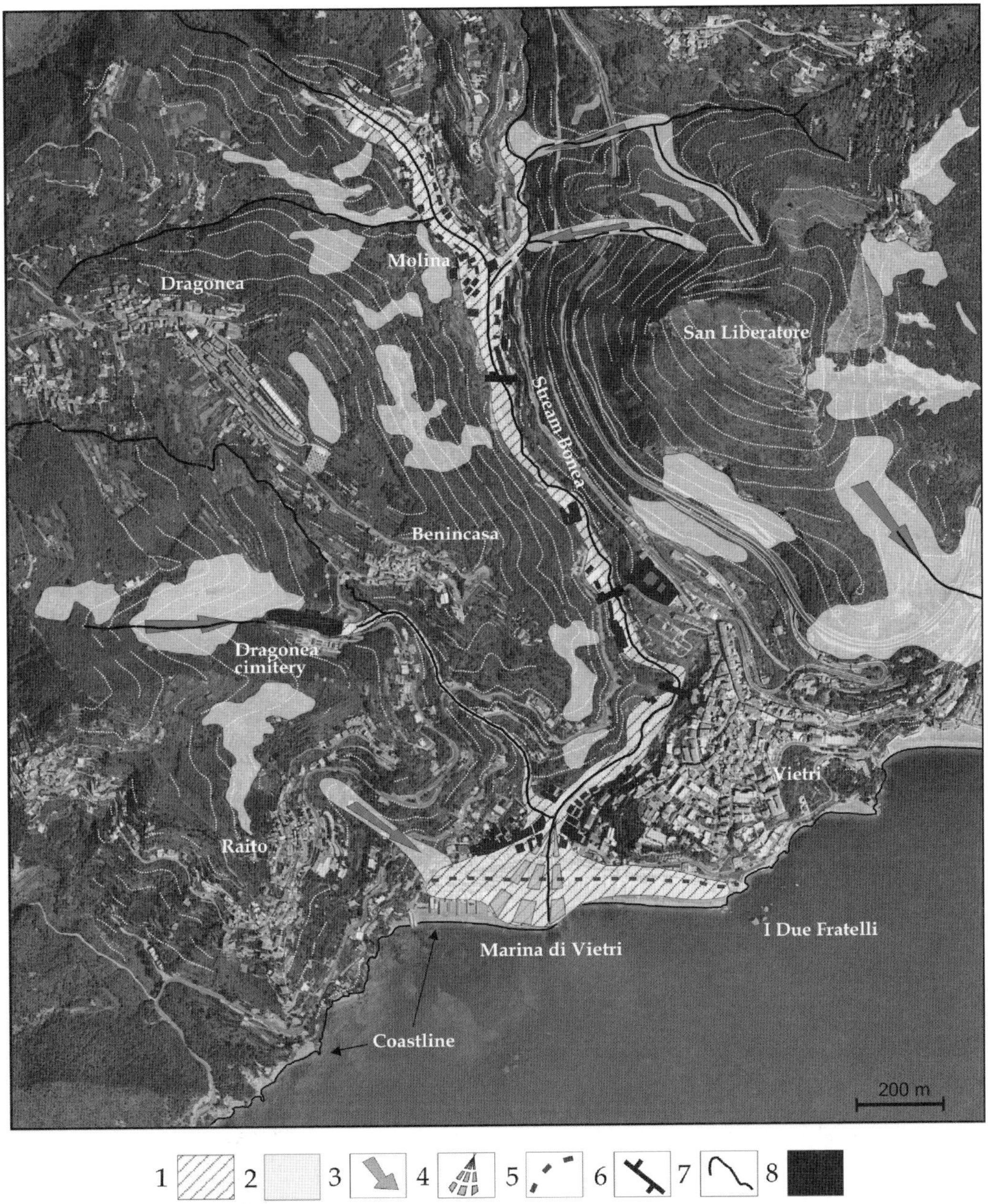

Fig. 22. Geological effects and damage pattern induced by the 1954 flood at Vietri sul Mare (eastern Tyrrhenian Sea). (1) Hyperconcentrated stream flow; (2) evidence of past denudations; (3) channelled debris flow; (4) alluvial fan; (5) shoreline before 1954 event; (6) temporary dam along the Stream Bonea; (7) drainage network; (8) medium to heavy damage. Modified from Esposito *et al.* (2004*b*). (See Fig. 4 for location.)

Conclusions

Geological constraints for hazard assessment in rocky coastal areas are intrinsically associated with mass-wasting phenomena involving rapid, catastrophic sediment transfers. Cliff recession, large slope failures and floods of steep coastal streams are part of the processes that naturally

Fig. 23. Stratigraphic section located within a narrow gorge reach along the Albori, stream Amalfi coast (eastern Tyrrhenian Sea), showing slack-water flood deposits passing upward into coarser debris-flow deposits. This sedimentary sequence formed during the catastrophic 1954 flood. The section is about 60 cm high.

drive the evolution of rocky coasts and significantly expose coastal communities and human activity to hazards. The processes that regulate rapid mass movements include hydrodynamic forces, volcanic activity, tectonics and slope-to-stream delivery, and their understanding is critical for assessment of hazard and related geological risk.

The marine areas of rocky coasts are typically characterized by high gradients, low width and abrupt continental margins, with submarine canyons close to the shoreline. In these environments, material produced by mass movements is delivered at intermittent time intervals, in the form of cliff debris, landslide deposits, coarse-grained deltas and sandy tabular bodies resulting from fluvial turbiditic flows. The combination of steep continental shelves, which are unable to dissipate wave energy, and episodic coastal supply prevents a stable beach profile being maintained, although significant beach areas can develop at the cliff toes owing to locally reduced bathymetric gradients. Lack of extensive coastal plains along rocky coasts is further due to the capture of sand at the heads of submarine canyons, so that it is carried out of the coastal system to the deep sea.

Low persistence of protective materials in the littoral environment exposes rocky coasts to erosion by hydrodynamic forces, thus producing an irreversible loss of land. However, long periods of no deposition and apparent stability induce people and even authorities to forget the nature and risks of events of rapid sediment transfers, particularly for poorly monitored coastal streams. For these reasons, the combination of marine and historical investigations is crucial to identify hazard-related geological processes and to recognize geologically sensitive rocky coastal areas (Fig. 24).

References

ALLISON, R. J. 1989. Rates and mechanisms of change on hard rock coastal cliffs. *Zeitschrift für Geomorphologie Neue Folge, Supplementband*, **73**, 125–138.

ANTHONY, E. J. & JULIAN, M. 1999. Source-to-sink sediment transfers, environmental engineering and hazard mitigation in the steep Var River catchment, French Riviera, southeastern France. *Geomorphology*, **31**, 337–354.

BAKER, V. 1994. Glacial to modern changes in global river fluxes. *In*: NATIONAL RESEARCH COUNCIL (ed.) *Material Fluxes on the Surface of the Earth.* National Academic Press, Washington, DC, 86–89.

BAKER, V. R. & KOCHEL, R. C. 1988. Flood sedimentation in bedrock fluvial systems. *In*: BAKER, V. R., PATTON, P. C. & KOCHEL, R. C. (eds) *Flood Geomorphology*. Wiley, New York, 123–137.

BALDWIN, J. E., DONLEY, H. F. & HOWARD, T. R. 1987. On debris flow/avalanche mitigation and control, San Francisco Bay area, California. *In*: COSTA, J. E. & WIECZOREK, G. F. (eds) *Debris Flow, Avalanches: Process, Recognition, and Mitigation.* Geological Society of America, Review in Engineering Geology, **7**, 223–236.

BEGET, J. E. & KIENLE, J. 1992. Cyclic formation of debris avalanches at Mount St. Augustine volcano. *Nature*, **356**, 701–704.

BELL, W. T. & OGILVIE, A. E. 1978. Weather compilations as a source of data for the reconstruction of European climate during Medieval period. *Climatic Change*, **1**, 331–348.

BELMONTE, A. C. & BELTRÀN, F. S. 2001. Flood events in Mediterranean ephemeral streams (ramblas) in Valencia region, Spain. *Catena*, **45**, 229–249.

BENDA, L. 1990. The influence of debris flows on channels and valley floors in the Oregon Coast Range, USA. *Earth Surface Processes and Landforms*, **15**, 457–466.

BENITO, G., SÁNCHEZ-MOYA, Y. & SOPEÑA, A. 2003. Sedimentology of high-stage flood deposits of the Tagus River, Central Spain. *Sedimentary Geology*, **157**, 107–132.

BHASIN, R. & KAYNIA, A. M. 2004. Static and dynamic simulation of a 700 m high rock slope in Western Norway. *International Journal of Engineering Geology*, **71**, 213–226.

BLAIR, M. L., VLASIC, T. C., COTTON, W. R. & FOWLER, W. 1985. *When the Ground Falls. Planning and Engineering Response to Debris Flows*. University of Colorado, Boulder.

BLIKRA, L. H., LONGVA, O., HARBITZ, C. & LØVHOLT, F. 2005. Quantification of rock-avalanche and tsunami hazard in Storfjorden, western Norway. *In*: SENNESET, K., FLAATE, K. & LARSEN, J. O. (eds) *Landslides and Avalanches. ICFL 2005 Norway*. Taylor & Francis, London, 57–64.

BLODGETT, T. A., BLIZARD, C. & ISACKS, B. L. 1998. Andean landslide hazards. *In*: KALVODA, J. & ROSENFELD, C. L. (eds) *Geomorphological Hazards in High Mountain Areas*. Kluwer, Dordrecht, 211–227.

BORGIA, A., FERRARI, L. & PASQUARÉ, G. 1992. Importance of lateral spreading in the tectonic and volcanic evolution of Mount Etna. *Nature*, **357**, 231–235.

BOULTBEE, N., STEAD, D., SCHWAB, J. W. & GEERTSEMA, M. 2006. The Zymoetz River rock avalanche, June 2002, British Columbia, Canada. *Engineering Geology*, **83**, 76–93.

BOZZANO, F., CHIOCCI, F. L. *ET AL*. 2006. Subaerial and submarine characterisation of the landslide responsible for the 1783 Scilla tsunami. EGU 2006. *Geophysical Research Abstracts*, **8**, 10422.

BRAY, M. J. & HOOKE, J. M. 1997. Prediction of soft-cliff retreat with accelerating sea-level rise. *Journal of Coastal Research*, **13**, 453–467.

BRUNNER, C. A., NORMARK, W. R., ZUFFA, G. G. & SERRA, F. 1999. Deep-sea sedimentary record in the late Wisconsin cataclysmic floods from the Columbia River. *Geology*, **27**, 463–466.

BUDILLION, F., VIOLANTE, C., CONFORTI, A., ESPOSITO, E., INSINGA, D., IORIO, M. & PORFIDO, S. 2005. Event beds in the recent prodelta stratigraphic record of the small flood-prone Bonea Stream (Amalfi Coast, Southern Italy). *Marine Geology*, **222–223**, 419–441.

CAMARASA, A. M. 1994. La intensitat de la precipitacion. *In*: PÉREZ CUEVA, A. (ed.) *Atlas climatic de la Comunitat Valenciana*. COPUT, Valencia, 100–101.

CAMARASA, A. M. & SEGURA, F. S. 2001. Flood events in Mediterranean ephemeral streams (ramblas) in Valencia Region, Spain. *Catena*, **45**, 229–249.

CARTER, R. W. G. 1988. *Coastal Environments: an Introduction to the Physical, Ecological and Cultural Systems of Coastlines*. Academic Press, London.

CHIOCCI, F. L. & DE ALTERIIS, G. 2006. The Ischia debris avalanche. First clear submarine evidence in the Mediterranean of a volcanic island pre-historic collapse. *Terra Nova*, **18**, 202–209.

CINQUE, A. & ROBUSTELLI, G. 2009. Alluvial and coastal hazards caused by long-range effects of Plinian eruptions: the case of the Lattari Mts. after the AD 79 eruption of Vesuvius. *In*: VIOLANTE, C. (ed.) *Geohazard in Rocky Coastal Areas*. Geological Society, London, Special Publications, **322**, 155–171.

CLAGUE, J. J. & EVANS, S. G. 1994. *Formation and failure of natural dams in the Canadian Cordillera*. Geological Survey of Canada Bulletin, **464**.

CROSTA, G. B., CHEN, H. & FRATTINI, P. 2006. Forecasting hazard scenarios and implication for the evaluation of countermeasure efficency for large debris avalanches. *Engineering Geology*, **83**, 236–253.

DE ALTERIIS, G. & VIOLANTE, C. 2009. Catastrophic landslides off Ischia volcanic island (Italy) during prehistory. *In*: VIOLANTE, C. (ed.) *Geohazard in Rocky Coastal Areas*. Geological Society, London, Special Publications, **322**, 73–104.

DEPLUS, C., LE FRIANT, A. *ET AL*. 2001. Submarine evidence for large-scale debris avalanches in the Lesser Antilles Arc. *Earth and Planetary Science Letters*, **192**, 145–157.

DUBAR, M. & ANTHONY, E. 1995. Holocene environmental change and river-mouth sedimentation in the Baie des Anges, French Riviera. *Quaternary Research*, **43**, 329–343.

ELIASON, J., KJARAN, S. P., HOLM, S. L., GUDMUSSON, M. T. & LARSEN, G. 2007. Large hazardous floods as translatory waves. *Environmental Modelling and Software*, **22**, 1392–1399.

EMERY, K. & KUHN, G. 1982. Sea cliffs: their processes, profiles, and classifications. *Geological Society of America Bulletin*, **93**, 644–654.

ESPOSITO, E., PORFIDO, S., VIOLANTE, C. & ALAIA, F. 2003. Disaster induced by historical floods in a selected coastal area (southern Italy). *In*: THORNDYCRAFT, V. R., BENITO, G., BARRIENDOS, M. & LLASAT, M. C. (eds) *Paleofloods, Historical Floods and Climatic Variability: Applications in Flood Risk Assessment, Proceedings of the PHEPRA Workshop, Barcelona, 16–19 October, 2002*, CSIC – Centro de Ciencias Medio ambientales, Madrid, 143–148.

ESPOSITO, E., PORFIDO, S. & VIOLANTE, C. (eds) 2004*a*. *Il Nubifragio dell'Ottobre 1954 a Vietri sul Mare, Costa d'Amalfi, Salerno. Scenario ed effetti di una piena fluviale catastrofica in un'area di costa rocciosa*. Pubblicazione Gruppo Nazionale per la Difesa dalle Catastrofi Idrogeologiche, **2870**.

ESPOSITO, E., PORFIDO, S., VIOLANTE, C., BISCARINI, C., ALAIA, F. & ESPOSITO, G. 2004*b*. Water events and historical flood recurrences in the Vietri sul Mare coastal area (Costiera Amalfitana, southern Italy). *In*: RODDA, G. & UBERTINI, L. (eds) *The Basis of Civilization—Water Science?* International Association of Hydrological Sciences Publication, **286**, 1–12.

FAURÉS, J. M., GOODRICH, D. C., WOOLHISER, D. A. & SOROOSHIAN, S. 1995. Impact of small-scale spatial rainfall variability on runoff modelling. *Journal of Hydrology*, **174**, 309–326.

FERENTINOS, G., PAPATHEODOROU, G. & COLLINS, M. B. 1988. Sediment transport processes on active submarine fault escarpment: Gulf of Corinth, Greece. *Marine Geology*, **83**, 43–61.

FRANCIS, P. W. & WELLS, G. 1988. Landsat Thematic Mapper observations of debris avalanche deposits in the central Andes. *Bulletin of Volcanology*, **50**, 258–278.

GARCIA, M. O. 1996. Turbidites from slope failures on Hawaiian volcanoes. *In*: MCGUIRE, W. J., JONES, A. P. & NEUBERG, J. (eds) *Volcano Instability on the Earth and other Planets*. Geological Society, London, Special Publications, **110**, 281–292.

GEE, M. J. R., MASSON, D. G., WATTS, A. B. & MITCHELL, N. C. 2001. Passage of debris flows and turbidity currents through a topographic constriction: seafloor erosion and deflection of flow pathways. *Sedimentology*, **48**, 1389–1409.

GEERTSEMA, M., HUNGR, O., SCHWAB, J. W. & EVANS, S. G. 2006. A large rockslide–debris avalanche in cohesive soil at Pink Mountain, northeastern British Columbia, Canada. *Engineering Geology*, **83**, 120–143.

GLICKEN, H. 1998. Rockslide–debris avalanche of May 18, 1980, Mount St. Helens volcano, Washington. *Geological Survey of Japan Bulletin*, **49**, 55–106.

GRAZIANI, L., MARAMAI, A. & TINTI, S. 2006. A revision of the 1783 Calabrian (southern Italy) tsunamis. *Natural Hazard and Earth System Sciences*, **6**, 1053–1060.

GRIGGS, G. B. 1994. California's coastal hazards. *Journal of Coastal Research, Special Issue*, **12**, 1–15.

HAMPTON, M. A., LEE, H. J. & LOCAT, J. 1996. Submarine landslides. *Reviews of Geophysics*, **34**, 33–59.

HAPKE, C., REID, D. & BORRELLI, M. 2007. *The National Assessment of Shoreline Change: A GIS Compilation of Vector Cliff Edges and Associated Cliff Erosion Data for the California Coast.* US Geological Survey, Open-File Report, **2007–1112**.

HASLETT, S. 2000. *Coastal Systems.* Routledge Introductions to Environmental Series II.

HERMANNS, R. L., BLIKRA, L. H., NAUMANN, M., NILSEN, B., PANTHI, K. K., STROMEYER, O. & LONGVA, O. 2006. Examples of multiple rock-slope collapses from Kofels (Otz valley, Austria) and western Norway. *Engineering Geology*, **83**, 94–108.

HOLCOMB, R. T. & SEARLE, R. C. 1991. Large landslides from oceanic volcanoes. *Marine Geotechnology*, **10**, 19–32.

HSÜ, K. J. 1975. On sturzstroms—catastrophic debris streams generated by rockfalls. *Geological Society of American Bulletin*, **86**, 129–140.

HUBBEL, D. M., LANENE, J. M. & MCKENZIE, S. V. 1983. *Characteristic of Columbia River sediment following the eruption of Mount St. Helens on May 18, 1890.* US Geological Survey Circular, **850 J. 21**.

HURLIMANN, M., LEDESMA, A. & MARTÌ, J. 2001. Characterization of a volcanic residual soil and its implications for large landslide phenomena: application to Tenerife, Canary Islands. *Engineering Geology*, **59**, 115–132.

INOKUCHI, T. 1988. Gigantic landslides and debris avalanches on volcanoes in Japan. *In*: *Proceeding of the Kagoshima International Conference on Volcanoes.* National Institute for Research Administration, Japan, 456–459.

JOHNSON, R. W. 1987. Large-scale volcanic cone collapse: the 1888 slope failure of Ritter volcano, and other examples from Papua New Guinea. *Bulletin of Volcanology*, **49**, 669–679.

JULIAN, M. & ANTHONY, E. J. 1997. Mouvements de masse et torrentialité: cas du Tuébi et de la haute Tinée, Alpes-Maritimes, France. *Géomorphologie: relief, processus, environnement*, **1**, 59–71.

KELLEY, J. T. 2004. *Coastal bluffs of New England.* US Geological Survey, Professional Papers, **1693**.

KILBURN, C. R. J. & PETLEY, D. N. 2003. Forecasting giant, catastrophic slope collapse: lessons from Vajont, Northern Italy. *Geomorphology*, **54**, 21–32.

KIM, S. B., CHOUGH, S. K. & CHUN, S. S. 1995. Bouldery deposits in the lowermost part of the Cretaceous Kyokpori Formation, SW Korea: Cohesionless debris flows and debris falls on a steep-gradient delta slope. *Sedimentary Geology*, **98**, 97–119.

KOKELAAR, P. & ROMAGNOLI, C. 1995. Sector collapse, sedimentation and clast population evolution at an active island-arc volcano: Stromboli, Italy. *Bulletin of Volcanology*, **57**, 240–262.

LABAZUY, P. 1996. Recurrent landslide events on the submarine flank of Piton de la Fournaise volcano (Réunion Island). *In*: MCGUIRE, W. J., JONES, A. P. & NEUBERG, J. (eds) *Volcano Instability on the Earth and other Planets.* Geological Society, London, Special Publications, **110**, 295–306.

LARSEN, M. C., WIECZOREK, G. F., EATON, L. S., MORGAN, B. A. & TORRES-SIERRA, H. 2001. *Natural hazards on alluvial fans; The Venezuela debris flow and flash flood disaster.* US Geological Survey, Fact Sheet, **103–01**.

LAWRENCE, P. L. 1994. Natural hazards of shoreline bluff erosion: a case study of Horizon View, Lake Huron. *Geomorphology*, **10**, 65–81.

LAZZARI, A. 1954. Aspetti geologici dei fenomeni verificatisi nel Salernitano in conseguenza del nubrifagio del 25–26 Ottobre 1954. *Bollettino Società dei Naturalisti in Napoli*, **68**, 131–142.

LIPPMAN, P. W. & MULLINEAUX, D. (eds) 1981. *The 1980 eruption of Mount St. Helens.* US Geological Survey, Professional Papers, **1250**.

LIPPMAN, P. W., NORMARK, W. R., MOORE, J. G., WILSON, J. B. & GUTMACHER, C. E. 1988. The giant Alika debris slide, Mauna Loa, Hawaii. *Journal of Geophysical Research*, **93**, 4279–4299.

LIU, K., CROOK, K. A. W., CLARKE, J. H. & WHITMORE, G. P. 1995. Submarine features of modern open-sea fan deltas, Huon peninsula, Papua New Guinea. *Sedimentary Geology*, **98**, 63–77.

LOBO, F. J., FERNANDEZ-SALAS, L. M., MORENO, I., SANZ, J. L. & MALDONADO, A. 2006. The sea-floor morphology of a Mediterranean shelf fed by small rivers, northern Alboran Sea margin. *Continental Shelf Research*, **26**, 2607–2628.

LOCAT, J. & SANFAÇON, R. 2000. *Multibeam surveys: a major tool for geosciences.* Proceedings of the Canadian Hydrographics Conference, Montréal 2000, Canada, 1–11.

LO GIUDICE, E. & RASA, R. 1992. Very shallow earthquakes and brittle deformation in active volcanic areas. The Etnean region as an example. *Tectonophysics*, **202**, 257–268.

MAJOR, J. J., PIERSON, T. C., DIEARTH, R. L. & COSTA, J. E. 2000. Sediment yield following severe volcanic disturbance—A two decade perspective from Mount St. Helens. *Geology*, **28**, 819–822.

MARTÌ, J., HURLIMANN, M., ABLAY, G. A. & GUDMUNDSSON, A. 1997. Vertical and lateral collapses on Tenerife (Canary Islands) and other volcanic ocean islands. *Geological Magazine*, **131**, 715–727.

MARTIN-VIDE, J. P., NINEROLA, D., BATEMAN, A., NAVARRO, E. & VELASCO, E. 1999. Runoff and sediment transport in a torrential ephemeral stream of the Mediterranean coast. *Journal of Hydrology*, **225**, 118–129.

MASSON, D. G., CANALS, M., ALONSO, B., URGELES, R. & HUHNERBACH, V. 1998. The Canary debris flow: source area morphology and failure mechanisms. *Sedimentology*, **45**, 411–432.

MASSON, D. G., WATTS, A. B., GEE, M. J. R., URGELESS, R., MITCHELL, N. C., LE BAS, T. P. & CANALS, M. 2002. Slope failures on the flanks of the western Canary Islands. *Earth-Science Reviews*, **57**, 1–35.

MCGREAL, W. S. 1979. Cliffline recession near Kilkeel, N. Ireland; an example of a dynamic coastal system. *Geografiska Annale, Series A, Physical Geography*, **61**, 211–219.

MCGUIRE, W. J., PULLEN, A. D. & SAUNDERS, S. J. 1990. Recent dyke-induced large-scale block movement at Mount Etna and potential slope failure. *Nature*, **343**, 357–359.

MCGUIRE, W. J., HOWARTH, R. J. *ET AL*. 1997. Correlation between rate of sea-level change and frequency of explosive volcanisms in the Mediterranean. *Nature*, **389**, 473–476.

MEADE, R. H. & PARKER, R. S. 1985. Sediment in rivers of the United States. *US Geological Survey, Water-Supply Paper*, **2275**, 46–60.

MEADE, R. H., YUYZK, T. R. & DAY, T. J. 1990. Movement and storage of sediments in rivers of the United States and Canada. *In*: RIGGS, H. C. & WOLMAN, M. G. (eds) *Surface Water Hydrology*. The Geology of North America, **50**, 255–280.

MELOSH, H. J. 1979. Acoustic fluidization—A new geologic process. *Journal of Geophysical Research*, **84**, 7513–7520.

MELOSH, H. J. 1987. The mechanics of large rock avalanches. *Review of Engineering Geology*, **7**, 41–49.

MERTES, L. A. K. & WARRICK, J. A. 2001. Measuring flood output from 110 coastal watersheds in California with field measurements and SeaWiFs. *Geology*, **29**, 659–662.

MILLIMAN, J. D. & SYVITSKI, J. P. M. 1992. Geomorphic/tectonic control of sediment discharge to the ocean: The importance of small mountainous rivers. *Journal of Geology*, **100**, 525–544.

MILLIMAN, J. D., SNOW, J., JAGER, J. & NITTROUER, C. A. 1996. Catastrophic discharge of fluvial sediment to the ocean: evidence of Jokulhlaups events in the Alsek Sea Valley, southeast Alaska (USA). *In*: *Erosion and Sediment Yield: Global and Regional Perspectives*. IAHS Publication, Proceedings of the Exeter Symposium, July 1996, **236**, 367–379.

MITCHELL, N. A., MASSON, D. G., WATTS, A. B., GEE, M. J. R. & URGELESS, R. 2002. The morphology of submarine flanks of volcanic ocean islands. A comparative study of the Canary and Hawaiian hotspot islands. *Journal of Volcanology and Geothermal Research*, **115**, 83–107.

MOHRIG, D., WHIPPLE, K. X., HONDZO, M., ELLIS, C. & PARKER, G. 1998. Hydroplaning of subaqueous debris flows. *Geological Society of America Bulletin*, **110**, 387–394.

MONTGOMERY, D. R., BALCO, G. & WILLET, S. D. 2001. Climate, tectonics, and the morphology of the Andes. *Geology*, **29**, 579–582.

MOORE, J. G. & MOORE, G. W. 1984. Deposit from a giant wave on the island of Lanai, Hawaii. *Science*, **226**, 1312–1315.

MOORE, J. G., CLAGUE, D. A., HOLCOMB, R. T., LIPMAN, P. V., NORMARK, W. R. & TORRESAN, M. E. 1989. Prodigious submarine landslides on the Hawaiian Ridge. *Journal of Geophysical Research*, **94**, 17465–17484.

MOORE, J. G., NORMARK, W. R. & HOLCOMB, R. T. 1994. Giant Hawaiian landslides. *Annual Review of Earth and Planetary Sciences*, **22**, 119–124.

MOORE, L. J., BENUMOF, B. & GRIGGS, G. B. 1999. Coastal erosion hazards in Santa Cruz and San Diego Counties, California. *Journal of Coastal Research*, **28**, 121–139.

MULDER, T., MIGEON, S., SAVOYE, B. & JOUANNEAU, J. M. 2001. Twentieth century floods recorded in the deep Mediterranean sediments. *Geology*, **29**, 1011–1014.

MULDER, T., SYVITSKY, P. M., MIGEON, S., FAUGÉRES, J. C. & SAVOYE, B. 2003. Marine hyperpycnal flows: initiation, behaviour and related deposits. A review. *Marine and Petroleum Geology*, **20**, 861–882.

MÜLLER, L. 1964. The rock slide in the Vaiont valley. *Felsmechanik und Ingenieurgeologie*, **2**, 148–212.

MURRAY, J. B. & VOIGHT, B. 1996. Slope stability and eruption prediction on the eastern flank of Mount Etna. *In*: MCGUIRE, W. J., JONES, A. P. & NEUBERG, J. (eds) *Volcano Instability on the Earth and other Planets*. Geological Society, London, Special Publications, **110**, 111–114.

MUTTI, E., DAVOLI, G., TINTERRI, R. & ZAVALA, C. 1996. The importance of ancient fluvio-deltaic systems dominated by catastrophic flooding in tectonically active basins. *Memorie della Società Geologica Italiana*, **48**, 233–291.

MUTTI, E., TINTERRI, R., BENEVELLI, G., DI BIASE, D. & CAVANNA, G. 2003. Deltaic, mixed turbidite sedimentation of ancient foreland basins. *Marine and Petroleum Geology*, **20**, 733–755.

NAVA-SANCHEZ, E. H., CRUZ-OROZOCO, R. & GORSLINE, D. S. 1995. Morphology and sedimentology of two contemporary fan deltas on the southeastern Baja California peninsula, Mexico. *Sedimentary Geology*, **98**, 45–62.

NAVA-SANCHEZ, E. H., GORSLINE, D. S., CRUZ-OROZOCO, R. & GODINEZ-ORTA, L. 1999. The El-Coyote fan delta: a wave dominated example from the Gulf of California, Mexico. *Quaternary International*, **56**, 129–140.

NEMEC, W. 1990. Deltas—remarks on terminology and classification. *In*: COLELLA, A. & PRIOR, D. B. (eds) *Coarse Grained Deltas*. International Association of Sedimentologists, Special Publications, **10**, 3–12.

NOUH, M. 1990. Flood hydrograph estimation from arid catchment morphology. *Hydrological Processes*, **4**, 103–120.

NUNN, P. D. & PASTORIZIO, M. A. 2007. Geological histories and geohazard potential of Pacific Islands illuminated by myths. *In*: PICCARDI, L. & MASSE, W. B. (eds) *Myths and Geology*. Geological Society, London, Special Publications, **273**, 143–163.

ORTON, G. J. & READING, H. G. 1993. Variability of deltas in terms of sediment supply, with particular emphasis on grain size. *Sedimentology*, **40**, 475–512.

PARESCHI, M. T., BOSCHI, E. & FAVALLI, M. 2006. Lost tsunami. *Geophysical Research Letters*, **33**, doi: 10.1029/2006GL027790.

PENTA, F., LUPINO, R., CAMOZZA, F. & ESU, F. 1954. Effetti dell'alluvione del 26 ottobre 1954 nel Salernitano. *Rivista Italiana di Geotecnica*, **6**, 245–257.

PEREZ, F. L. 2001. Matrix granulometry of catastrophic debris flows (December 1999) in central coastal Venezuela. *Catena*, **45**, 163–183.

PORFIDO, S., ESPOSITO, E., ALAIA, F., MOLISSO, F. & SACCHI, M. 2009. The use of documentary sources for reconstructing flood chronologies on the Amalfi rocky coast (southern Italy). *In*: VIOLANTE, C. (ed.) *Geohazard in Rocky Coastal Areas*. Geological Society, London, Special Publications, **322**, 173–187.

PRIOR, D. B. & BORNHOLD, B. D. 1988. Submarine morphology and processes of fjord fan deltas and related high gradient sytems: modern examples from British Columbia. *In*: NEMEC, W. & STEEL, R. J. (eds) *Fan Deltas: Sedimentology and Tectonic Settings*. Blackie, Glasgow, 125–143.

PRIOR, D. B. & BORNHOLD, B. D. 1989. Submarine sedimentation on a developing Holocene fan delta. *Sedimentology*, **36**, 1053–1076.

PRIOR, D. B. & BORNHOLD, B. D. 1990. The underwater development of Holocene fan deltas. *In*: COLELLA, A. & PRIOR, D. B. (eds) *Coarse Grained Deltas*. International Association of Sedimentologists, Special Publications, **10**, 75–89.

PRIOR, D. B., BORNHOLD, B. D., WISEMAN, W. J. & LOWE, D. R. 1981. Turbidity current activity in a British fjord. *Science*, **237**, 1330–1333.

RICHARDS, K. S. & LORRIMAN, N. R. 1987. Basal erosion and mass movement. *In*: ANDERSON, M. G. & RICHARDS, K. S. (eds) *Slope Stability: Geotechnical Engineering and Geomorphology*. Wiley, New York, 331–357.

ROTHWELL, R. G., PEARCE, T. J. & WEAVER, P. P. E. 1992. Late Quaternary evolution of the Madeira Abyssal Plain, Canary Basin, NE Atlantic. *Basin Research*, **4**, 103–131.

ROTHWELL, R. G., THOMSON, J. & KÄHLER, G. 1998. Low sea-level emplacement of a very large late Pleistocene 'megaturbidite' in the western Mediterranean Sea. *Nature*, **392**, 377–380.

SABATO, L. & TROPEANO, M. 2004. Fiumara: a kind of high hazard river. *Physics and Chemistry of the Earth*, **29**, 707–715.

SACCHI, M., MOLISSO, F. *ET AL*. 2009. Insights into flood-dominated fan deltas: very high-resolution seismic examples off the Amalfi cliffed coasts, eastern Tyrrhenian Sea. *In*: VIOLANTE, C. (ed.) *Geohazard in Rocky Coastal Areas*. Geological Society, London, Special Publications, **322**, 33–72.

SATAKE, K. & KATO, Y. 2001. The 1741 Oshima-Oshima eruption: extent and volume of submarine debris avalanche. *Geophysical Research Letters*, **28**, 427–430.

SCHUMM, S. A. 1977. *The Fluvial System*. Wiley, New York.

SCHUSTER, R. L. & HIGHLAND, L. M. 2001. *Socioeconomic and environmental impacts of landslides in the western hemisphere*. US Geological Survey, Open-File Report, **01-276**.

SHREVE, R. L. 1968. Leakage and fluidization in air-layer lubricated landslides. *Geological Society of America Bulletin*, **79**, 653–658.

SIEBERT, L. 1984. Large volcanic debris avalanches: characteristics of source areas, deposits, and associated eruptions. *Journal of Volcanology and Geothermal Research*, **22**, 163–197.

SIEBERT, L., GLICKEN, H. X. & UI, T. 1987. Volcanic hazards from Bezymianny- and Bandai-type eruptions. *Bulletin of Volcanology*, **49**, 435–459.

SMITH, M. S. & SHEPHERD, J. B. 1996. Tsunami waves generated by volcanic landslides: an assessment of the hazard associated with Kick'em Jenny. *In*: MCGUIRE, W. J., JONES, A. P. & NEUBERG, J. (eds) *Volcano Instability on the Earth and other Planets*. Geological Society, London, Special Publications, **110**, 115–123.

SOH, W., TANAKA, T. & TAIRA, A. 1995. Geomorphology and sedimentary processes of a modern slope-type fan delta (Fujikawa fan delta), Suruga Trough, Japan. *Sedimentary Geology*, **98**, 79–95.

SUNAMURA, T. 1983. Processes of sea cliff and platform erosion. *In*: KOMAR, P. D. (ed.) *CRC Handbook of Coastal Processes and Erosion*. CRC Press, Boca Raton, FL, 233–265.

SUNAMURA, T. 1992. *Geomorphology of Rocky Coasts*. Wiley, New York.

SUNAMURA, T. 1997. A relationship between wave-induced cliff erosion and erosive force of waves. *Journal of Geology*, **85**, 613–618.

SYVITSKI, J. P. M. & FARROW, G. E. 1989. Fjord sedimentation as an analogue for small hydrocarbon-bearing submarine fans. *In*: WHATELEY, M. K. G. & PICKERING, K. T. (eds) *Deltas: Sites and Traps for Fossil Fuels*. Geological Society, London, Special Pubications, **41**, 21–43.

TALLING, P. J., WYNN, R. B. *ET AL*. 2007. Onset of submarine debris flow deposition far from original giant landslide. *Nature*, **450**, 541–544.

TAPPIN, D. R., WATTS, P., MCMURTRY, G. M., LAFOY, Y. & MATSUMOTO, T. 2001. The Sissano, Papua New Guinea tsunami of July 1998—offshore evidence on the source. *Marine Geology*, **175**, 1–23.

TINTI, S., BORTOLUCCI, E. & ROMAGNOLI, C. 2000. Computer simulations of tsunamis due to flank collapse at Stromboli, Italy. *Journal of Volcanology and Geothermal Research*, **96**, 103–128.

TREHAILE, A. S. 1987. *The Geomorphology of Rock Coasts*. Oxford University Press, Oxford.

TREHAILE, A. S. 2002. Rock coasts, with particular emphasis on shore platforms. *Geomorphology*, **48**, 7–22.

TREHAILE, A. S. 2004. Modelling the accumulation and dynamics of beach onshore platforms. *Marine Geology*, **206**, 55–72.

UI, T., YAMAMOTO, H. & SUZUKI-KAMATA, K. 1986. Characterization of debris avalanche deposits in Japan. *Journal of Volcanology and Geothermal Research*, **29**, 231–243.

VIOLANTE, C., BUDILLON, F., ESPOSITO, E., PORFIDO, S. & VITTORI, E. 2004. Submerged hummocky topographfies and relations with landslides on the northwestern flank of Ischia island, southern Italy. *In*: PICARELLI, L. (ed.) *Proceedings of International Conference on Fast slope Movements, Prediction and Prevention for Risk Mitigation, Sorrento, Italy*. Patron Editore, Bologna, 309–315.

VIOLANTE, C., DE ALTERIIS, G. & ESPOSITO, E. 2006. Seafloor base maps, geological features and hazard assessment in marine coastal areas. Example from the Bay of Naples, Campania, Southern Italy. *In*: Institut Geologic de Calalunya. *Proceedings of the 5th European Congress on Geoscientific Cartography and Information Systems, Barcelona*. Institute of Geology Catalunia, **1**, 534–537.

VIOLANTE, C., BISCARINI, C., ESPOSITO, E., MOLISSO, F., PORFIDO, S. & SACCHI, M. 2009. The consequences of hydrologic events on steep coastal watersheds: the Costa d'Amalfi, eastern Tyrrhenian sea. *In*: LIEBSHER, H. J. *ET AL.* (eds) *The Role of Hydrology in Water Resource Management. Capri Italy, 2008*. IAHS Publication, **327**, 102–113.

VOIGHT, B. 1978. *Rockslides and Avalanches: 1. Natural Phenomena*. Elsevier, Amsterdam.

VOIGHT, B., JANDA, R. J., GLICKEN, H. & DOUGLAS, P. M. 1983. Nature and mechanics of the Mount St Helens rockslide–avalanche of 18 May 1980. *Géotechnique*, **33**, 243–273.

WARD, S. N. 2001. Landslide tsunami. *Journal of Geophysical Research*, **106**, 11201–11215.

WATTS, A. B. & MASSON, D. G. 2001. New sonar evidence for recent catastrophic collapses of the north flank of Tenerife, Canary Islands. *Bulletin of Volcanology*, **63**, 8–19.

WEAVER, P. P. E., MASSON, D. G. & KIDD, R. B. 1994. Slumps, slides, and turbidity currents—sea level change and sedimentation in the Canary Basin. *Geoscientist*, **4**, 14–16.

WEAVER, P. P. E., MASSON, D. G., GUNN, D. E., KIDD, R. B. & MADDISON, D. A. 1995. Sediment mass wasting in the Canary basin. *In*: PICKERING, K. T., HISCOTT, R. N., KENYON, N. H., RICCI-LUCCHI, F. & SMITH, R. D. A. (eds) *Atlas of Deep-Water Environments: Architectural Style in Turbidite Systems*. Chapman & Hall, London, 287–296.

WHEACROFT, R. A., SOMMERFIELD, C. K., DRAKE, D. E., BORGELD, J. C. & NITTROUER, C. A. 1997. Rapid and widespread dispersal of flood sediment on the northern California margin. *Geology*, **25**, 163–166.

WOOLHISER, D. A. & GOODRICH, D. C. 1988. Effect of storm rainfall intensity patterns on surface runoff. *Journal of Hydrology*, **102**, 335–354.

Insights into flood-dominated fan-deltas: very high-resolution seismic examples off the Amalfi cliffed coasts, eastern Tyrrhenian Sea

M. SACCHI[1]*, F. MOLISSO[1], C. VIOLANTE[1], E. ESPOSITO[1], D. INSINGA[1], C. LUBRITTO[2], S. PORFIDO[1] & T. TÓTH[3,4]

[1]*Istituto per l'Ambiente Marino Costiero (IAMC)—CNR, Napoli, Calata P.ta di Massa, Porto di Napoli, 80133—Napoli, Italy*

[2]*Dipartimento di Scienze Ambientali, Seconda Università di Napoli, Via Vivaldi, 43—81100 Caserta, Italy*

[3]*Department of Geophysics, Eötvös Loránd University, Pázmány Péter sétány 1/C, H-1117 Budapest, Hungary*

[4]*Present address: Geomega Ltd., Mester u. 4, 1095 Budapest, Hungary*

**Corresponding author (e-mail: marco.sacchi@iamc.cnr.it)*

Abstract: A high-resolution (IKB-Seistec) seismic survey calibrated with gravity-core data, off the Amalfi coast, a rocky coastal area on the southern side of the Sorrento Peninsula (Italy), documents the internal stratigraphic architecture of a series of small fan-deltas that develop at the mouth of major bedrock streams. The fan-delta system mostly postdates the Plinian eruption of Vesuvius of AD 79 and displays various phases of development associated with periods of high sediment supply from the adjacent river basins. During these periods landscape-mantling loose pyroclastic deposits (mostly air-fall tephra from Vesuvius) were quickly eroded and delivered to the continental shelf by sheet wash and flash flood events. Depositional processes on the foresets were dominated by sediment gravity flows originating from hyperpycnal river flow and pyroclastic fall deposits. This in turn created favourable conditions for sea-floor instability, soft sediment failure, slumping and sliding, which characterize the deltaic stratigraphic architecture. The intermittently increased sediment yield during the various phases of the evolution of the fan-delta system was probably influenced also by the morphoclimatic regime. This may have resulted in varying rates of progradation of the delta foresets, tentatively correlated with the main climatic oscillations of the last 2000 years. The Amalfi fan-delta system represents a small-scale analogue for larger flood-dominated fan-deltas of the world and may be regarded as a useful example for a better understanding of inner-shelf, mixed siliciclastic–volcaniclastic fan-delta systems in the stratigraphic record.

In recent years, deltaic depositional settings at the mouth of small rivers of the Mediterranean and other temperate regions have received growing attention, because of the relevance of these facies associations in the understanding of the late Quaternary evolution of inner-shelf depositional systems and their interaction with fluvio-deltaic processes, sea-floor instability of delta slopes, coastal volcanism, active tectonics, and the climatic regime (Nava-Sanchez *et al.* 1999; Sacchi *et al.* 2005; Trincardi & Syvitski 2005; Lobo *et al.* 2006; McConnico & Bassett 2007).

The renewed interest in fan-deltas has partly derived from the recognition that deltas are unstable sedimentary systems prone to severe physiographic changes (e.g. modification of the coastline or sea-floor instability) that may occur on human time scales. Recent research has been particularly focused on sea-floor failures that are common both on continental slopes and in some deltaic settings; they can represent a major threat not only to oil and other offshore installations but also to the marine environment and coastal facilities. Moreover, in recent years sedimentologists have returned to modern fan-deltas as a possible source of reliable criteria for the recognition of fan-delta deposits in the stratigraphic record (Nemec & Steel 1988; Colella & Prior 1990; Nemec 1990*a*; Wescott & Ethridge 1990).

There is a large variability of underwater fan-delta architecture, depending on the process combinations and the relative dominance of each. Differences in types and rates of sediment supply and the onshore morphology influence underwater delta growth. Sediment dispersal underwater is directly related to supply by rivers. In the case of

From: VIOLANTE, C. (ed.) *Geohazard in Rocky Coastal Areas*. The Geological Society, London, Special Publications, **322**, 33–71. DOI: 10.1144/SP322.2

bedrock rivers and streams of temperate regions that form fan-deltas along high-relief sea-cliffed coasts (Nava-Sanchez *et al.* 1999; Fernández-Salas *et al.* 2003; Hasiotis *et al.* 2006; Lobo *et al.* 2006; Violante 2009; Violante *et al.* 2009), the fluvial regime is basically controlled by episodic, but sometimes catastrophic discharges that cause flooding on the fans. Long-term development of fan-deltas obviously involves a wide range of possible processes but variations in sediment supply and in the morphoclimatic regime appear to be major controls (e.g. Colella & Prior 1990; Reading 1996; Einsele 2000).

It has long been recognized that the fronts of marine deltas are prone to failure of unconsolidated sediments, and in some cases are dominated by products of mass movements deriving from instability of the delta front (Coleman & Prior 1982; Lindsay *et al.* 1984; Coleman 1988). The main reasons for sediment-induced deformation include: (1) the relatively high sedimentation rate on the delta front, which causes undercompaction and high pore-fluid pressures or liquefaction, leading in turn to loss of shear strength within the deposits; (2) biodegradation of organic debris and associated free methane gas, which weakens the sediment stability; (3) shocking of accumulated sediment by storm and wave action; (4) sediment instability induced by earthquakes.

Among the factors that may have a significant impact on fan-delta construction, and eventually sea-floor instability at the delta front and slopes are, hence, the frequency of recurrence of exceptional river floods, mudflows and explosive eruptions (involving pyroclastic falls, surges and flows) of coastal volcanoes. These processes can all induce the supply of large volumes of loose or poorly consolidated sediment into the delta system and over vast areas of the continental shelf (Cinque *et al.* 1997; Major *et al.* 2000; Lirer *et al.* 2001; Sacchi *et al.* 2005; Sulpizio *et al.* 2006; Bisson *et al.* 2007; Cinque & Robustelli 2009).

This paper focuses on the sequence stratigraphic interpretation of high-resolution seismic and gravity-core data acquired on a series of small fan-deltas that have developed on the inner shelf of the Amalfi coast. The aim of the study is the detailed reconstruction of stratal architecture of the fan-deltas and the interpretation of seismic facies in terms of depositional processes and environmental setting. The application of ultra high-resolution seismic data to inner-shelf and small-scale fan-deltas in shallow water provides unprecedented detailed insights into shallow-water depositional systems, documenting sea-floor morphology and areas of active erosion or deposition, allowing for bed-to-bed correlation in the gravity-core calibration of the seismic record.

Geological setting

The eastern Tyrrhenian margin is characterized by a number of peripheral basins that evolved during the latest Neogene–Quaternary across the hinge zone between the southern Apennines fold and thrust belt and the Tyrrhenian back-arc extensional area (Fig. 1). These basins, which include the present-day Bay of Salerno, formed in response to large-scale orogen-parallel extension and associated transtensional tectonics that accompanied the anti-clockwise rotation of the Apennine belt and lithospheric stretching in the central Tyrrhenian basin (Malinverno & Ryan 1986; Oldow *et al.* 1993; Sacchi *et al.* 1994; Ferranti *et al.* 1996; Patacca & Scandone 2007). Quaternary orogen-parallel extension caused the formation of half-graben systems (e.g. Bay of Naples, Bay of Salerno) and intervening structural highs (e.g. Sorrento Peninsula) that are perpendicular to the main axis of the Apennine thrust belt (Bartole *et al.* 1984; Mariani & Prato 1988; Sacchi *et al.* 1994; Milia & Torrente 1999; Acocella & Funiciello 2006). The extensional processes caused in turn the onset of intense volcanism. Active volcanic centres are represented by Somma–Vesuvius, the island of Ischia and the district of the Campi Flegrei, with its numerous vents both onshore and offshore the Bay of Naples (Rosi & Sbrana 1987; Santacroce 1987; Vezzoli 1988; Milia *et al.* 1998; Santacroce & Sbrana 2003).

The Sorrento Peninsula

The study area is located on the southern slope of the Sorrento Peninsula (Amalfi coast). The peninsula is a major Quaternary morpho-structural unit of the western flank of the Southern Apennines and forms a narrow and elevated mountain range (up to 1444 m) that separates two major embayments of the eastern Tyrrhenian margin, namely the Bay of Naples and the Bay of Salerno. It is mostly formed by a pile of Mesozoic carbonate rocks, covered by Tertiary to Quaternary siliciclastic and pyroclastic units, and is deeply cut by a complex of bedrock rivers and channels characterized by relatively small catchment areas and pronounced disequilibrium of the stream profiles (Fig. 2). These rivers have flow that shows a distinct seasonality and a torrential behaviour (Esposito *et al.* 2004*a*, *b*; Budillon *et al.* 2005; Liquete *et al.* 2005; Violante *et al.* 2009); their source is very high relative to the base level, and erosion processes have proceeded rapidly and generated a rugged morphology (Reineck & Singh 1975; Einsele 2000).

Being a horst-like structure in a half-graben basin setting the Sorrento Peninsula displays a remarkable asymmetry in the morphology of its two flanks, the southern one (Amalfi coast) being

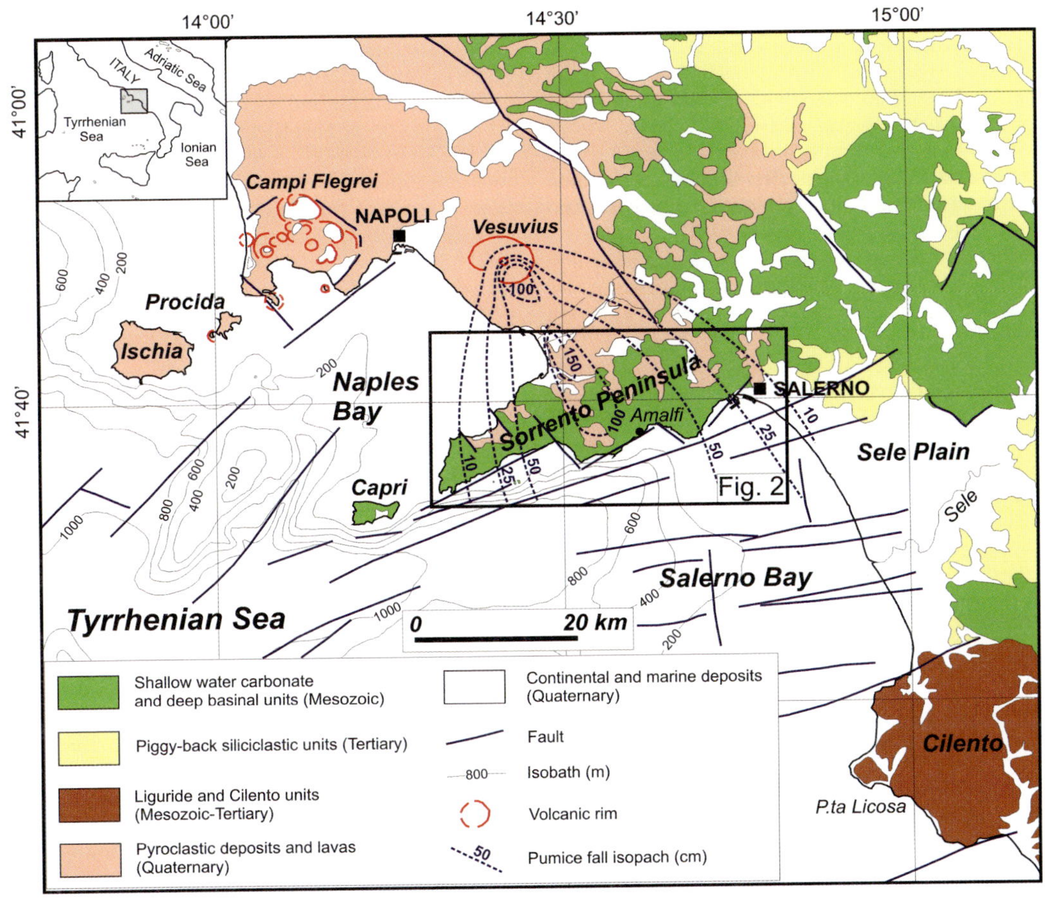

Fig. 1. Tectonic sketch-map of the Campania continental margin (eastern Tyrrhenian Sea) and location of the study area (Amalfi coast of the Sorrento Peninsula).

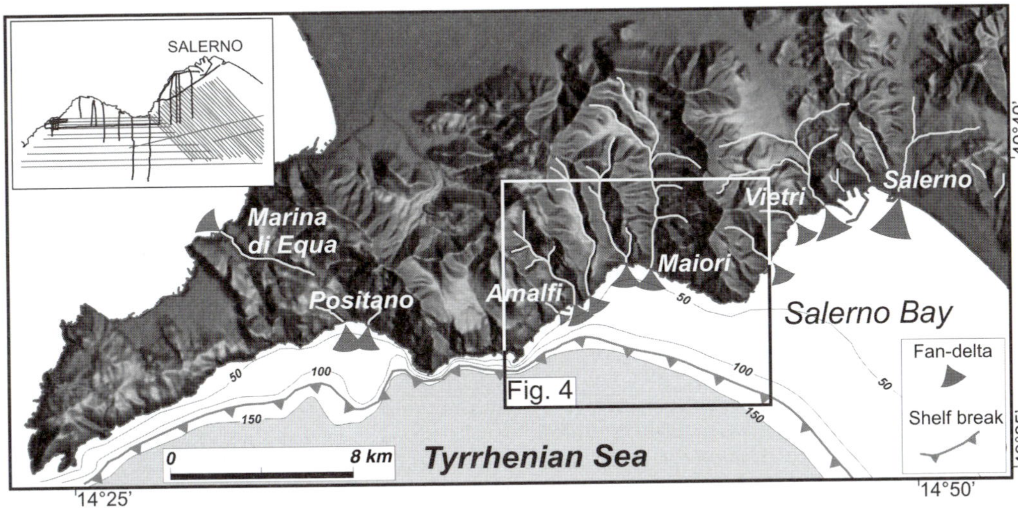

Fig. 2. Shaded relief map of the Sorrento Peninsula and location of the fan-delta systems fed by the main streams of the Amalfi coast. Box shows location of Figure 4. Inset shows the full seismic database used for this research.

steeper and narrower than the northern one (Sorrento coast). This asymmetry can also be observed offshore, where the narrow continental shelf on the Amalfi side contrasts with a wider shelf in the southern part of the Bay of Naples. As a consequence the Amalfi flank of the peninsula is characterized by tectonically uplifted rocky and steep backdrops, deeply incised gorges, and coastal cliffs (Brancaccio *et al.* 1991).

Coarse-grained coastal alluvial fans confined by narrow valleys at the mouth of the major streams are relatively common in this setting (Fig. 2). They are formed by deposition from flash floods, during periods of heavy rainfall. The delivery of sediments towards the coastal fans is favoured by the steep slopes and the loose source-area material of a wide size range (diameters vary from 2 m or more down to clay size) that includes bedrock river gravel, slope-weathering products, soil and unconsolidated volcaniclastic deposits.

Landscape-mantling pyroclastic deposits and the recent volcanic activity of Somma–Vesuvius

The Sorrento Peninsula, located about 20 km south of Somma–Vesuvius, has been repeatedly mantled during recent millennia by the pyroclastic products of the volcano, which has alternated intense phases of explosive activity accompanied by emission of fine ashes with periods of quiescence (Santacroce 1987). The Late Holocene activity of Somma–Vesuvius includes the Plinian event of the 'Avellino Pumice' and six inter-Plinian events (AP1–AP6; Andronico & Cioni 2002) during the nearly 1600 year period preceding another Plinian event ('Pompeii Pumice'), which destroyed the Roman cities of Pompeii, Stabiae and Herculaneum in AD 79 (Fig. 3).

The volume of the Pompeii fall products has been estimated between 4 km^3 (Sigurdsson *et al.* 1985) and 1–1.5 km^3 (Cioni *et al.* 1999) of Dense Rock Equivalent (DRE). Following the eruptive event, the Sorrento Peninsula was covered by up to 2 m of pyroclastic air-fall tephra (Lirer *et al.* 1973; Sigurdsson *et al.* 1985; Carey & Sigurdsson 1987; Cioni *et al.* 1992). This exceptionally high thickness of pyroclastic fall deposits was due to both the large volume of the erupted products and the main direction of dispersal of the air-fall tephra towards the SW, possibly as a consequence of the direction of dominant winds during the eruptive event (see Fig. 1). Deposits related to the so-called 'Pompeii' eruption have been recognized offshore (Carbone *et al.* 1984; Buccheri *et al.* 2002; Insinga 2003; Munno & Petrosino 2004). More recently a detailed description of the proximal deposits in the Bays of Naples and Salerno was reported by Sacchi *et al.* (2005), Insinga *et al.* (2008) and Milia *et al.* (2008).

The three centuries of activity following the AD 79 eruption were characterized by the emission of ash ('S. Maria Cycle', Andronico *et al.* 1995) until two sub-Plinian events that occurred in AD 472 (the so-called 'Pollena' eruption of Rosi & Santacroce 1983; Santacroce 1987; Rolandi *et al.* 2004; Sulpizio *et al.* 2005) and in AD 512. Intense activity, including four major eruptions the ('Medieval' events of Rolandi *et al.* 1998), characterizes the history of Somma–Vesuvius up to the 12th century. Products related to this period cover a wide area in the eastern and southeastern sector of the volcano both on land (Rolandi *et al.* 1998; Santacroce & Sbrana 2003) and offshore in the Bay of Naples (Insinga *et al.* 2008).

Following the AD 1631 sub-Plinian eruption (Rolandi *et al.* 1993; Rosi *et al.* 1993), the volcano experienced a 300 year period of semi-persistent, mild activity with a total volume of erupted material of the order of 10^7 m^3 (Arrighi *et al.* 2001). During this period only a few eruptions reached sub-Plinian intensity (e.g. in 1822 and 1906) and only five events were accompanied by a violent phreatomagmatic phase (i.e. in 1779, 1794, 1822, 1906 and 1944; Arrighi *et al.* 2001).

The explosive eruptions of Somma–Vesuvius in recent millennia have repeatedly accumulated loose pyroclastic material over large areas of the Campania region, thus creating favourable conditions for volcaniclastic debris to generate mass flows on slopes (Sulpizio *et al.* 2006; Bisson *et al.* 2007).

The Amalfi continental shelf

The continental shelf in the Bay of Salerno extends for about 70 km from the island of Capri in the north to Punta Licosa in the south. It is characterized by a varied morphology, controlled by both the tectonic activity and uplift of the Sorrento Peninsula and the associated high sediment supply during the Late Quaternary. The width of the continental shelf also varies significantly, from a maximum of 20–25 km in the southern Bay of Salerno to 0.5–4 km in the northern part of the Bay, off the Amalfi coast, and the shelf break is typically located at a water depth of *c.* 120 m (Fig. 4). Here the continental shelf is dominated by a thick prograding succession of Late Pleistocene age (Conforti 2003; Sacchi *et al.* 2004) that is truncated by a major erosional unconformity (ES). The latter is due to subaerial exposure during the sea-level fall and lowstand of the last glacial age and/or marine erosion associated with a transgressive landward shift of the fair-weather wave base during the rapid sea-level rise that accompanied the Holocene

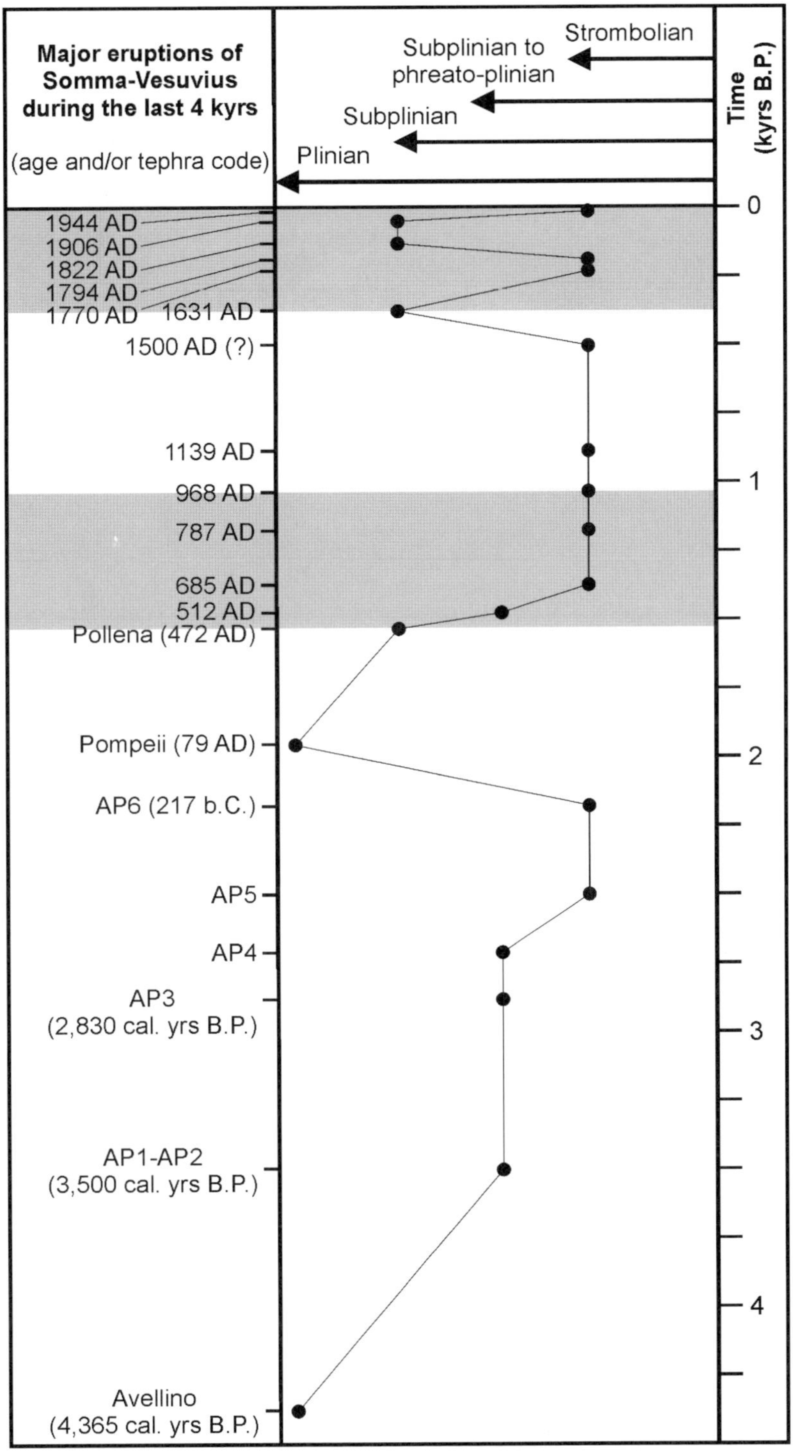

Fig. 3. List of the major eruptions of Somma–Vesuvius during the last 4000 years, with indication of the approximate explosive potential of events (after Santacroce *et al.* 2008). Time slices in grey are periods of semi-persistent volcanic activity.

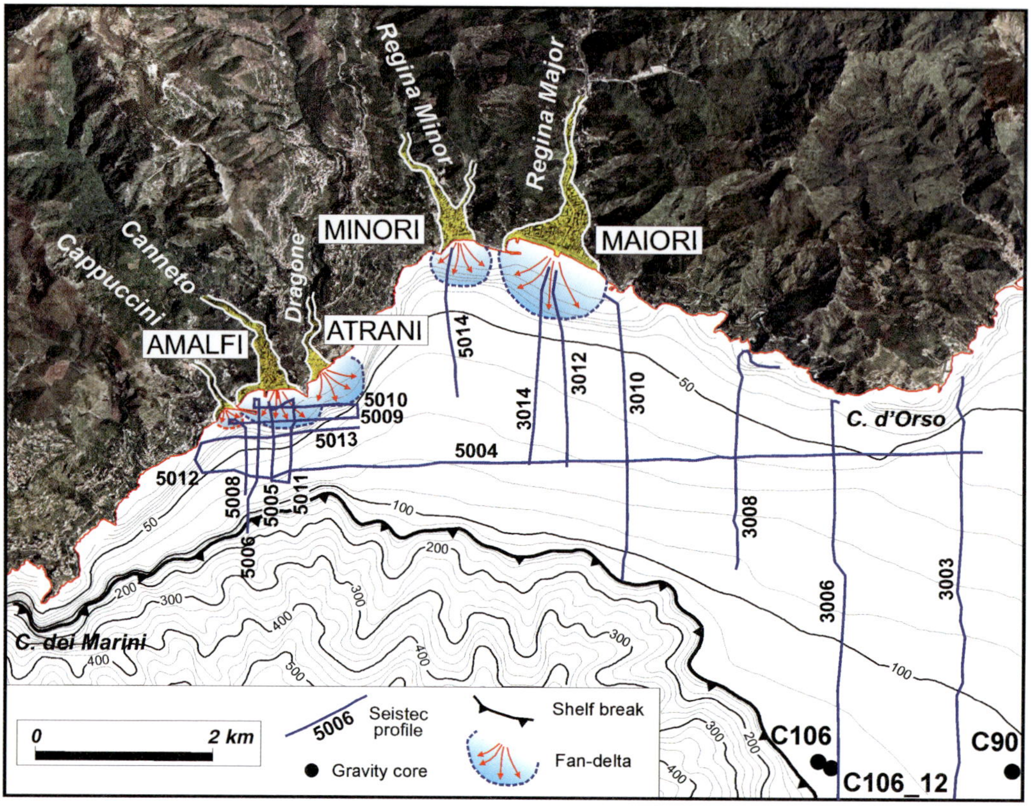

Fig. 4. Sketch-map of the alluvial fan-delta system of the Amalfi cliffed coast between Conca dei Marini and Capo d'Orso, with location of Seistec profiles and gravity cores illustrated in this study. (See Fig. 2 for location.)

deglaciation. The unconformity marks a stratigraphic gap ranging from a few thousand years to more than 70 ka. The maximum thickness of the post-glacial succession of the continental shelf in the northern Bay of Salerno may exceed 40 m.

Material and methods

This study is based on a very high-resolution (IKB-Seistec), single-channel reflection seismic survey carried out on the Amalfi inner shelf, between Salerno and Amalfi, in July 2004 (Fig. 4). The overall control for the stratigraphy and depositional setting of the late Quaternary depositional sequence comes from an extended dataset that includes multibeam bathymetry, side-scan sonar imagery, single-channel sparker and chirp-sonar profiles, sediment cores, and integrated biostratigraphic and chronological data acquired by the IAMC-CNR between 1997 and 2004. The sequence stratigraphic nomenclature adopted for seismic interpretation is after Hunt & Tucker (1992).

Seismic data acquisition

The seismic data presented in this paper include a grid of more than 100 km of very high-resolution single channel (uniboom) reflection profiles acquired using the IKB-Seistec profiler. This profiler has been designed specifically for collecting very high-resolution data in shallow-water environments but can also be used in water depths >200 m (Simpkin & Davis 1993; Mosher & Simpkin 1999). The Seistec system comprises a 2.5 m long catamaran supporting both the boomer source and receiver. The source is an IKB model B3 wide-band electrodynamic 'boomer' producing a single positive peak pressure impulse with a primary pulse width of 120 μs. The receiving system is a line-in-cone receiver located adjacent to the boomer plate (70 cm). The source emits useful frequencies in the range 1–20 kHz and, because of this wide frequency band, allows resolution of reflectors spaced 20 cm apart. Penetration is up to 100 m in soft sediments, and 200 m in deep-water soft sediments. During the survey, on board a small boat at a speed of about 3 knots, a STEP power supply was used with a power of 150 J and a shot rate of 4–6 shots per second. Seismic data were recorded with a PreSeis digital acquisition system with 16 bits per sample, a sampling frequency of 100 kHz and a recording length of 60–100 ms. The position during navigation was determined by a differential global positioning system (GPS) directly mounted at the common depth point (CDP) of the IKB-Seistec profiler.

The exceptional time resolution and the fixed source–receiver geometry of the Seistec profiling system, together with its high sub-bottom penetration, allow for a quantitative analysis of the different seismic-signature shapes and geometries and signal amplitudes.

Gravity coring and laboratory analysis

Ground truthing of seismic records was provided by the detailed analysis of three gravity-cores (C90, C106, C106_12) collected on the outer shelf off the Amalfi coast (Fig. 4), as well as by the general description of a number of supplementary cores recovered from the mid–outer shelf of the northern Bay of Salerno. Sedimentological analysis included the recognition of major lithofacies associations, sedimentary structures, and grain-size analysis of selected stratigraphic intervals by laser diffractometry. The grade of bioturbation was estimated according to Droser & Bottjer (1991). Textural analysis of cores C90 and C106_12 was conducted at an average sampling rate of 3 cm, throughout the entire core length. Denser sampling was conducted for detailed study of transitional zones between different lithologies and/or sedimentary structures. Grain-size statistical parameters have been calculated following the classic graphical equations developed by Folk & Ward (1957).

Quantitative micropalaeontological analysis was conducted on 3 cm thick samples spaced every 10 cm. A minimum count of 100 specimens of benthic foraminifers for each sample was observed following the methods described by Fatela & Taborda (2002). The generic identification of foraminifers was made following Loeblich & Tappan (1988) and Sgarrella & Moncharmont-Zei (1993).

The carbonate content was measured on 10 g dry sediment samples, spaced every 10 cm. Mollusc fragments that could affect $CaCO_3$ analyses were handpicked and removed. Therefore, variations in carbonate content of the samples through the core record the occurrence of calcareous microfossils, in particular foraminifers and ostracodes, and nannofossils.

^{14}C accelerating mass spectrometry (AMS) measurements were performed with a system based on a tandem accelerator (9SDH-2, National Electrostatics Corporation) with a maximum terminal voltage of 3 MV. The $\delta^{13}C$ of each sample was also measured using an elemental analyser (ThermoFinnigan EA 1112) coupled with an IR mass spectrometer (ThermoFinnigan Deltaplus). Samples were pre-treated in accordance with the procedures outlined by Hoefs (1987), Vogel *et al.* (1984) and Passariello *et al.* (2007). Radiocarbon ages were calibrated by using the CALPAL-2007 software (Weninger & Jöris 2008; Weninger *et al.* 2008).

Data and results

Gravity-core stratigraphy

Gravity-cores C90, C106, C106_12 were collected on the outer shelf of the Bay of Salerno, between Capo d'Orso and Amalfi, at water depths of 103–116 m and provide a calibration of the entire post-glacial succession. The length of recovered core sections ranges from 485 cm in cores C90 and C106 to 687 cm in core C106_12 (Figs 5 and 6).

Stratigraphic correlation of the gravity-cores is clear from the sedimentological analysis, thus allowing the construction of a composite stratigraphic section that represented the basis for the geological calibration adopted in the seismic stratigraphic interpretation. The cored succession consists of a *c.* 5 m thick transgressive sequence that overlies the major erosional surface (ES) associated with the sea-level fall and lowstand of the last glacial maximum. Below this unconformity, core C106_12 sampled a sandy silt succession older than 50 ka, within Upper Pleistocene deposits (Fig. 6).

The entire cored sequence is punctuated by at least 10 tephras, which we labelled from top to bottom as tS1, tS1-α, tS1-β, tS1-γ, tS2, tS2-α, tS3, tS3-α and tS4 to tS6. Of these, tS1–tS4 are interbedded within the post-glacial succession, whereas tS5 and tS6 are interbedded within the Upper Pleistocene deposits underlying unconformity ES. The tephra layers typically are several centimetres to a few decimetres thick. Most of them have a sharp base, normal or inverse grading, poor sorting and typically a gradual transition to the overlying deposits. An exceptionally thick tephra is represented by the 80–100 cm thick pyroclastic bed deposited during the AD 79 Plinian eruption of Vesuvius (tephra tS2; Sacchi *et al.* 2005; Insinga *et al.* 2008).

On the basis of sedimentological analysis and quantitative changes in benthic foraminiferal assemblages of the core samples, three main lithofacies associations can be recognized (Figs 7 and 8). A composite stratigraphic section, from bottom to top, consists of: facies A, shelf mud with volcaniclasts

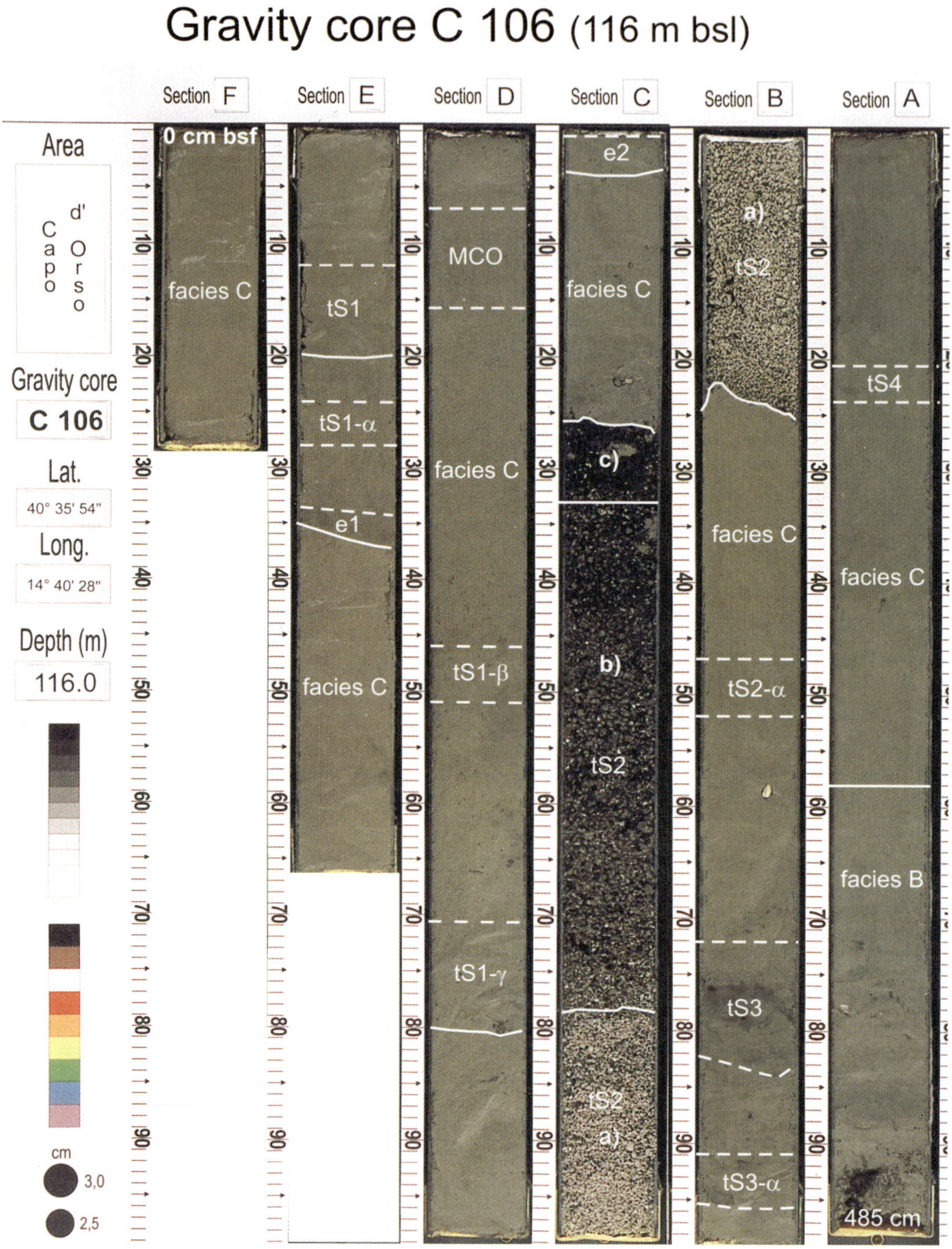

Fig. 5. Photograph of C106 core splits. Labels F to A denote core sections from top (0 cm bsf) to bottom (485 cm bsf). Total length of the core is 485 cm. MCO, Medieval Climatic Optimum; e1, suspension plume deposit; facies B, shoreface sand and pebble; facies C, bioturbated prodelta mud. Tephra layers (tSn) are also indicated. Key to sub-units of tS2 ('Pompeii', AD 79) fallout deposits: (a) white pumice; (b) grey pumice; (c) fine-grained lapilli. (For this and following figures, see Tables 1–3 for further information on chronology and nature of event beds.)

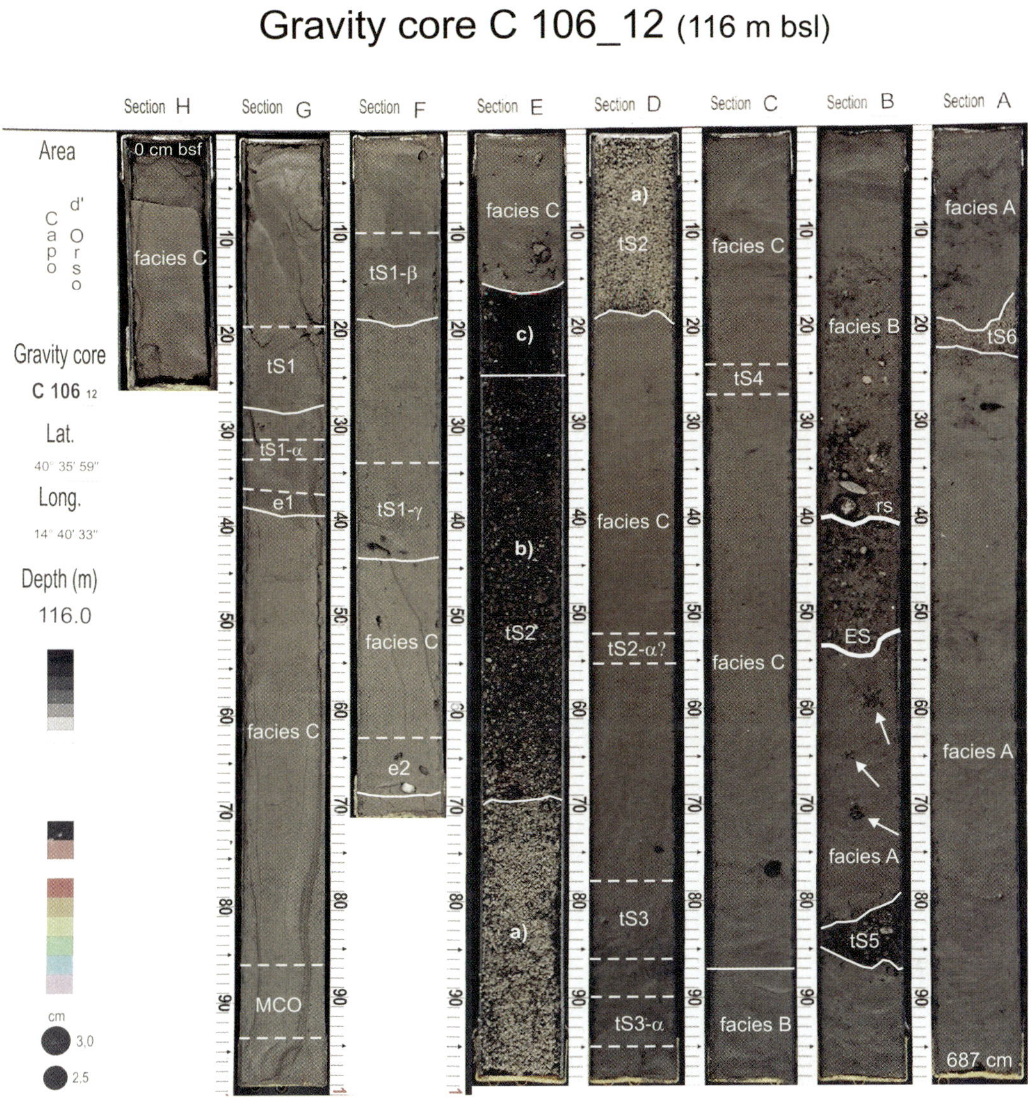

Fig. 6. Photograph of C106_12 core splits. Labels H to A denote core sections from top (0 cm bsf) to bottom (687 cm bsf). Total length of the core is 687 cm. Abbreviations as in Figure 5 and: ES, base of post-glacial succession; rs, ravinement surface; facies A, shelf mud with volcaniclasts and bioclasts. Arrows indicate *Thalassinoides* burrows (*Glossifungites* ichnofacies).

and bioclasts; facies B, shoreface sand and pebbles; facies C, bioturbated prodelta mud.

Facies A, shelf mud with volcaniclasts and bioclasts

This facies is represented only in core C106_12 and has a thickness of *c.* 110 cm. From base to top, it consists of poorly sorted olive grey (5Y4/2) clayey sandy silt and dark grey (5Y4/1) sandy clayey silt interbedded with very thin volcaniclastic lenses. At 612 cm and 580 cm below sea floor (bsf) two pyroclastic layers occur: a 4 cm thick tephra (tS6) and a 9 cm thick tephra (tS5). The sandy fraction is represented by volcaniclasts and bioclasts and displays a gradual increase, especially in the volcaniclastic component, towards the top (Figs 6 and 7).

The benthic foraminiferal assemblage is mainly dominated by *Uvigerina peregrina*, *Cassidulina laevigata carinata*, *Hyalinea baltica*, *Melonis*

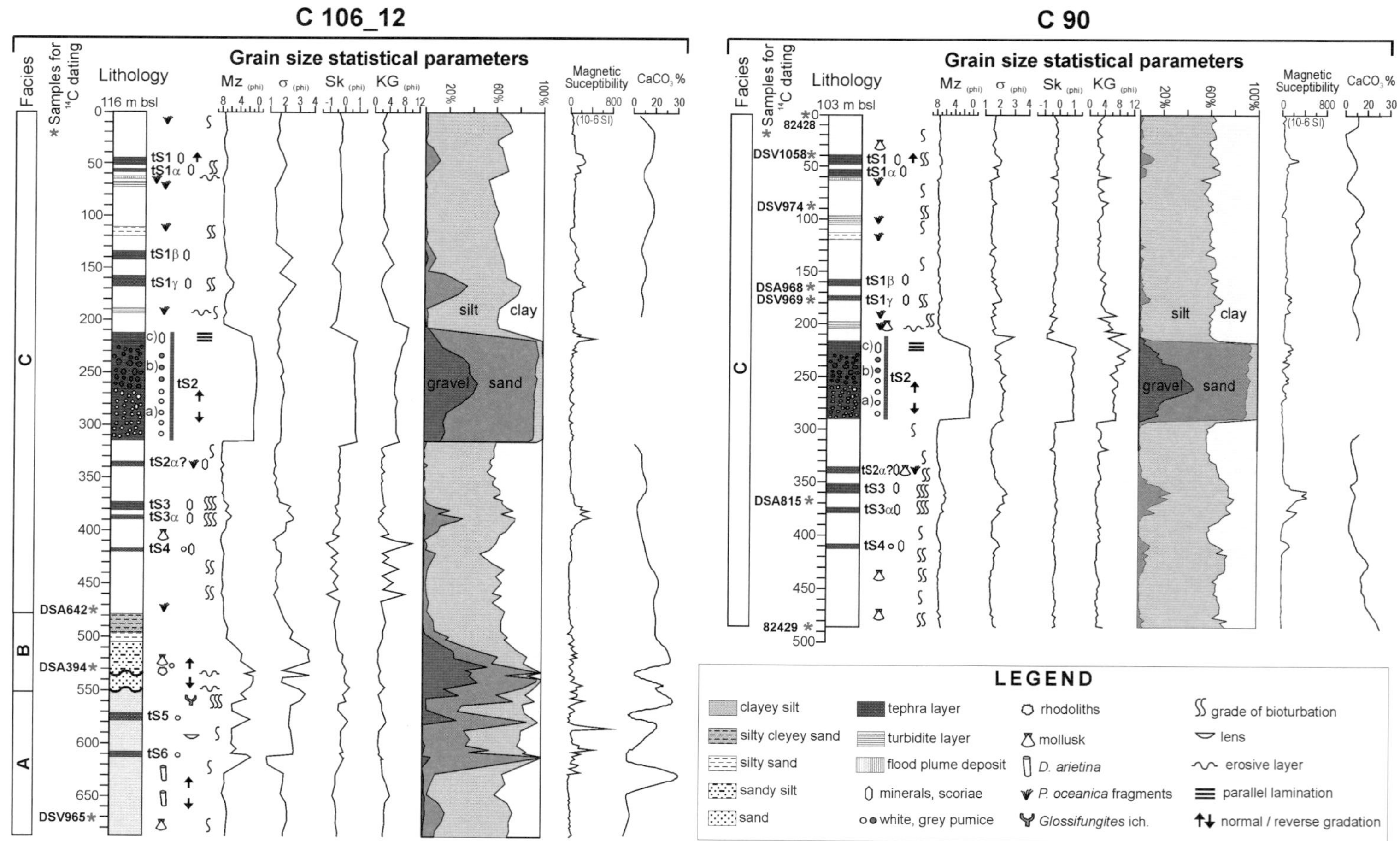

Fig. 7. Lithology, textures, sedimentary structures, magnetic susceptibility, calcium carbonate content and facies associations of gravity cores C90 and C106_12. Grain-size statistical parameters (Folk & Ward 1957): Mz, mean; σ, sorting; Sk, skewness; KG, kurtosis. Facies and tephra units as in Figure 5.

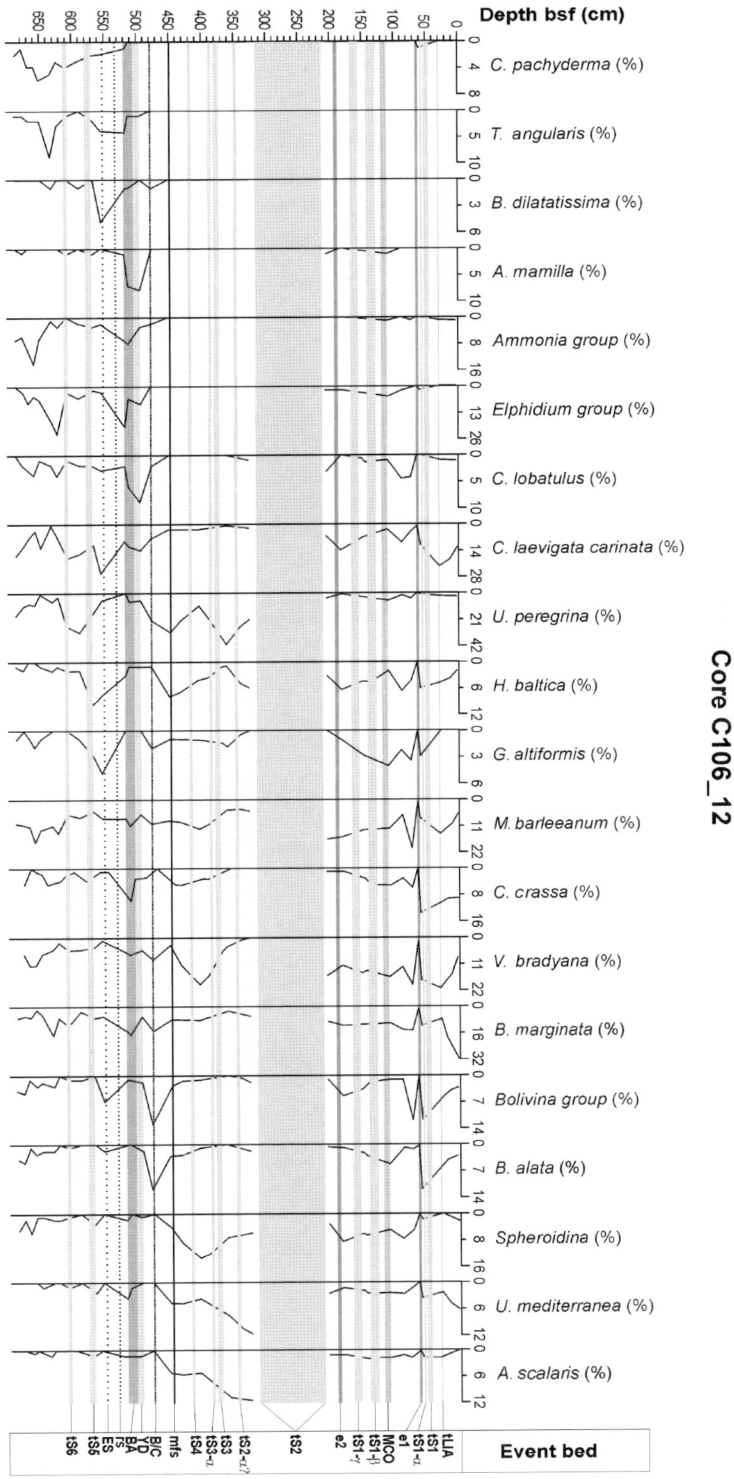

Fig. 8. Relative frequency (%) of selected benthic foraminifers and event beds of core C106_12.

barleeanum and *B. marginata*. These forms tend to increase in number towards the top of the unit, where they occur associated with rare *Valvulineria bradyana*, and the *Elphidium* and *Ammonia* groups. This association is typical of the lower circalittoral zone (e.g. Sgarrella & Barra 1984; Sgarrella & Moncharmont-Zei 1993). Specimens of broken pteropods are found (*Limacina retroversa*, *L. inflata* and *Clio pyramidata*). Fragments of polychaetes (*Ditrupa arietina*), bivalves, gastropods, bryozoans and echinoderms are present at 670 and 620 cm.

The occurrence in the foraminiferal assemblages of species typical of cold and productive waters, such as *C. laevigata carinata*, *Trifarina angulosa* and *H. baltica* (Murray 1991; Asioli 1996), of the *Elphidium* and *Ammonia* groups, as well as of *M. barleeanum* and *B. marginata*, coupled with the presence of *D. arietina*, suggests cold waters and stagnation episodes, possibly related to freshwater runoff and relatively high concentration of fine-grained sediments and organic matter (Jorissen 1988; Grémare *et al.* 1998). The occurrence of the epipelagic boreal guest *Limacina retroversa*, from the northern cold-water regions, at present absent in the Mediterranean, suggests a cold period of the glacial Pleistocene (Buccheri 1984; Buccheri *et al.* 2002; Fig. 8).

The uppermost 20 cm of this unit show sparsely branching burrows of ichnogenus *Thalassinoides*. The burrows have well-defined circular boundaries, with diameter ranging from 1 to 2 cm, and are passively infilled with lithofacies B. This ichnofabric corresponds to the *Glossifungites* ichnofacies (Frey & Seilacher 1980; Pemberton & Frey 1985; MacEachern *et al.* 1992; Gingras *et al.* 2001; see Fig. 6). The unit is bounded at the top by an erosional surface that correlates with the unconformity ES of seismic profiles, and is characterized by the occurrence of shell debris. *Glossifungites* ichnofabric is commonly taken to indicate a firmground (see Gingras *et al.* 2001, and references therein). In this context it probably represents colonization of the Pleistocene eroded substrate during minor breaks in sedimentation following storm events, below storm wave base in the foreshore–offshore transition zone, during the early transgressive systems tract (TST). The lithofacies assemblage of this succession, coupled with the seismic evidence of thick Upper Pleistocene forestepping parasequences beneath unconformity ES, suggests that this unit may be interpreted as a progradational deltaic sequence characterized by shelf mud deposit with volcaniclasts and bioclasts.

Facies B, shoreface sand and pebble

These deposits directly overlies facies A and consist of a 70 cm thick unit in core C106_12, whereas they represent the lowermost 40 cm of succession sampled at the C106 core site (Figs 5–7). The grain sizes of this unit range from silty sand to pebble, with very poor sorting. The lowermost 10 cm are represented by a medium sandy pebble sequence with inverse gradation that is bounded at the top by an erosive surface that correlates with the ravinement surface recognized on seismic profiles. Towards its top, the deposit is characterized by coarse-grained constituents represented by volcaniclasts, lithoclasts and bioclasts, often fragmented. The matrix is formed by the same constituents. Lithoclasts are rounded, often reddened, and include carbonate rock fragments (calcilutite) at places encrusted by red coralline algae (rhodoliths). Volcaniclasts are sub-rounded to sub-angular, locally reddened, and consist of light grey and green pumice, mineral grains (mostly sanidine, pyroxene and rare magnetite crystals) that are often rounded and abraded, scoriae and glass. The bioclasts include bivalve and gastropod shell fragments, which are sometimes abraded, bioeroded and encrusted, solitary corals, bryozoans and echinoid fragments. Plant debris (*Posidonia oceanica*) is also found (Fig. 7).

In the lower part of this unit, the benthic assemblage is represented by the *Elphidium* group, *B. marginata*, *C. laevigata carinata*, *M. barleeanum* and the *Ammonia* group, which are partly replaced towards the top by epiphytic species typical of detrital sediments (*Cibicides lobatulus*, *Asterigerinata mamilla*, *Planorbulina mediterranensis*) (Jorissen 1987; Sgarrella & Moncharmont-Zei 1993; Fig. 8). The occurrence of *C. laevigata carinata* and *M. barleeanum* in the benthic assemblage at the bottom of the unit may indicate an increased delivery of organic matter but minor oxygen depletion (Caralp 1988; Asioli *et al.* 2001). The benthic assemblage at the top of the interval indicates that water depth is somewhat shallower than for the underlying facies A, and the occurrence of *C. laevigata carinata* suggests cold and productive waters (e.g. Asioli 1996). The *Elphidium* group and *C. lobatulus* tolerate hyposaline conditions and may resist severe salinity fluctuations (Murray 1973, 1976; Boltovskoy 1976). These observations, along with the occurrence of rhodolith-bearing pebbles and shells, indicate a infralittoral to circalittoral zone associated with relatively low-salinity conditions. On the basis of lithofacies assemblages and the seismic-stratigraphic architecture, this unit can be interpreted as a transgressive lag containing reworked material from the substrate and/or the early shoreface deposits ('healing phase deposits') (e.g. Posamentier & Allen 1993).

A negative peak of the $CaCO_3$ curve in the interval between 495 and 505 cm (C106_12), which suggests a relatively low water temperature, may be tentatively correlated with the Younger Dryas (Kallel *et al.* 1997). Analogously, the underlying interval between 505 and 525 cm (C106_12), marked by a positive peak of the $CaCO_3$, may be taken as corresponding to the Bölling–Allerød event (Fig. 7).

Facies C, bioturbated prodelta mud

Facies C is represented in all the study cores and has a thickness of 212 cm (C106_12) to 118 cm (C90 and C106). It consists of a mud-supported lithofacies characterized by a grey to olive grey homogeneous, bioturbated clayey silt with at least eight tephra (or cryptothepra) interbedded with a few thin layers of fine-grained turbidites (Figs 5–7).

The sandy fraction is rare and consists of fine to very fine volcaniclasts, sub-rounded to sub-angular grey pumice, mineral grains, scoriae and glass, and bioclasts (represented by fragments of gastropods, bivalves, bryozoans and echinoids) and very rare fragments of pteropods (*C. pyramidata*, *Creises* sp., *L. inflata*). Turbidite layers are a few centimetres thick and consist of volcaniclasts (rounded pumice, locally reddened scoriae, fragments of minerals), sub-rounded lithic fragments and reworked bioclasts (fragments of bivalves, gastropods, bryozoans, phanerogamous seagrass remains) (Fig. 7).

In the upper part of the succession, a few centimetres below tephra ts1-α, a 2–3 cm thick layer of grey homogeneous bioturbated clayey silt, characterized by an extremely reduced faunal content and rare *Posidonia oceanica* fragments, occurs (layer e1 in Figs 5–8). This layer can be correlated across all the study cores and may be interpreted as a suspension plume deposit (Nemec 1995; Parsons *et al.* 2001), probably associated with the exceptional flood events that affected the Amalfi coast in 1581 and 1588 (Figs 8–11; Porfido *et al.* 2009).

Facies C marks a significant change in the relative abundance of many benthic taxa, accompanied by the disappearance and/or abrupt decrease of

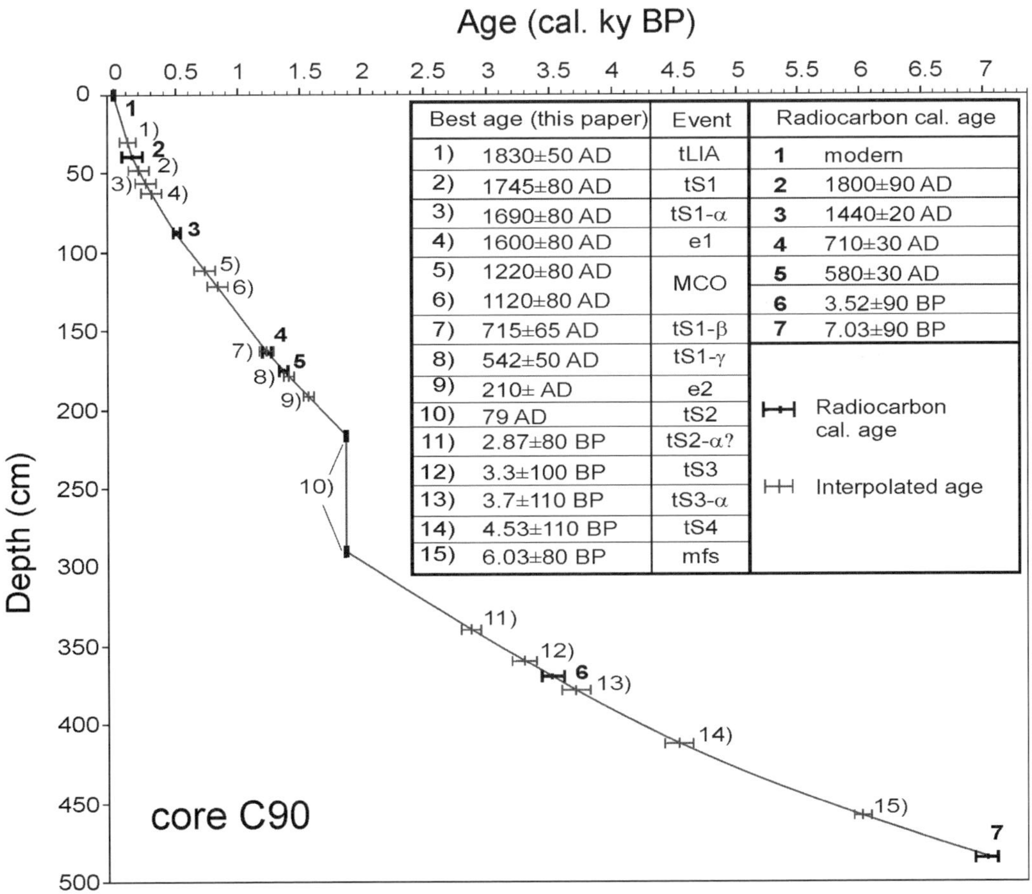

Fig. 9. Age–depth model of gravity core C90 based on calibrated AMS ^{14}C data and tied to the age of the tephra bed deposited by Vesuvius during the Plinian eruption of Pompeii (AD 79).

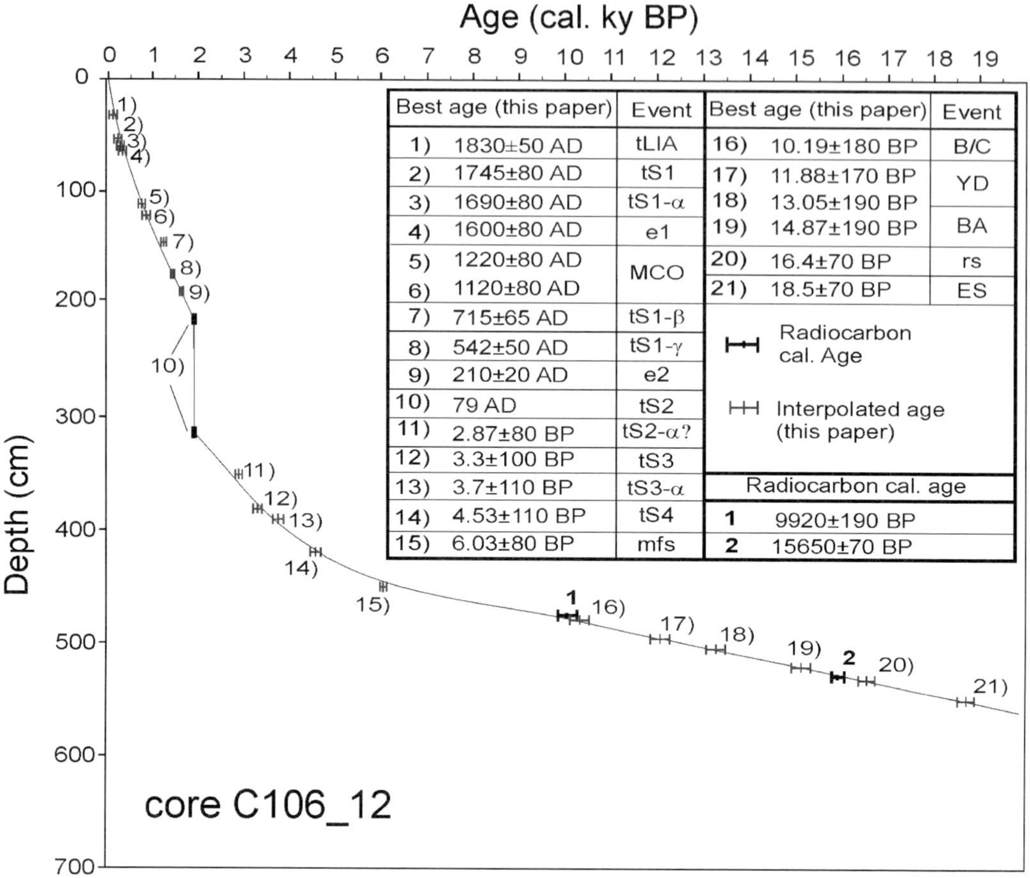

Fig. 10. Age–depth model of gravity core C106_12 based on calibrated AMS ^{14}C data and tied to the tephra bed deposited by Vesuvius during the Plinian eruption of Pompeii (AD 79).

some species (*Cibicidoides pachyderma*, *T. angulosa*, *Bolivina dilatatissima*, *Ammonia* group) and the parallel increase of others (e.g. *U. peregrina*, *U. mediterranea*, *Amphycorina scalaris*, *M. barleeanum*, *B. marginata*, *H. baltica*). Peaks of maximum abundance of some taxa, along with a relative maximum in the $CaCO_3$ curve, seem to concentrate around 450 cm, at a stratigraphic horizon we interpret as the maximum flooding surface (mfs) (Figs 7 and 8).

The AD 79 Pompeii tephra bed (tS2) corresponds to a marked environmental change in the faunistic assemblage (Fig. 8). Some taxa associated with the shallow infaunal microhabitat (*U. peregrina*, *U. mediterranea*, *H. baltica*, *Gyroidina* sp.) that had proliferated at the top of facies B, display, after tephra tS2, a sudden decrease in relative abundance, whereas opportunistic species (*V. bradyana*), intermediate infaunal (*M. barleeanum*) and deep infaunal species (*B. marginata*) that could probably better survive the increased sedimentation rates are recorded particularly after this event (Jorissen 1987; Linke & Lutze 1993; Jorissen & Wittling 1999; Schmiedl *et al.* 2000). Further upward the gradual decrease in abundance of taxa typical of cold and productive waters (*C. laevigata carinata*, *H. baltica*) (Murray 1991; Asioli 1996) probably indicates a relative maximum in the warming of bottom waters and culminates in a period that can be correlated with the Medieval Climatic Optimum (MCO). At 30 cm bsf the same species display a relative increase and are associated with opportunistic species (*V. bradyana*, *M. barleeanum*), thus suggesting relatively cooler waters and possibly increased freshwater input (Asioli 1996). According to our results this event has a probable age of AD 1830 $\pm/-$50 and may be tentatively correlated with the termination of the Little Ice Age (tLIA) (Figs 8–10).

In general, the benthic assemblage of this unit indicates a rapid increase in sedimentation rate relative to the underlying facies B and a predominantly

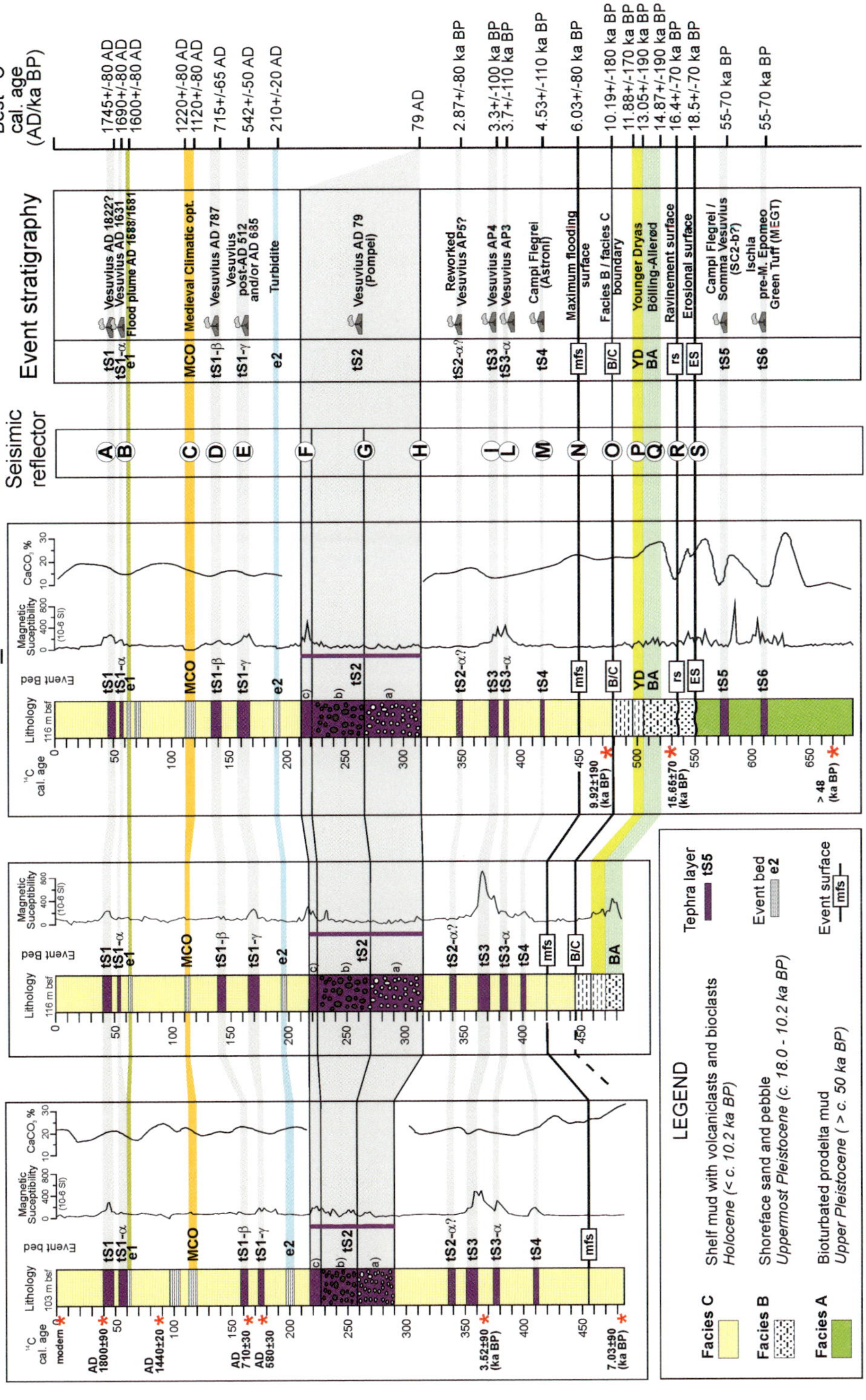

Fig. 11. Correlation of gravity cores used in this study and interpolated ^{14}C chronology of major tephra layers and other event beds. Key to sub-units of tS2 as in Figure 5.

muddy setting, characterized by relatively high organic matter and low oxygen concentration, typical of the modern 'mud belt' (Jorissen 1988). Correlation with seismic profiles suggests that this lithofacies corresponds to prodelta mud deposits associated with the modern fan-delta system of the Amalfi shelf.

^{14}C chronology and tephra layers

The AMS radiocarbon ages obtained from sampled marine material are reported in Table 1. The calibrated ages obtained are in a general agreement with the stratigraphic position of the samples and provide an absolute age constraint for the last *c.* 16 ka BP as well as a potential cross-check for chronological attribution of tephra layers. A modern age (i.e. AD 1950) resulted from the top of core C90 (sample 82428), whereas age determination for sample DSV965 was not possible as it was outside the scale of the adopted AMS system (i.e. >48 ka BP). The ^{14}C calibrated age models obtained for cores C90 and C106_12 are illustrated in Figures 9 and 10. According to the ^{14}C calibrated age–depth plots we obtain interpolated ages that provide a chronological reference for the recognized event beds and can be used for tentative correlation of the studied tephra layers with age-dated volcanic events onland. Correlation between the study cores is illustrated in Figure 11. The preferred ages and proposed interpretation of tephra layers and other event beds are reported in Tables 2 and 3.

Tephras tS1, tS2, tS3 and tS3-α have been recently described by Insinga *et al.* (2008). These pyroclastic layers have been interpreted, on the basis of stratigraphic and chemical analysis, as the products of the Vesuvian activity of the last 3 kyrs and have been correlated with the eruptions of AD 1822 or 1631 (tS1), the eruption of AD 79 (tS2), and the inter-Plinian events (tS3-α and tS3) referred to in the literature as AP3 and AP4, respectively (Andronico & Cioni 2002). Interpolated ages allow us to include tephra tS1 and tS1-α in the time span AD 1825–1665 and 1770–1610, respectively. During these time intervals only the AD 1631 and 1822 events reached sub-Plinian intensity and only the AD 1779 and 1794 eruptions were accompanied by a phreatomagmatic phase. This evidence suggests a possible correlation of tephra tS1 with the 1822 event (Insinga *et al.* 2008, and references therein) and tephra tS1-α with the 1631 eruption (Rolandi *et al.* 1993; Rosi *et al.* 1993). Sedimentological analysis, however, indicates that an eventual amalgamation of more tephra layers deposited by different eruptions occurring during the semi-persistent activity of the Vesuviust hat followed the AD 1631 event cannot be excluded.

Table 1. *Summary of the AMS ^{14}C results on sediment core samples from the Bay of Salerno*

Core	Lab. code	Sample number	Depth bsf (cm)	Material	^{14}C age (years BP)	δ^{13}C	$\Delta(\delta^{13}$C)	Cal. age (years BP)
C90	LLL	82428	0–3	Planktonic foraminifers	Modern	–	–	Modern
C90	CIRCE	DSV1058	39	Shell	484 ± 21	2.95	0.42	AD 1800 ± 90
C90	CIRCE	DSV974	87	Marine plant remains	826 ± 25	−24.49	1.19	AD 1440 ± 20
C90	CIRCE	DSA968	164	Shell	1680 ± 20	1.06	0.22	AD 710 ± 40
C90	CIRCE	DSV969	175	Bryozoan	1872 ± 25	6.82	0.63	AD 580 ± 30
C90	CIRCE	DSA815	370	Shell	3678 ± 84	3.24	1.16	3520 ± 90 years BP
C90	LLL	82429	480–485	Planktonic foraminifers	6515 ± 45	–	–	7030 ± 90 years BP
C106_12	CIRCE	DSA642	475	Shell	9239 ± 106	7.80	1.78	9920 ± 190 years BP
C106_12	CIRCE	DSA394	529	Shell	13 676 ± 45	5.33	0.49	15 650 ± 70 years BP
C106_12	CIRCE	DSV965	672	Shell	>48 ka	−10.25	0.48	>48 ka

LLL, Lawrence Livermore Laboratory, CA, USA; CIRCE, Centre for Isotopic Research for Cultural and Environmental Heritage, Caserta, Italy.

Table 2. *Interpolated ages of post-glacial event beds at C90 and C106_12 core sites, according to the* ^{14}*C cal. age model presented in this study (Figs 9 and 10)*

Bed code	Nature of event	Gravity core	Depth to base (cm)	Depth to top (cm)	Thickness (cm)	Correlative sesimic reflector	^{14}C cal. age
tS1	Tephra layer	C90	48	39	9	A (top + bottom)	AD 1745 ± 80
		C106_12	53	45	8		
tS1-α	Tephra layer	C90	60	53	7	B (top + bottom)	AD 1690 ± 80
		C106_12	59	57	2		
e1	Flood plume deposit AD 1588–1581	C90	63	60	3		AD 1600 ± 80
		C106_12	65	63	2		
MCO	Medieval Climatic Optimum	C106_12	120	113	7	C (top + bottom)	AD 1220 ± 80 to 1120 ± 80
		C106_12	120	111	9		
tS1-β	Tephra layer	C90	163	157	6	D (top + bottom)	AD 715 ± 65
		C106_12	143	134	9		
tS1-γ	Tephra layer	C90	178	172	6	E (top + bottom)	AD 542 ± 50
		C106_12	168	158	10		
e2	Turbidite layer	C90	204	197	7	–	AD 210 ± 20
		C106_12	194	187	7		
tS2	'Pompeii' pumice fall AD 79	C90	290	217	73	F (top) G (white–grey pumice boundary) H (bottom)	AD 79
		C106_12	314	213	101		
tS2-α?	Reworked tephra layer	C90	342	336	6	–	2.87 ± 80 ka BP
		C106_12	350	345	5		
tS3	Tephra layer	C90	361	351	10	I (top + bottom)	3.3 ± 100 ka BP
		C106_12	381	373	8		
tS3α	Tephra layer	C90	379	375	4	L (top + bottom)	3.7 ± 110 ka BP
		C106_12	390	385	5		
tS4	Tephra layer	C90	412	408	4	M (top + bottom)	4.53 ± 110 ka BP
		C106_12	420	417	3		
mfs	Maximum flooding surface	C90	455	–	–	N	6030 ± 80 ka BP
		C106_12	450	–	–		
B/C	Facies B–C boundary	C106_12	479	–	–	O	10.19 ± 180 ka BP
YD	Younger Dryas	C106_12	505	497	8	P (top + bottom)	11.88 ± 170 to 13.05 ± 190 ka BP
BA	Bölling–Allerød	C106_12	521	505	16	Q (top + bottom)	13.05 ± 190 to 14.87 ± 190 ka BP
rs	Ravinement surface	C106_12	531	–	–	R	16.4 ± 70 ka BP
ES	Erosional surface	C106_12	550	–	–	–	18.5 ± 70 ka BP

See also Figures 11 and 13 for further detail on correlation between event beds and seismic reflectors.

Table 3. *Marine tephra layers described in this study and proposed correlation with proximal pyroclastic deposits documented onland*

Tephra code	Lithology and texture of marine tephra	Suggested correlation with proximal tephra (age expressed as ^{14}C cal. years BP)	References
tS1	Fine- to very fine-grained dark grey ash (5y3/2) with clayey matrix. Ash made of leucite-bearing scoria (mostly analcimized), scarce pumiceous glass shards, lithic fragments, loose crystals. Sharp base	AD 1822? (Vesuvius)	Arrighi *et al.* 2001 Budillon *et al.* 2005 Insinga *et al.* 2008
tS1-α	Fine- to very fine-grained dark grey ash (5y3/2) with clayey matrix. Ash made of scoria, abundant lithic fragments and very rare glass shards (pumiceous type), loose crystals (clinopyroxene and feldspar). Sharp base in core C106	AD 1631 (Vesuvius)	Rolandi *et al.* 1993 Rosi *et al.* 1993
tS1-β	Fine- to very fine-grained bioturbated ash dispersed in clayey matrix (cryptotephra). Ash made of brown pumiceous glass shards with tubular and ovoid vesicles, leucite-bearing lithic fragments, minor scoria and obsidian. Loose crystals (mostly K-feldspar, clinopyroxene and biotite)	AD 787 (Vesuvius)	Rolandi *et al.* 1998
tS1-γ	Fine- to medium-grained dark grey (5y3/2) ash made of dark and light grey, leucite-bearing scoria, pumiceous glass shards, leucite-bearing lithic fragments and loose crystals (K-feldspar, clinopyroxene and biotite). Sharp base	Post AD 512 activity and/or AD 685 (Somma–Vesuvius)	Rolandi *et al.* 1998 Iorio *et al.* 2004
tS2	Coarse- to fine-grained white phonolitic pumice lapilli overlain by coarse-grained grey tephriphonolitic pumice lapilli and coarse-grained dark grey ash. Scoria, carbonate lithic fragments and loose crystals are diffuse throughout the entire tephra. Sharp base and top	AD 79 ('Pompeii') (Somma–Vesuvius)	Sacchi *et al.* 2005 Insinga *et al.* 2008
tS3	Fine- to medium-grained dark grey (5y3/2) bioturbated ash in silty matrix. Ash made of leucite-bearing phonotephritic scoria and scarce pumice. Loose crystals (mostly K-feldspar and biotite)	AP4 (Somma–Vesuvius)	Rolandi *et al.* 1998 Andronico & Cioni 2002 Iorio *et al.* 2004 Insinga *et al.* 2008
tS3α	Fine- to medium-grained dark grey (5y3/2) bioturbated ash in silty matrix. Ash made of leucite-bearing phonotephritic scoria and scarce pumice. Loose crystas (mostly K-feldspar and biotite)	AP3 (Somma–Vesuvius) 2830 ± 50 years BP	Rolandi *et al.* 1998 Andronico & Cioni 2002 Iorio *et al.* 2004 Insinga *et al.* 2008
tS4	Fine- to very fine-grained ash dispersed in clayey silty matrix. Ash made mostly of loose K-feldspars, white pumice with tubular vesicles, glass shards, minor scoria and lithic fragments. Glass shards show blocky, platy, tubular and rare thick-walled morphologies. Loose crystals of clinopyroxene	Astroni (Campi Flegrei) 4244 ± 91 years BP	Di Vito *et al.* 1999
tS5	Coarse- to medium-grained ash made of reddish sub-angular, almost aphyric pumice with tubular and ovoidal vesicles, brown and yellow glass shards and dark grey scoria. Scarce lithic fragments. Glass shards show blocky, tubular and thin-walled bubble wall morphologies	SC2-b (?) (Campi Flegrei or Somma–Vesuvius) *c.* 55–65 ka BP	Di Vito *et al.* 2008
tS6	Coarse- to medium-grained white ash made of sub-angular, almost aphyric pumice, scarce blocky shards and lithic fragments. Loose crystals (mostly K-feldspar and pyroxene). Sharp base	pre-'M. Epomeo Green Tuff' events (Ischia) *c.* 55–65 ka BP	Vezzoli 1988 Brown *et al.* 2008 Di Vito *et al.* 2008

Cryptotephra tS1-β and tephra tS1-γ can be correlated with the Medieval activity of Vesuvius between the end of the fifth century and the eighth century. Their phonotephritic composition (unpublished data) suggests that they might be the product of the explosive activity that postdate the AD 512 eruption. Historical documents (Figliolo & Marturano 1994; Lirer *et al.* 2005) reported the AD 685 and 787 eruptions as the major events during this period. More recently, Cioni *et al.* (2008) described onland some ash layers referred to volcanic activity that took place soon after AD 512. According to these considerations we propose a possible correlation of tephra tS1-β with the AD 787 eruption, whereas tephra tS1-γ may represent the products of the activity that occurred after AD 512, including the AD 685 event. Tephra layers associated with this chronological interval occur as well-preserved marker horizons within the Upper Holocene succession and have been recently recognized offshore the Campania region (Insinga *et al.* 2008, and references therein) as well as in distal settings (i.e. Monticchio Lake, Wulf *et al.* 2008).

Tephra tS2 is a *c.* 1 m thick, distinctive pyroclastic bed formed by the air-fall deposits associated with the Plinian eruption of Vesuvius of AD 79 (Sigurdsson *et al.* 1985; Cioni *et al.* 1992; Sacchi *et al.* 2005). At the core C106_12 site, it consists of three major horizons represented from bottom to top by: (1) 48 cm thick, inversely graded, subrounded, white pumice and lapilli; (2) 44 cm thick, normally graded, sub-angular, coarse grey pumice lapilli and lithic fragments; (3) 9 cm thick, parallel-laminated coarse pumiceous and scoriaceous lapilli, including loose crystals (biotite and pyroxenes), abundant small mollusc fragments, carbonate lithoclasts, and other lithic fragments (see Fig. 6). The tephra bed corresponds to a pronounced seismic horizon that can be traced all across the Bay of Salerno (Sacchi *et al.* 2005).

Tephras tS3 and tS3-α have been correlated by Insinga *et al.* (2008) with the AP inter-Plinian activity (Andronico & Cioni 2002), on the basis of their lithological and chemical composition. In particular, these deposits have been interpreted as the products of the AP4 and AP3 (2830 $\pm$ 50 ^{14}C cal. years; Rolandi *et al.* 1998) events. The age–depth plots we have obtained here, however, point to significantly older ^{14}C interpolated ages, namely 3300 $\pm$ 100 years (cal. BP) for tephra tS3 and 3700 $\pm$ 110 years (cal. BP) for tephra tS3-α. On the basis of this interpretation, the volcaniclastic layer that lies above ts3 (ts2-α in Table 2) may be tentatively correlated with the younger AP eruptions (Andronico & Cioni 2002).

The interpolated ^{14}C age for tephra tS4 of 4350 $\pm$ 90 years BP, along with its homogeneous trachyphonolitic composition (unpublished data), suggests a correlation of this deposit with the explosive Campi Flegrei events of Astroni dated at 4244 $\pm$ 91 ^{14}C cal. years BP (Di Vito *et al.* 1999). Products related to these eruptions have been reported elsewhere in the Bay of Naples (Insinga 2003) and more distal settings (Siani *et al.* 2004; Wulf *et al.* 2004).

Tephras tS5 and tS6 have been cored within the Upper Pleistocene deposits, beneath the unconformity ES (Figs 6 and 12–14) and are characterized by a trachytic and trachyphonolitic composition respectively (unpublished data). On the basis of the AMS radiocarbon data that indicate ages older than 48 ka BP, the expected ages of these tephras are between *c.* 60 ka and 80 ka BP. The attribution of tephra tS5 to a specific source is not an easy task because of the large number of volcanic events with similar geochemical signatures that occurred in the region between *c.* 44 ka and 75 ka BP (Pappalardo *et al.* 1999; Di Vito *et al.* 2008). However, the lithology and stratigraphic position of tephra tS5 suggests a correlation with the subaerial products of the activity of Somma–Vesuvius or Campi Flegrei between 55 and 70 ka BP (e.g. tephra SC2-b of Di Vito *et al.* (2008)). The very evolved trachyphonolitic composition of pumice suggests the island of Ischia as a possible source area for tephra tS6 (Insinga 2003). High-intensity explosive eruptions took place on the island during that time interval (Brown *et al.* 2008) and caused the widespread deposition of several tephra layers over the Campania Plain (Di Vito *et al.* 2008) and at more distal sites (Paterne *et al.* 1988; Wulf *et al.* 2004). In particular, tephra tS6 might be correlated with one of the several explosive events characterized by the deposition of pumice lapilli that predate the Monte Epomeo Green Tuff (MEGT) eruption (*c.* 55 ka BP; Vezzoli 1988), such as the Mago and/or Mt. S. Angelo eruptions (55–74 ka BP) described by Brown *et al.* (2008).

Core and seismic data correlation

Integrated stratigraphy and correlation of gravity-cores allowed for a bed-to-bed calibration of seismic reflectors. All the studied cores are located on the outer shelf, where the Holocene wedge is relatively thin and the typical range of penetration of standard gravity-cores ensured sampling of the entire post-glacial succession, down to the upper Pleistocene prograding units. In particular, core–seismic correlation for the entire Holocene section was established at C106_12 core site, located at *c.* 500 m from the southern edge of the track of Seistec profile 3006 (Figs 12 and 13). Depth (m) to time (ms) conversion of the C106_12 core log was calculated using an average velocity of seismic waves of *c.* 1600 m s^{-1} within the cored succession.

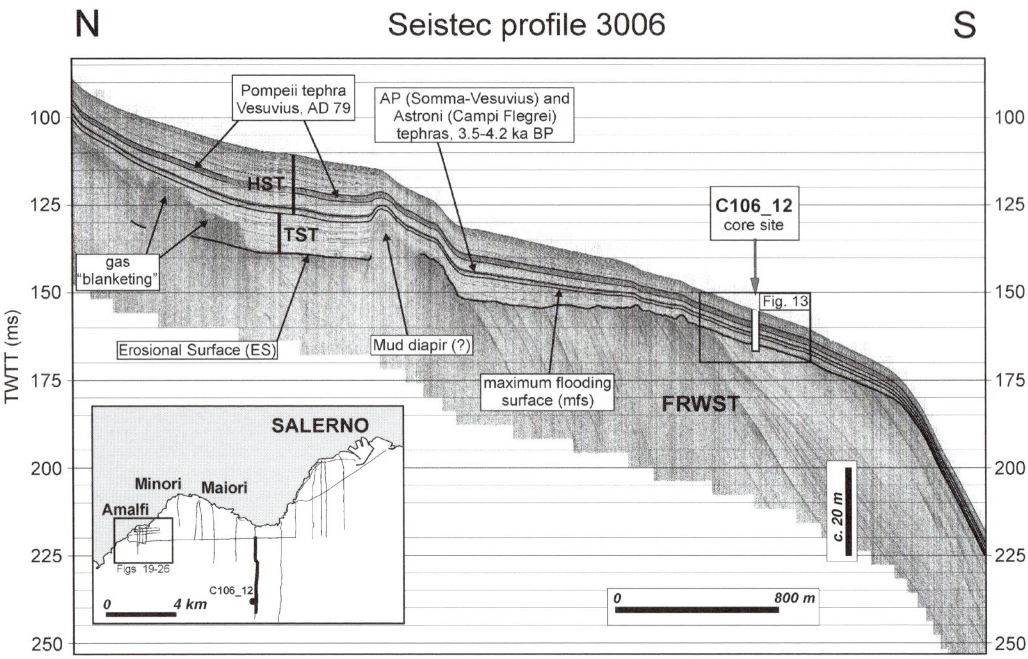

Fig. 12. Seistec profile 3006 showing the general architecture of the continental shelf off the Amalfi coast and the location of C106_12 core site. Erosional surface (ES) separates the Upper Pleistocene forced regression wedge prograding units (FRWST) from the above Uppermost Pleistocene–Holocene, post-glacial, transgressive systems tract (TST) and highstand systems tract (HST) deposits. TWTT, two-way travel time. Sequence stratigraphic nomenclature is after Hunt & Tucker (1992). Inset shows location of profile.

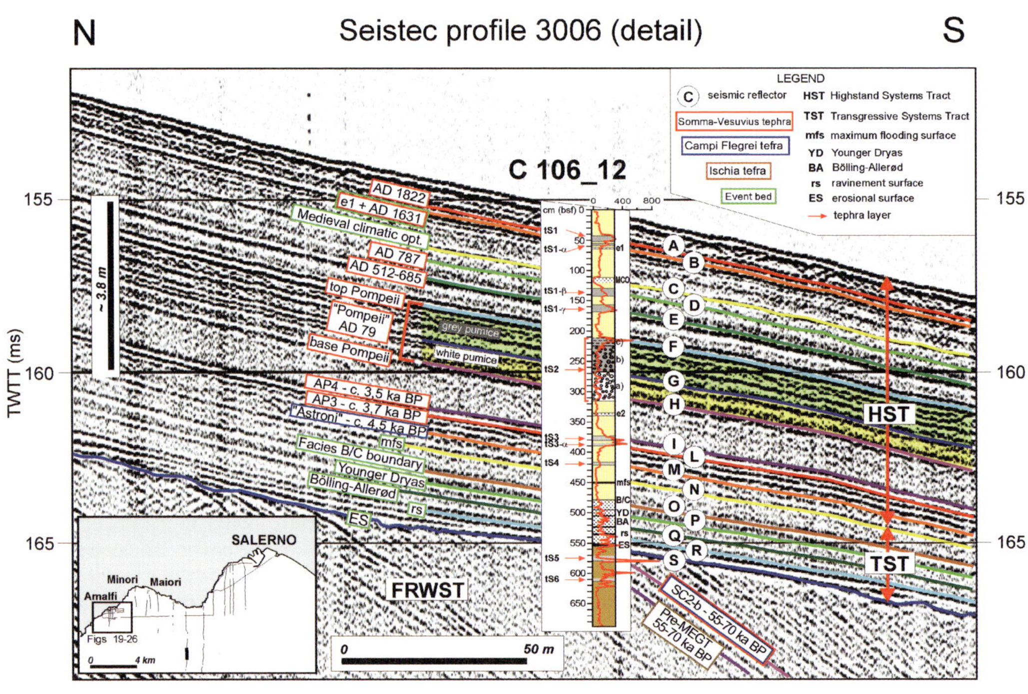

Fig. 13. Geological calibration and labelling of seismic reflectors of Seistec profile 3006 at C106_12 core site (see Fig. 12 and inset for location of profile). Key to sub-units of tS2 as in Figure 5.

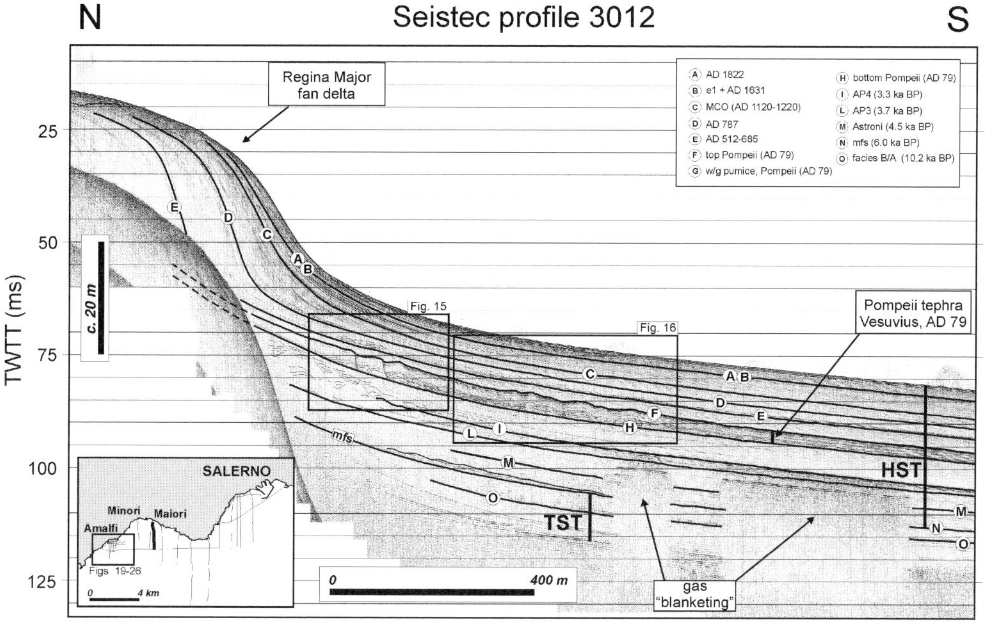

Fig. 14. Seistec profile 3012 offshore Maiori and its interpretation showing the fan-delta fed by the Regina Major stream and sedimentary structures associated with gravity-driven instability on delta slopes. Inset shows location of profile. Boxes indicate location of Figures 15 and 16.

Correction for sediment compaction was obtained by using the sea floor, tephra tS2 and unconformity ES as stratigraphic tie points (Fig. 13).

Core–seismic correlation allowed for calibration of the major sequence stratigraphic surfaces on seismic profiles (e.g. transgressive surface, ravinement surface, maximum flooding surface). Moreover, it indicates that tephra layers tend to correspond to high-amplitude, well-developed seismic reflectors. This is clearly the case, for instance, of the Vesuvius early Medieval tephra tS1-γ, the tS2 (AD 79) pumice fall layer, the tS3 and tS3-α (AP inter-Plinian deposits) and tephra tS4 (Astroni–Averno).

Seismic interpretation

The IKB–Seistec seismic profiles imaged the seafloor subsurface down to *c.* 60 ms two-way travel time (*c.* 100 m) beneath the sea floor. Maximum penetration of the seismic signal was recorded in the Capo d'Orso area, where the Upper Pleistocene prograding units can be thoroughly traced beneath unconformity ES (Figs 12 and 13). In the Amalfi area acoustic penetration was somewhat less, because of significant attenuation of seismic waves at depth or blanketing of the seismic signal by gas-saturated layers. Maximum vertical resolution is of the order of 20 cm.

Seistec profiles show that unconformity ES separates two main seismic stratigraphic units (Figs 12 and 13). The lower one is represented by a prograding succession truncated at the top by a dramatic erosional surface, and mostly consists of Upper Pleistocene forced regressive wedge systems tracts (FRWST) deposits (Hunt & Tucker 1992).

Above unconformity ES, seismic profiles show relatively continuous, parallel and subparallel reflectors, gently inclined towards the SE as a result of low-angle backstepping and aggrading of layers. This unit is represented from bottom to top by the transgressive systems tract (TST) and highstand systems tract (HST) deposits that formed in response to the transgressive landward shift of the coastline during the rapid sea-level rise that accompanied the last deglaciation (*c.* 18–6 ka). The thickness of the uppermost Pleistocene–Holocene shelf wedge varies in the study area from 35–40 m in the inner–mid-shelf, to a minimum of 4–2 m at the shelf edge.

The HST deposits of the southern shelf of the Sorrento Peninsula between Amalfi and Capo d'Orso are characterized by the occurrence of a number of remarkably developed reflectors that can be correlated with different pyroclastic layers interbedded mostly within the upper Holocene succession. According to recent tephrostratigraphic (Buccheri *et al.* 2002; Insinga 2003; Sacchi *et al.*

2005; Insinga *et al.* 2008) and sequence stratigraphic research (Conforti 2003; Sacchi *et al.* 2004, 2005) conducted on the eastern Tyrrhenian margin off the Campania region these beds may be correlated to major tephra layers that originated from explosive eruptions of Somma–Vesuvius, the Campi Flegrei and Ischia. In particular, the seismic signature of the late HST deposits in the entire northern Bay of Salerno is characterized by a remarkable seismic horizon bounded by two high-amplitude reflectors that can be correlated with tephra layer tS2 deposited offshore during the AD 79 eruption of Vesuvius (Sacchi *et al.* 2005; Insinga *et al.* 2008) (Figs 12–26). The thickness of the tS2 tephra horizon ranges from a maximum of about 200 cm in the shoreface area off the Amalfi coast to some 10 cm at the shelf edge (Fig. 12).

Seismic profiles showed that the late HST (Upper Holocene) succession of the Amalfi shelf is characterized by a number of small prograding deltas that develop at the mouth of the small bedrock rivers with torrential regime. The best developed deltaic wedges in the study area occur at the mouth of the Regina Major torrent, offshore Maiori; at the mouth of the Regina Minor, off the village of Minori; and at the mouth of Canneto, Dragone and Cappuccini torrents offshore Amalfi and Atrani. These deltaic bodies represent the subaqueous components of the confined alluvial fans that developed in the narrow coastal plain and pocket beaches of the Amalfi coast (see Fig. 4).

The Regina Major fan-delta

Offshore Maiori the thickness of the post-glacial succession is of the order of 50 ms. In particular, the HST deposits on the Maiori mid-shelf are significantly thicker than the correlative layers off the cliffed coast of Capo d'Orso and may exceed 35 ms (Fig. 14). The best throughgoing reflectors are represented by the *c.* 2 m thick seismic horizon bounded by reflectors F–H that is readily correlated with the pyroclastic fall of the AD 79 eruption of Vesuvius (tS2), and reflectors E (tephra tS1-γ), I (tephra tS3) and L (tephra tS3-α). Towards the inner shelf, a well-developed fan-delta is imaged by seismic profiles, which extends for *c.* 800 m from the mouth of the Regina Major torrent (Figs 15–19). The delta is characterized by sigmoid prograding foresets and a bottomset

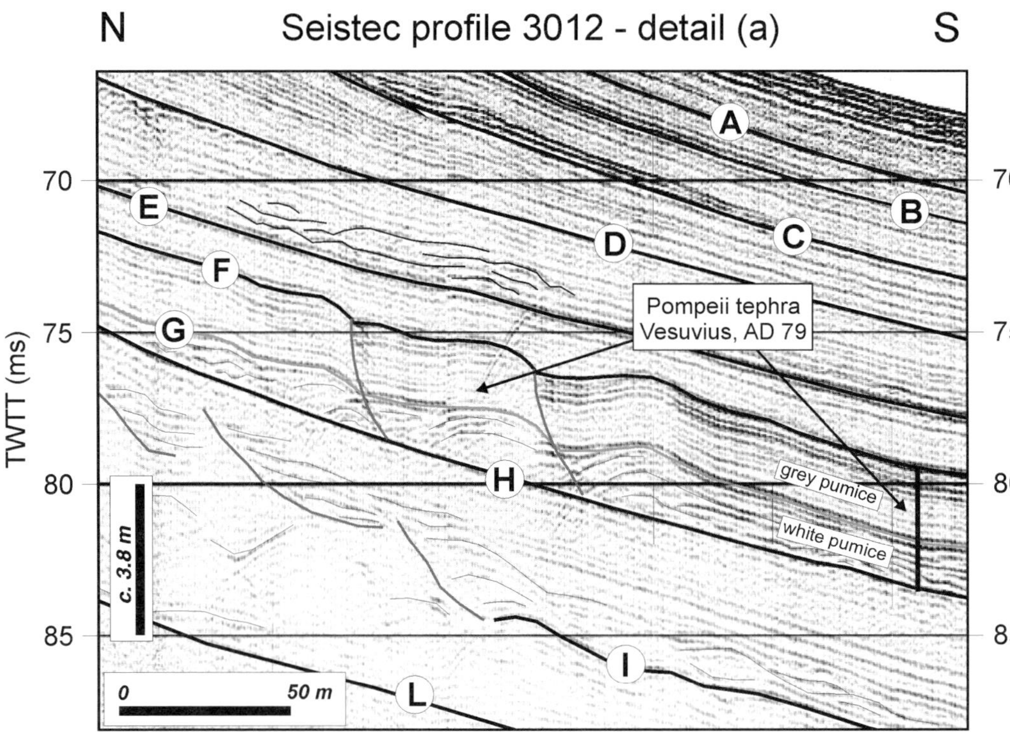

Fig. 15. Detail of Seistec profile 3012 showing the failure of the tephra bed deposited by Vesuvius 25 August, AD 79 ('Pompeii' eruption). (Note the fault-slumping of the tephra layer in the evacuation zone.) Figure 14 shows location and key to reflector labels.

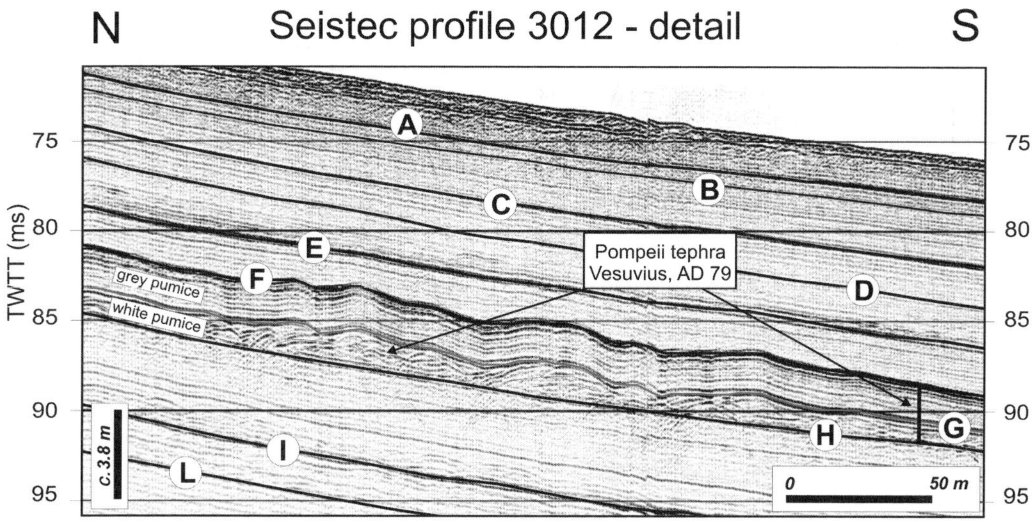

Fig. 16. Detail of Seistec profile 3012 showing the slumping of the Pompeii tephra bed (AD 79). (Note the soft-sediment deformation within the lower part of the tephra layer (i.e. white pumice), associated with detachment and sliding of the bed above its basal surface soon after its deposition.) Figure 14 shows location and key to reflector labels.

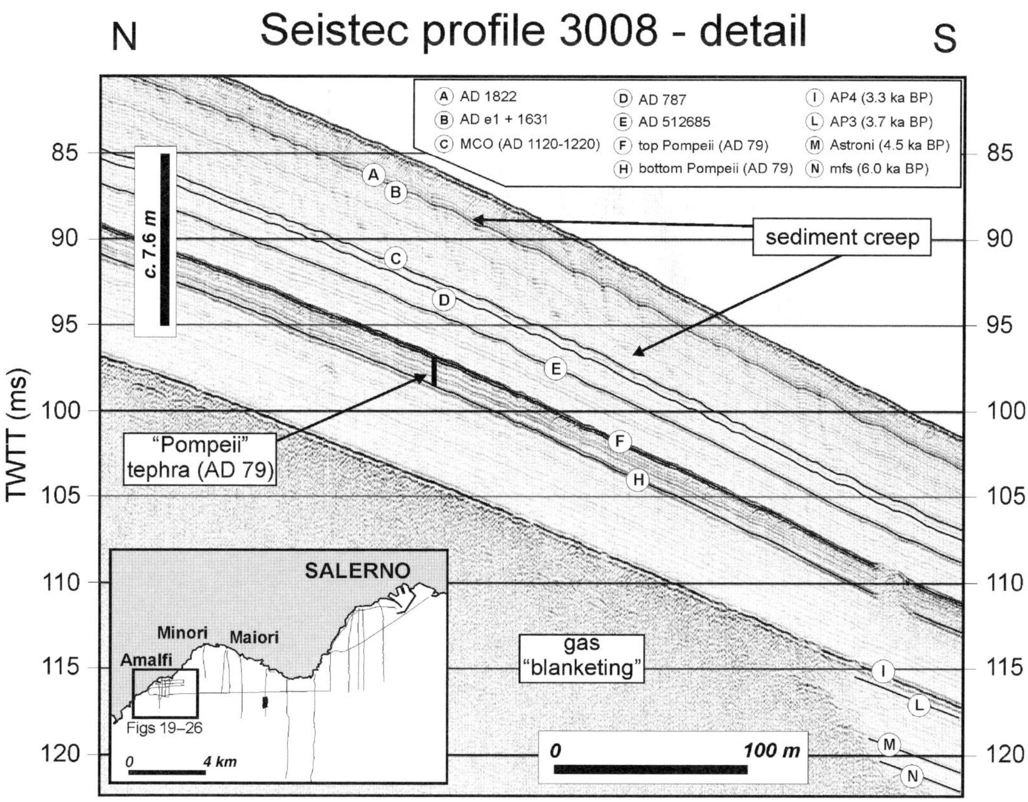

Fig. 17. Detail of Seistec profile 3008 showing crenulation of reflectors as a result of creep within the upper HST deposits of the prodelta area, mostly above the AD 79 ('Pompeii') tephra bed. Inset shows location of profile.

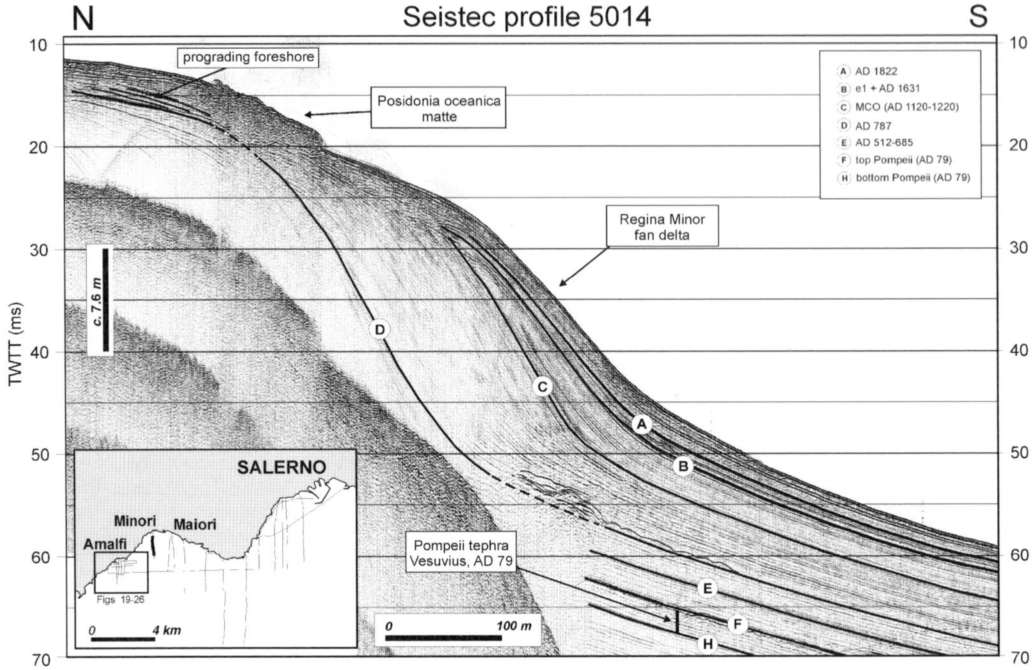

Fig. 18. Seistec profile 5014 off the Regina Minor stream mouth and its interpretation. (Note gravity-flow deposits above reflector D.) Inset shows location of profile.

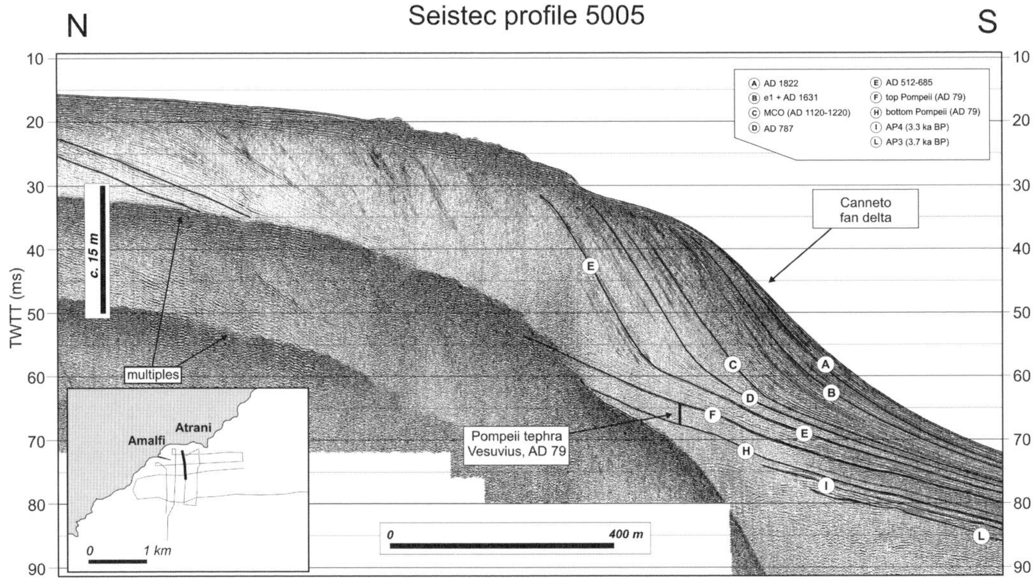

Fig. 19. Seistec profile 5005 offshore the Canneto stream, Amalfi, and its interpretation. Reflector E separates two different phases of development of the Canneto fan-delta. Inset shows location of profile.

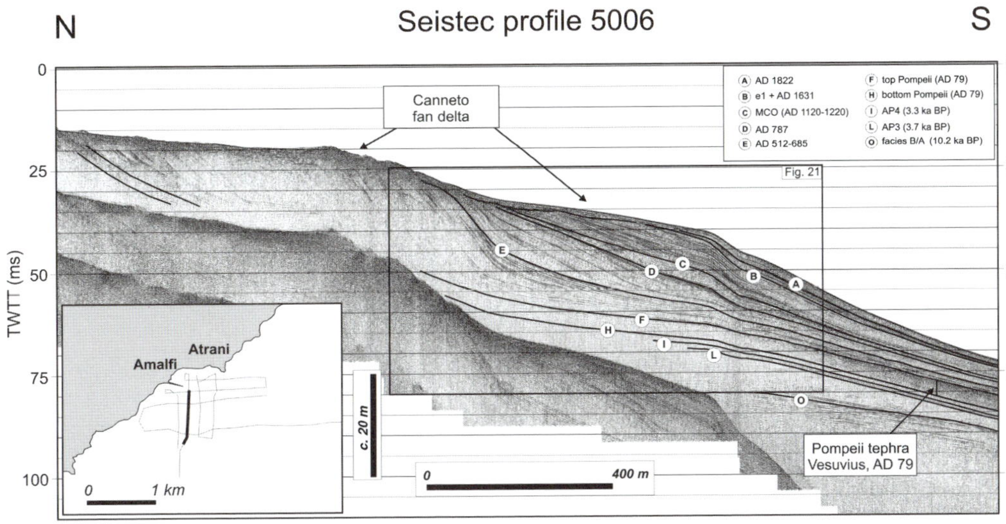

Fig. 20. Seistec profile 5006 offshore the Canneto stream, Amalfi, and its interpretation. (Note the significant growth of the fan-delta between reflectors F and E, together with the difference in the stratal architecture between the two parts of the deltaic body separated by horizon E.) Inset shows location of profile.

represented by the top of the AD 79 tephra layer. Inclination of foresets is between 10° in the inner fan-delta and 30° towards the delta front, whereas the present-day delta front slope is *c.* 20°.

Seistec profile 3012 documents the occurrence of sedimentary structures at the base of the Regina Major delta front that may be associated with a general gravity-driven instability and soft-sediment deformation above distinct stratigraphic surfaces, represented by reflector H (base of AD 79 pumice layer) and reflector I. Minor, but still clear evidence of sea-floor instability can be recognized above reflectors L and E (Fig. 14). Seismic interpretation suggests that soft-sediment deformation above the base of tephra tS2 mostly involves the pyroclastic layer itself and consists of slump–slide folding

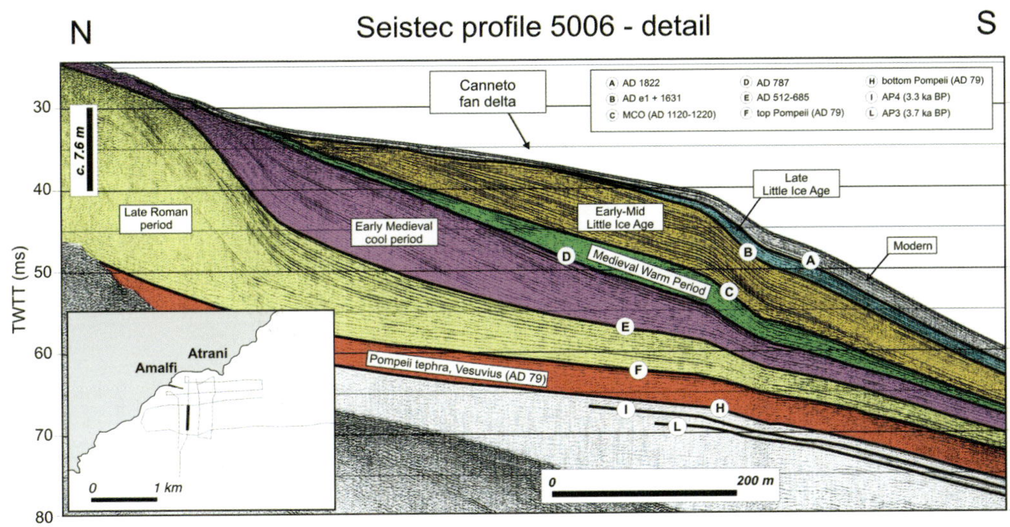

Fig. 21. Detail of Seistec profile 5006 off the Canneto stream, Amalfi, showing approximate correlation of prograding delta units with major climatic changes of the last 2000 years. (Note the backstepping and aggradation of strata within the stratigraphic unit corresponding to the apex of the Medieval Warm Period, which represents an exception to the generally prograding units of the fan delta.) Inset and Figure 20 show location.

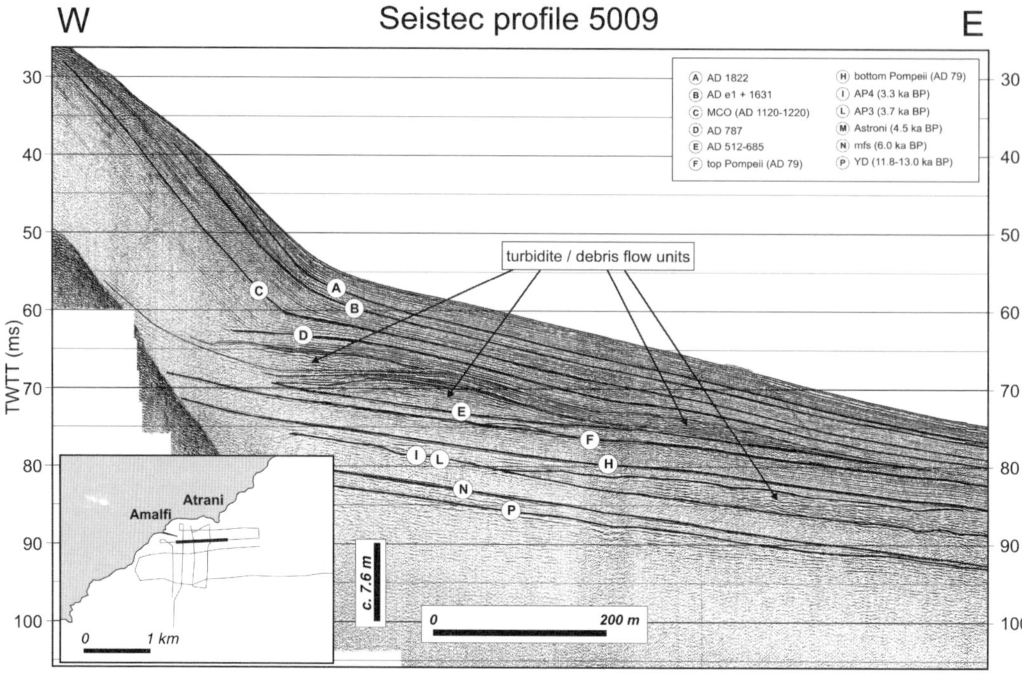

Fig. 22. Seistec profile 5009 offshore the Canneto stream, Amalfi, showing turbidite lobes or debris flow units occurring above different tephra layers (reflectors I, F and E). Inset shows location of profile.

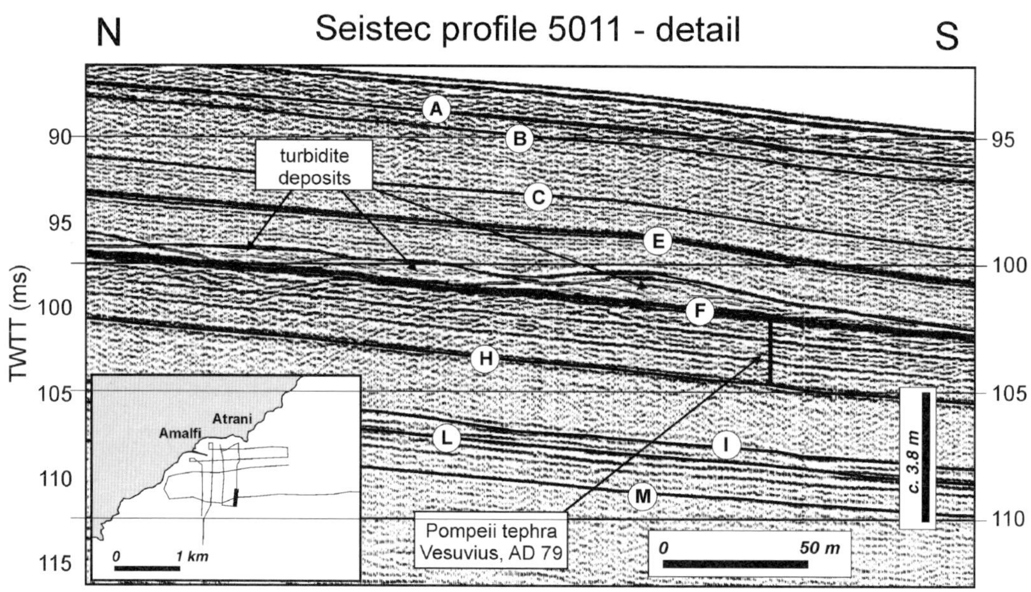

Fig. 23. Detail of Seistec profile 5011 off the Canneto fan prodelta showing distal fine-grained turbidites above the Pompeii (AD 79) tephra bed. Inset shows location of profile, and Tables 2 and 3 give legend for event beds.

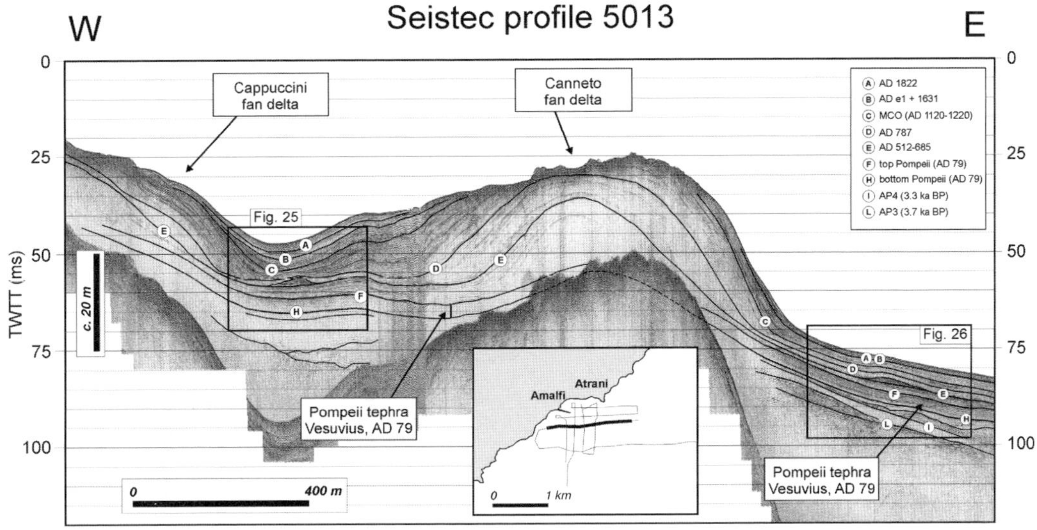

Fig. 24. Seistec profile 5013 offshore Amalfi showing a transverse section of the Cappuccini–Canneto delta system. (Note the development of turbidite and debris-flow units above different tephra layers (reflectors E, I and L).) Inset shows location of profile.

and slump–fault rupture of the tephra layer (Figs 14–16). The AD 79 pyroclastic unit is sandwiched between normal flat-lying dipping layers. The wavelength of the slump-folds is 20–60 m, and fold amplitude is of the order of 0.5–1.0 m (Figs 15 and 16).

At a lower stratigraphic level other evidence exists of sediment failure, represented by hummocky

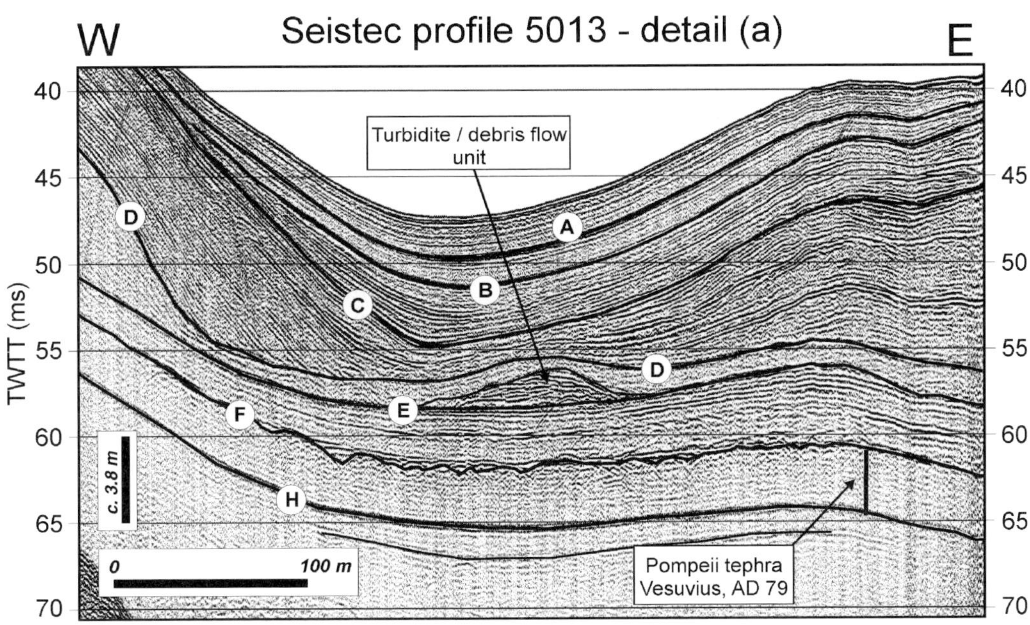

Fig. 25. Detail of Seistec profile 5013 showing turbidite and debris-flow deposits above seismic horizon E (tephra layer tS-1γ) in the channel between the Cappuccini and Canneto fan-deltas. Figure 24 shows location and key to reflector labels.

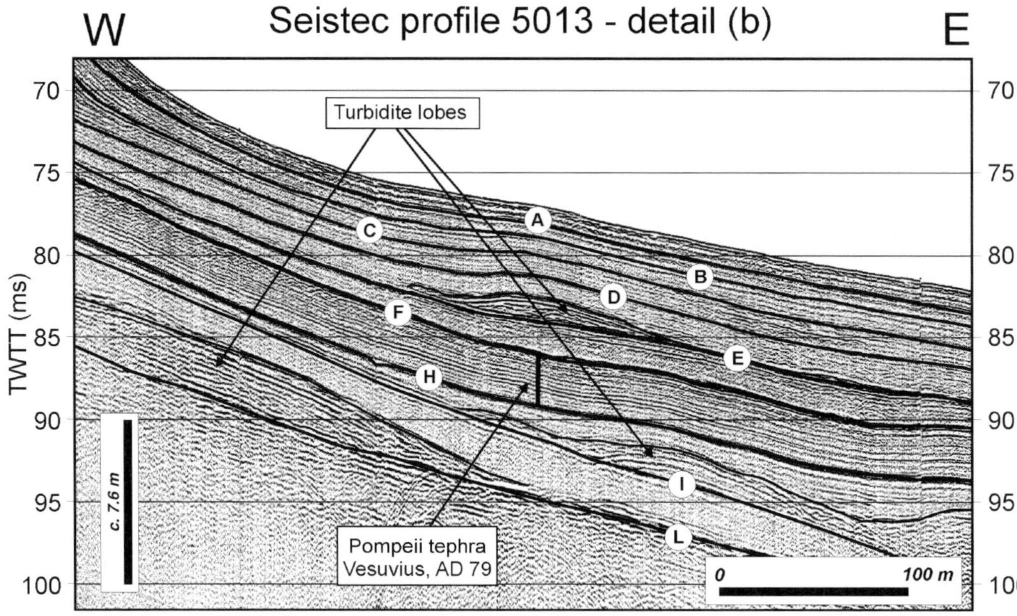

Fig. 26. Detail of Seistec profile 5013 off the Canneto fan-delta showing turbidite units deposited above different tephra layers (seismic horizons E, I and L). Figure 24 shows location and key to reflector labels.

bedforms with internal chaotic seismic facies that can be interpreted as slide or debris-flow deposits accumulated at the base of the delta front (Fig. 14). The sediment slide accumulation is sandwiched between seismic horizons H and I. Single hummocks are 50–100 m wide and up to 3 m high. Further evidence of sediment failure is also found above reflector N (maximum flooding surface), before the onset of the Holocene fan-delta *sensu stricto* (Fig. 15). Seismic profiles also indicate the occurrence of irregular geometry and crenulation of reflectors within the upper HST deposits of the prodelta area, mostly above reflector F (Fig. 17).

The Regina Minor fan-delta (Minori)

Adjacent to Maiori, off the village of Minori there is a major embayment of the Amalfi coast that hosts the westernmost part of the relatively large continental shelf of the northern Bay of Salerno, with a shelf break located at *c.* 3.5 km from the coast. Seistec profile 5014 shows a *c.* 50 ms (*c.* 40 m) thick post-glacial succession over the mid-shelf, and a well-developed fan-delta towards the inner shelf, as the underwater counterpart of the confined alluvial fan fed by the Regina Minor stream (Fig. 18).

The Regina Minor fan-delta is *c.* 800 m wide and the base of the foresets is about 500 m from the Minori seashore. The pattern of seismic reflectors indicates that the stratigraphic architecture of the deltaic wedge is characterized, mostly above reflector F, by an average increase coupled with significant variations of progradation rates in the delta foresets. These variations are particularly evident in the delta front area (Fig. 18).

A pronounced step in the sea-bed morphology associated with chaotic internal reflections in the fan-delta topset area between 15 and 20 ms (*c.* 11–15 m water depth) is visible from seismic profile 5014 and can be interpreted as the result of the occurrence of a *Posidonia oceanica* 'meadow' that colonizes the sea floor (Fig. 18). It may also be observed from the seismic record that the meadow exerts a 'damming effect' and protects the modern foreshore from erosion by storm waves, thus allowing for the development of the foreshore sediment fill landward of the *Posidonia* meadow. The small prograding unit downlapping above reflector D landward of the *Posidonia* meadow may be thus interpreted as the product of a temporary seaward migration of the outer foreshore deposits during a period of coastline progradation.

Minor evidence of gravity-driven deformation is found at several stratigraphic levels towards the base of the delta front, above reflector D, and small-scale, creep-like, slope instability may be recognized at the top of the AD 79 tephra layer (reflector F) (Fig. 18).

The Canneto–Dragone fan-delta system (Amalfi–Atrani)

Off Amalfi and Atrani the width of the continental shelf is reduced to less than 2 km. The post-glacial succession offshore Amalfi–Atrani reaches a maximum thickness of *c.* 25 ms (*c.* 20 m) and is significantly thinner than in the adjacent sector off Minori and Maiori. Seismic correlation between the two sectors was ensured by a cross-check of the dip seismic profiles with the tie section 5004, which allowed tracing of the main Holocene reflectors across the entire Amalfi shelf. In this area seismic interpretation shows the occurrence of a sedimentary apron formed by the coalescence of two fan-deltas developing at the mouth of Torrente Canneto (Amalfi) and Torrente Dragone (Atrani), and a smaller delta representing the offshore extension of a small alluvial fan at the mouth of the Cappuccini torrent, immediately west of Amalfi harbour. The fan-delta system extends for about 1.5 km, from the mouth of the Dragone to that of the Cappuccini torrent, with delta-front foresets developing as far as 700 m from the coastline (Figs 19–26).

As in the cases of the Regina Major and Regina Minor fan-deltas, the AD 79 tephra horizon (tS2) marks a significant increase in the average prograding rates of foresets, which may be associated with a coarsening of the sediments entering the fan-delta system. Dip sections across the delta system show that the Canneto–Dragone prograding wedge may be divided into two sub-units by a downlap surface represented by reflector E. The two sub-units of the prograding wedge probably represent different phases in the development of the fan-delta system (Figs 20, 21 and 24–26).

Evidence of gravity-driven instability is common at various stratigraphic levels in the front of the Canneto–Dragone–Cappuccini delta system, particularly at the base of the prograding foresets, above horizons E, F and I–L. Seistec profiles show that gravity flows above horizon E locally form debris-flow units and/or turbidite lobes with a thickness of a few metres and width of a few tens of metres (Figs 22 and 24–26). Fine-grained turbidites are found above the AD 79 tephra layer, as suggested by the internal seismic facies and the external geometry of the deposits (Fig. 23).

Profile 5013 displays the stratigraphic architecture of the Canneto–Dragone–Cappuccini fan-delta system along strike and illustrates the occurrence of gravity-flow deposits above reflector E (Figs 24–26).

Discussion

Fan-deltas that develop along tectonically elevated, cliffed coasts at the mouth of ephemeral streams of temperate regions are typically fed by small alluvial fans, which are often confined in narrow, deeply incised V-shaped coastal valleys (Dabrio 1990; Nava-Sanchez *et al.* 1999; Fernández-Salas *et al.* 2003; Lobo *et al.* 2006). In these morphological settings the development and progradation of such small deltas generally have little or no subaerial expression. Furthermore, the recognition of their underwater component has often been hampered by the intrinsic resolution limits of the standard sub-bottom profiling, where coarse-grained deposits, shallow water and often gas saturation prevail. This is probably the reason why, until recently, these deltas have been mostly neglected by researchers in the outcrop and rarely described in the Holocene marine record (e.g. Prior & Bornhold 1990). Consequently, fan-deltas similar to those described in this study may be much more common than previously believed along cliffed coasts or even in deep inland lakes.

The IKB-Seistec seismic reflection profiles and gravity-core data used in this study have revealed unprecedented detailed views of the inner–mid-shelf depositional system of the northern Bay of Salerno, allowing the recognition of a number of fan-deltas that developed mostly during the last 2000 years, at the mouth of small rivers of the Amalfi cliffed coast. The deltaic bodies described in this study are small-scale fan-deltas, of about 1 km^2 area and a few tens of metres thick. They display a general conical morphology with a delta-front slope of *c.* 20° and foreset inclination between 15° and 30°. These bodies represent the underwater counterparts of coastal alluvial fans fed by small bedrock rivers with torrential regime that are part of the hydrographic network of the Sorrento Peninsula. These coastal alluvial fans are grouped at the seaward termination of the coastal gorges that dissect the tectonically elevated mountain range of the Lattari Mts., where virtually all land-derived sediment is deposited below sea level. In this context subaerial delta-plain components are practically absent and the narrow space at the exit of the valleys is filled with a coarse-grained alluvial prism up to a few tens of metres thick, whereas at the seashore the alluvial deposits are reworked into pocket-beach settings. Most of the modern settlements of the coastal villages of Maiori, Minori, Atrani and Amalfi are built on such deposits.

There is evidence in the literature suggesting that the dynamic regime of the alluvial fans of the Sorrento Peninsula–Amalfi coast is controlled by episodic, but often catastrophic sediment and water discharges that have caused repeated flooding of the fans in recent millennia (Esposito *et al.* 2004*a*, *b*, and references therein; Porfido *et al.* 2009; Violante *et al.* 2009). The ephemeral streams that feed the alluvial fan systems are subject to episodic sheet wash and flash floods that may cause accumulation of large volumes of sediment. This is caused by heavy rains generated by the seasonal change of the atmospheric circulation in the Mediterranean region. The episodes of catastrophic rain typically bring very large sediment volumes to the fan-delta and offshore. This was the case for the catastrophic floods that struck the Amalfi coast between 25 and 26 October 1954, causing over 300 deaths (Esposito *et al.* 2004*a*). Following this dramatic event the coastline at the mouth of Bonea Stream, in Vietri, prograded seaward for more than 100 m (Esposito *et al.* 2004*a*, *b*, and references therein; Violante *et al.* 2009).

Seismic interpretation suggests that the general stratigraphic architecture of the studied fan-deltas is of two types. The deltaic bodies off Maiori and Minori display relatively steep and long sigmoidal foresets, commonly associated with topset layers. The fan-deltas off the Amalfi–Atrani coast show a clear variation in the dip angle of foresets and a very reduced or absent topset towards the delta fronts.

Another interesting observation on the overall stratal architecture of the fan-deltas is the remarkable regularity and lateral continuity of layers within the prograding foresets, so that the general 3D image of the deltaic bodies may be simply represented by a number of smooth, conical to hemispherical, onion-like layered underwater fans. Apparently, seismic profiles show no significant evidence of the various features or morphological zones typically associated with many coarse-grained marine deltas, such as distributary channels, gravel chutes or swales, flutes, stacked or switched sandy lobes, etc. In other words, seismic interpretation suggests that there is a lack of a system for delivering sediments towards deeper environments. Gravity-flow (inertia and/or turbidity flow) dominated transport of sediment proceeding directly from the coastal edge may be the primary cause for a substantial bypass of the sediment load to the lower slope segment, rather than segregation into distributary subsystems (Postma *et al.* 1988; Postma 1990, 1995; Parsons *et al.* 2001; McConnico & Bassett 2007).

The results of the seismic survey and the stratigraphic analysis on gravity-core samples suggest that the major changes in the stratal patterns within the fan-deltas of the Amalfi coast are often associated with the occurrence of single or clustered tephra layers interbedded within the Holocene record. In fact, all the fan-deltas described in this study started to develop above the pyroclastic bed deposited by Vesuvius during the ‘Pompeii’ Plinian eruption of AD 79. A significant change in the stratal architecture of the fan-deltas occurred after another eruption of Vesuvius, during the early Medieval period (*c.* AD 512–685). This is documented by the development of a downlap surface that can be correlated with the oldest Medieval products of Vesuvius preserved in the Bay of Salerno (tephra layer tS1-γ). The seismic reflector (E) that correlates with these products consistently separates the fan-deltas of all the study areas into two sub-units showing distinct stratal patterns (see Figs 19–21). Furthermore, minor changes in the stratal patterns of the fan-delta system may be recognized within the upper (early Medieval) sub-unit, as illustrated in Figure 21.

Seistec profiles reveal evidence of gravity-driven instability at various stratigraphic levels within the fan-deltas. All the observed features are not randomly distributed in the stratigraphic record but appear concentrated along distinct stratigraphic horizons or intervals. Probably, more than one mechanism of sediment deformation or failure could explain the variety of the features described in the fan-deltas of the Amalfi coast and a thorough discussion of such mechanisms would require a better control on mechanical properties of the sediment involved (e.g. Sultan *et al.* 2004). The available data allow for the recognition of (1) crenulation in mud-dominated prodelta slopes, possibly associated with shear deformation of sediments by creeping; (2) slide or slump deformation of the AD 79 pyroclastic deposits; (3) gravity (inertia, turbidity or debris) flow deposits associated with selected stratigraphic intervals.

Crenulation in mud-dominated prodelta slopes

Irregular geometry and crenulation of reflectors within horizons or packages above reflector H, in the prodelta area, have been detected on seismic profiles off the Regina Major fan-delta. This is particularly clear in the upper HST deposits, above a basal surface represented by the Pompeii tephra of AD 79 (Fig. 17).

According to several reports, sea-floor crenulations are common in mud-dominated prodelta slopes offshore river mouths all around the Mediterranean and elsewhere (Correggiari *et al.* 2001; Lee *et al.* 2002). In some cases the stratal geometry of such crenulations may resemble typical sand-wave morphology, and the origin of these features has been regarded as controversial. They may occur under sea-floor gradients of tenths of a degree, display a variety of internal geometries and seem to be associated with high sedimentation rates (Correggiari *et al.* 2001; Canals *et al.* 2004; Cattaneo *et al.* 2004; Trincardi *et al.* 2004).

In the case of the Regina Major prodelta the lack of conspicuous sandy facies in any of the cores collected through the basal surface does not provide evidence in favour of the hypothesis of a buried field of sandy bedforms. Hence it may be proposed that the crenulations imaged by Seistec profiles represent the effect of gravity-driven deformation as a result of creep of the upper HST above a basal surface represented by the Vesuvius AD 79 tephra.

Slide or slump deformation of the AD *79 pyroclastic deposits*

Offshore Maiori, seismic profiles have revealed significant gravity-driven deformation involving AD 79 tephra layer. In particular, stratal geometries highlight evidence for the slumping of the entire 2 m thick tephra layer by extensive shear deformation as a result of partial detachment from its base, along with internal deformation of the lower part of the tephra bed (white pumice), probably caused by the load exerted by the upper part (grey pumice) soon after its rapid deposition (Figs 14–16). The seismic records clearly image the slumped AD 79 pyroclastic unit sandwiched between normal flat-lying dipping layers. The longer wavelength of the slump-folds is 20–50 m, and fold amplitude is of the order of 1–2 m. Similar examples of slumping of volcaniclastic beds deposited underwater have been described, for instance, in the pyroclastic deposits that crop out at Lake Mono (Miller *et al.* 1982; Miller 1989).

It appears that the slump-fold deformation of the Pompeii tephra off the Regina Major stream was caused by the loading associated with the highly water-saturated, 2 m thick pyroclastic layer on the fan-delta slope. According to recent reconstructions, the pyroclastic bed was deposited within less than 12 h between the early afternoon and the night of 24 August, AD 79 (Dal Maso *et al.* 1999). Such extremely rapid, largely undrained loading over the underwater delta slope was probably responsible for a rapid decrease in the shear strength at the base of the layer and the consequent slump-fold deformation of the whole unit.

Gravity (inertia, turbidity or debris) flow deposits associated with selected stratigraphic intervals

Slope instability is also documented on seismic profiles by the repeated occurrence at different stratigraphic levels of sediment prisms probably corresponding to accumulation of gravity flow-type deposits. These deposits are typically found at the base of the delta front and may range in size from a few cubic metres to several hundreds of cubic metres. According to seismic interpretation they do not occur randomly but, again, are mostly concentrated along specific stratigraphic levels. From bottom to top, gravity-flow deposits tend to occur above tephra layers; namely, tS4 (Astroni) tS3-α (AP3), tS3 (AP4), tS2 ('Pompeii' tephra bed) and tS1-γ (early Medieval eruption) (Figs 18 and 24–26). There does not seem to be evidence of significant evacuation zones of sediments that eventually collapsed (or were eroded) from the delta front and were redeposited downslope by gravity flows. This may reinforce the previous suggestion that the studied fan-deltas lack major sediment segregation zones in the delta front and are dominated by effective bypass of alluvial sediments directly from the foreshore area to the lower segments of the fan-delta slopes.

Clues to mechanisms controlling underwater sediment dispersal

Notwithstanding the recent progress in the study of modern subaqueous coarse-grained, steep-faced deltas the general understanding of the sediment transport mechanisms operating on steep subaqueous slopes of coarse-grained deltas is still relatively limited (Nemec 1990*b*, 1995; Postma 1990; Mulder & Syvitski 1995; Mulder *et al.* 2003; Mutti *et al.* 2003). This is probably due, on the one hand, to the virtual inaccessibility of the subaqueous delta slopes to direct sedimentological observation and, on the other, to the limited examples of surveyed subaqueous deltas documented in the literature.

It is generally accepted that medium–coarse sediments may be transported for long distances away from nearshore fan-deltas. Side-scan sonar data have shown, for example, that during exceptional river floods the riverborne load appears to have sufficient energy to overcome buoyancy and frictional effects at the river mouth (Prior & Bornhold 1990). The near-bed concentrations of sediments appears to be largely unaffected by seawater density and may proceed across the subaqueous slopes as hyperpycnal flows. Where the bottom gradient of the nearshore fan is greater than the slope of the river thalweg, the offshore transport of gravel and sand is enhanced. In the foreshore area river-driven transport is in some places replaced by gravity-driven transfer of sediment along the underwater slopes. Hence coarse-grained sediment transport downslope over the fan may be significantly controlled by high-density, pseudolaminar inertia flows, and/or turbidity flows, as described experimentally by Postma *et al.* (1988) and proposed on the basis of modelling (Mulder & Syvitski 1995; Mulder *et al.* 2003) and outcrop studies on ancient deltas (Mutti *et al.* 2003; Plink-Bjorklund & Steel 2004).

Gravity flows usually originate in one of two ways: as surge-type sediment gravity flows generated by localized slope failure (Prior *et al.* 1987; Bornhold & Prior 1990; Nemec 1990*b*) or by more sustained currents related to sediment-laden hyperpycnal effluent (Gilbert 1975; Bornhold & Prior 1990; Chikita 1990; Mutti *et al.* 2003). In the case of the fan-deltas of the Amalfi coast, the absence of slide scars and evacuation zones landward of the flow deposits suggests that the principal mechanism for dispersal on the steep delta slopes is by sediment gravity flows, or highly concentrated mixtures of sediment and water moving close to the bed that probably evolve into turbulent flows during downslope motion. These may include the whole spectrum of gravity-driven processes from debris flows to grain flows and even high-density turbidity currents as they build momentum down the foreset slope.

Effects of landscape-mantling volcaniclastic deposits on the river basin erosional dynamics

Seismic stratigraphic interpretation showed that most of the gravity-driven instability processes are not diffused across the fan-delta but are concentrated on a few stratigraphic horizons that invariably correspond to major tephra layers or tephra clusters. This observation, coupled with the recognition that the sediment supply to the fan-delta system is largely affected by high-energy river floods, suggests a direct relationship between the rates of erosion of the river basin slopes that follow deposition of landscape-mantling volcaniclastic deposits and the rates of underwater sediment that are delivered to the fan-deltas.

In fact, during and following subaerial eruptions, sediment erosion and delivery rates are typically high, not only because of the large volumes of unconsolidated material, but also because sediment-stabilizing vegetation is often destroyed or damaged. Landscape-mantling deposits, mostly airfall tephra, are deposited in comparatively thin sheets of unwelded pyroclastic deposits that mantle topography and/or cap sequences. These deposits are quickly eroded by sheetwash or rill and gully erosion and the sediment yield from these sources declines rapidly, within a few years, as the gullies reach permeable, less erodible substrates and resistant surface crust develops (Collins & Dunne 1986; Cinque & Robustelli 2009). However, it may be expected that the effects of accelerated slope erosion may last for several decades, or even longer. Valley-fill deposits, such as debris avalanche and volcaniclastic deposits, typically erode not from the surface but also from the side, thus causing channel widening or bank erosion and collapse. As a consequence, the newly formed rock debris may further enhance slope erosion rates and keep providing high sediment yields to the seashore for a relatively long time.

Role of climatic forcing on the development of fan-deltas

It is widely documented that climatic changes significantly affect the stratal patterns of inner-shelf systems in general, and particularly deltaic settings, worldwide. In recent millennia, periods of cooler climate in the Mediterranean region were typically accompanied by periods of higher rainfall that may have resulted, in turn, in enhancing erosion on slopes. For instance, abrupt shifts between arid and humid phases are known to create temporary disequilibrium between climate, biotic processes and geomorphological processes, when rainfall erosivity is high and protective vegetation cover not fully established (e.g. Dabrio 1990; Mulder & Syvitski 1995; Fernández-Salas *et al.* 2003; Lobo *et al.* 2006).

The interpretation proposed in this study implies that the growth rates of the fan-deltas of the Amalfi coast were primarily controlled by the average recurrence interval and magnitude of river flooding episodes that have provided high sediment yields to the delta system, concomitant with periods during which abundant, erosion-prone (volcani)clastic material was available on the slopes of the river basins. Accordingly, it may be proposed that the amount of sediments delivered to the coastline and hence the rates of development of the Amalfi fan-deltas in the last 2000 years were possibly dictated by the interplay of the availability of loose pyroclastic cover on the slopes of the alluvial basins and the varying erosional rates on the slopes caused by the climatic oscillations that have occurred during recent millennia (Fig. 27).

In particular, the major change detectable in the stratal geometries of the fan-deltas occurring in the early Medieval time (tephra tS1-γ) may be associated with the onset of a period of climatic cooling, known as Early Medieval Cool Period (*c.* AD 500–800), that developed immediately after the Roman Warm Period. Further minor changes in the stratal patterns of the delta foresets, which are consistently imaged by the seismic record in all the fan-deltas of the Amalfi coast, may be tentatively correlated with the Medieval Warm Period (*c.* AD 900–1100) and the Little Ice Age (*c.* AD 1400–1850) (Fig. 21).

Conclusion

The interpretation of the very high-resolution (IKB-Seistec) seismic survey, calibrated with gravity-core data, showed that the fan-delta system imaged off the Amalfi rocky coast developed after the AD 79 Plinian eruption of Vesuvius. During this time interval of *c.* 2000 years, both sea-level oscillation

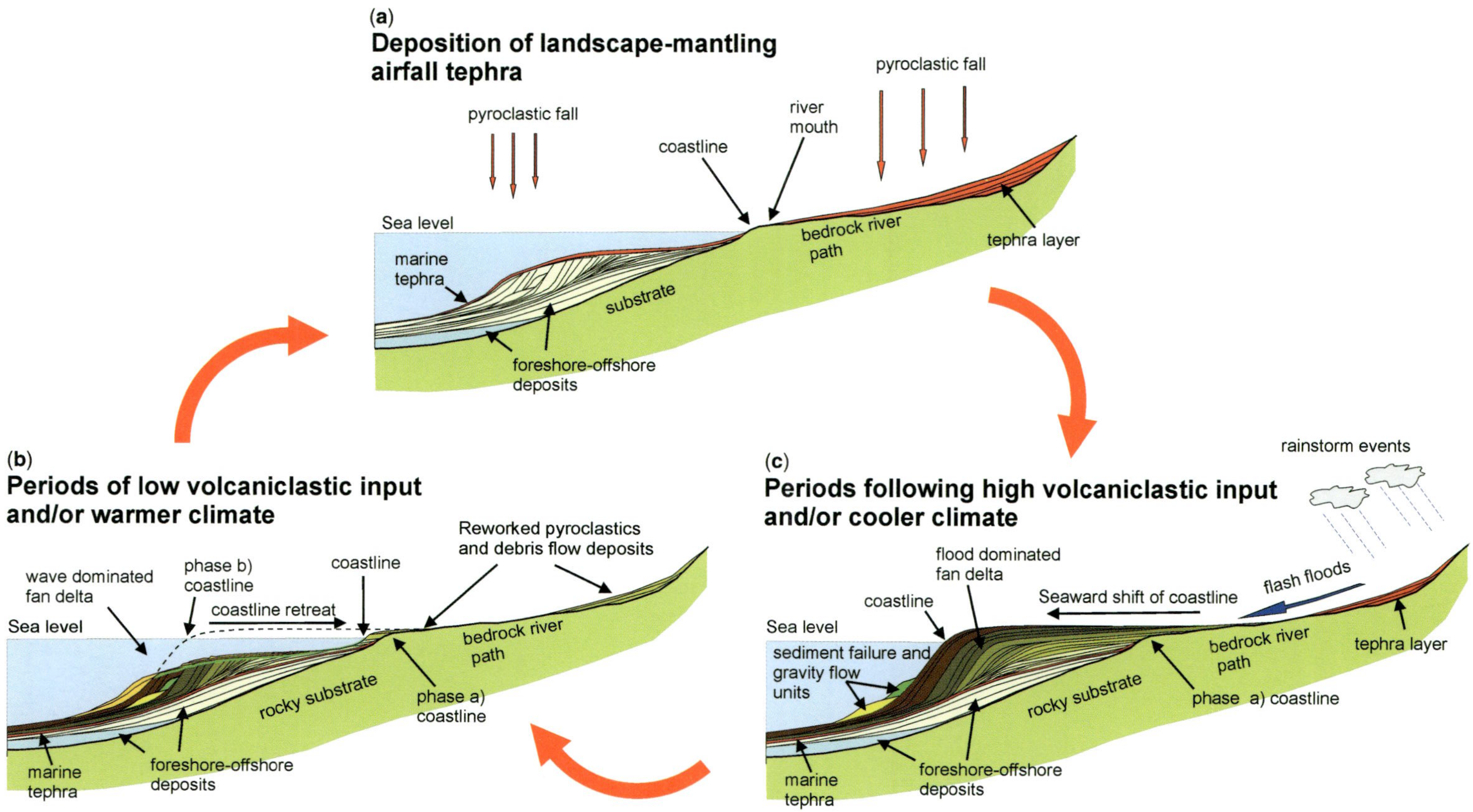

Fig. 27. Concept of Holocene delta growth cycles under varying climatic conditions and sediment input rates for mixed siliciclastic–volcaniclastic fan-deltas associated with Mediterranean-type cliffed coasts.

and tectonic subsidence or uplift were practically negligible in terms of influence on the overall stratigraphic architecture of the inner-shelf system, and the main factor controlling stratal geometries and patterns was the rate and mode of sediment supply.

The prominent gravity-driven instability and deformation of sediments detected at various stratigraphic levels within the delta slopes, along with the substantial lack of subaerial delta-plain components or other subaqueous segregation zones for sediments (e.g. distributary channels, levee complexes), suggest that the stratal geometry of the fan-deltas was dominantly dictated by the effective transfer of sediments by hyperpycnal (e.g. inertia, turbidity) flows directly fed by river flooding to the lower segments of the delta slopes.

The various phases of subaqueous delta growth were also controlled by the interplay of two main factors: (1) the accelerated erosion of slopes of the alluvial basins by sheetwash and flashfloods during and/or following periods of intense volcanism, which resulted in the delivery of considerable volumes of volcaniclastic debris and loose sediments to the shoreline; (2) the varying erosive potential of the river basin slopes under the changing morphoclimatic regimes over the last 2000 years.

IKB-Seistec seismic profiles were acquired in July 2004 and processed at various stages within the CNR–MTA (Hungarian Academy of Sciences) bilateral co-operation project, during the period 2004–2006, and processed at Geomega Ltd., Budapest. Gravity-cores and other geophysical data used in this study, including magnetic susceptibility logs, were acquired by the IAMC-CNR, Naples, between 1997 and 2004 for the geological mapping of the Italian coastal zone at 1:50 000 scale (CARG project). The ^{14}C AMS analyses were conducted at the Centre for Isotopic Research for Cultural and Environmental Heritage (CIRCE) Laboratory in Caserta and partly at the Lawrence Livermore Laboratory (LLL), US Geological Survey, CA, USA. Sincere thanks are due to M. Capodanno for her work in the sedimentological laboratory, and to P. Sclafani for proofreading of the English manuscript.

References

ACOCELLA, V. & FUNICIELLO, R. 2006. Transverse systems along the extensional Tyrrhenian margin of central Italy and their influence on volcanism. *Tectonics*, **25**, TC2003, doi:10.1029/2005TC001845.

ANDRONICO, D. & CIONI, R. 2002. Contrasting styles on Mount Vesuvius activity in the period between the Avellino and Pompeii Plinian eruptions, and some implications for assessment of future hazards. *Bulletin of Volcanology*, **64**, 372–391.

ANDRONICO, D., CALDERONI, G., CIONI, R., SBRANA, A., SULPIZIO, R. & SANTACROCE, R. 1995. Geological map of Somma–Vesuvius volcano. *Periodico di Mineralogia*, **64**, 77–78.

ARRIGHI, S., PRINCIPE, C. & ROSI, M. 2001. Violent strombolian and sub-Plinian eruptions at Vesuvius during post-1631 activity. *Bulletin of Volcanology*, **63**, 126–150.

ASIOLI, A. 1996. High resolution foraminifera biostratigraphy in the Central Adriatic basin during the last deglaciation: a contribution to the PALICLAS Project. *In*: OLDFIELD, F. & GUILIZZONI, P. (eds) *Palaeoenvironmental Analysis of Italian Crater Lake and Adriatic Sediments*. Memorie dell'Istituto Italiano di Idrobiologia, **55**, 197–218.

ASIOLI, A., TRINCARDI, F., LOWE, J. J., ARIZTEGUI, D., LANGONE, L. & OLDFIELD, F. 2001. Sub-millennial scale climatic oscillations in the central Adriatic during the Lateglacial: paleoceanographic implications. *Quaternary Science Reviews*, **20**, 1201–1221.

BARTOLE, R., SAVELLI, D., TRAMONTANA, M. & WEZEL, F. C. 1984. Structural and sedimentary features in the Tyrrhenian margin off Campania, Southern Italy. *Marine Geology*, **55**, 163–180.

BISSON, M., PARESCHI, M. T., ZANCHETTA, G., SULPIZIO, R. & SANTACROCE, R. 2007. Volcaniclastic debris-flow occurrences in the Campania region (Southern Italy) and their relation to Holocene–Late Pleistocene pyroclastic fall deposits: implications for large-scale hazard mapping. *Bulletin of Volcanology*, **70**, 157–167.

BOLTOVSKOY, E. 1976. Distribution of Recent foraminifera in the South American region. *In*: HEDLEY, R. H. & ADAMS, C. G. (eds) *Foraminifera, 2*. Academic Press, London, 171–236.

BORNHOLD, B. D. & PRIOR, D. B. 1990. Morphology and sedimentary processes on the subaqueous Noeick River delta, British Columbia, Canada. *In*: COLELLA, A. & PRIOR, D. B. (eds) *Coarse-Grained Deltas*. International Association of Sedimentologists, Special Publications, **10**, 169–181.

BRANCACCIO, L., CINQUE, A., ROMANO, P., RUSSO, F., SANTANGELO, N. & SANTO, A. 1991. Geomorphology and neotectonic evolution of a sector of the Tyrrhenian flank of the southern Apennines (Region of Naples, Italy). *Zeitschsift für Geomorphologie*, **82**, 47–58.

BROWN, R. J., ORSI, G. & DE VITA, S. 2008. New insights into Late Pleistocene explosive volcanic activity and caldera formation on Ischia (southern Italy). *Bulletin of Volcanology*, **70**, 583–603.

BUCCHERI, G. 1984. Pteropods as climatic indicators in Quaternary sequences. A lower–middle Pleistocene sequence outcropping in Cava Puleo (Ficarazzi, Palermo, Italy). *Palaeogeography, Palaeoclimatology, Palaeoecology*, **45**, 75–86.

BUCCHERI, G., CAPRETTO, G. *ET AL*. 2002. A high resolution record of the last deglaciation in the southern Tyrrhenian Sea: environmental and climatic evolution. *Marine Geology*, **186**, 447–470.

BUDILLON, F., VIOLANTE, C., CONFORTI, A., ESPOSITO, E., INSINGA, D., IORIO, M. & PORFIDO, S. 2005. Event beds in the recent prodelta stratigraphic record of the small flood-prone Bonea Stream (Amalfi Coast, Southern Italy). *Marine Geology*, **222–223**, 419–441.

CANALS, M., LASTRAS, G. *ET AL*. 2004. Slope failure dynamics and impacts from seafloor and shallow sub-seafloor geophysical data: case studies from the COSTA project. *Marine Geology*, **213**, 9–72.

CARALP, M. H. 1988. Size and morphology of the benthic foraminifera *Melonis barleeanum*: relationships with marine organic matter. *Journal of Foraminiferal Research*, **19**, 235–245.

CARBONE, A., LIRER, L. & MUNNO, R. 1984. Caratteri petrografici dei livelli piroclastici rinvenuti in alcuni gravity-cores nel Golfo di Pozzuoli e di Napoli. *Memorie della Società Geologica Italiana*, **27**, 195–204.

CAREY, S. & SIGURDSSON, H. 1987. Temporal variations in column height and magma discharge rate during the AD 79 eruption of Vesuvius. *Geological Society of America Bulletin*, **99**, 303–314.

CATTANEO, A., CORREGIARI, A., MARSSET, T., THOMAS, Y., MARSSET, B. & TRINCARDI, F. 2004. Seafloor undulation pattern on the Adriatic shelf and comparison to deep-water sediment waves. *Marine Geology*, **213**, 121–148.

CHIKITA, K. 1990. Sedimentation by river-induced turbidity currents: field measurements and interpretations. *Sedimentology*, **37**, 891–905.

CINQUE, A. & ROBUSTELLI, G. 2009. Alluvial and coastal hazards caused by long-range effect of Plinian eruptions: the case of the Lattari Mts. after the AD 79 eruption of Vesuvius. *In*: VIOLANTE, C. (ed.) *Geohazard in Rocky Coastal Areas*. Geological Society, London, Special Publications, **322**, 155–171.

CINQUE, A., AUGELLI, P. P. C. *ET AL*. 1997. Volcanism, tectonics and recent geomorphological change in the Bay of Napoli. *Geografia Fisica e Dinamica Quaternaria, Supplement*, **3**, 123–141.

CIONI, R., MARIANELLI, P. & SBRANA, A. 1992. Dynamics of the A.D. 79 eruption: stratigraphic, sedimentological and geochemical data on the successions from the Somma–Vesuvius southern and eastern sectors. *Acta Vulcanologica*, **2**, 109–123.

CIONI, R., SANTACROCE, R. & SBRANA, A. 1999. Pyroclastic deposits as a guide for reconstructing the multi-stage evolution of the Somma–Vesuvius caldera. *Bulletin of Volcanology*, **61**, 207–222.

CIONI, R., BERTAGNINI, A., SANTACROCE, R. & ANDRONICO, D. 2008. Explosive activity and eruption scenarios at Somma–Vesuvius (Italy): Towards a new classification scheme. *Journal of Volcanology and Geothermal Research*, **177**, 277–287.

COLELLA, A. & PRIOR, B. D. (eds) 1990. *Coarse-Grained Deltas*. International Association of Sedimentologists, Special Publications, **10**.

COLEMAN, J. M. 1988. Dynamic changes and processes in the Mississippi River Delta. *Geological Society of America Bulletin*, **100**, 999–1015.

COLEMAN, J. M. & PRIOR, D. B. 1982. Deltaic environments of deposition. *In*: SCHOLLE, P. A. & SPEARING, D. (eds) *Sandstone Depositional Environments*. American Association of Petroleum Geologists, Memoirs, **31**, 139–178.

COLLINS, B. D. & DUNNE, T. 1986. Erosion of tephra from the 1980 eruption of Mount St. Helens. *Geological Society of America Bulletin*, **97**, 896–905.

CONFORTI, A. 2003. *Stratigrafia integrata della sequenza Tardo-Quaternaria del settore settentrionale del Golfo di Salerno e di quello meridionale del Golfo di Napoli*. PhD thesis, University of Naples Federico II.

CORREGGIARI, A., TRINCARDI, F., LANGONE, L. & ROVERI, M. 2001. Styles of failure in late Holocene highstand prodelta wedges on the Adriatic shelf. *Journal of Sedimentary Research*, **71**, 218–236.

DABRIO, C. J. 1990. Fan-delta facies associations in late Neogene and Quaternary basins of southern Spain. *In*: COLELLA, A. & PRIOR, D. B. (eds) *Coarse-Grained Deltas*. International Association of Sedimentologists, Special Publications, **10**, 91–111.

DAL MASO, C., MATURANO, A. & VARONE, A. 1999. Pompei, il racconto dell'eruzione. *Le Scienze*, **371**, 58–65.

DI VITO, M., ISAIA, R. *ET AL*. 1999. Volcanism and deformation since 12,000 years at the Campi Flegrei caldera (Italy). *Journal of Volcanology and Geothermal Research*, **91**, 221–246.

DI VITO, M. A., SULPIZIO, R., ZANCHETTA, G. & D'ORAZIO, M. 2008. The late Pleistocene pyroclastic deposits of the Campanian Plain: new insights into the explosive activity of Neapolitan volcanoes. *Journal of Volcanology and Geothermal Research*, **177**, 19–48.

DROSER, M. J. & BOTTJER, D. J. 1991. Trace fossils and ichnofabric in Leg 119 cores. *In*: BARRON, J., LARSEN, B. *ET AL*. (eds) *Proceedings of the Ocean Drilling Program, Scientific Results, 119*. Ocean Drilling Program, College Station, TX, 635–641.

EINSELE, G. 2000. *Sedimentary Basins: Evolution, Facies and Sedimentary Budget*. Springer, Berlin.

ESPOSITO, E., PORFIDO, S. & VIOLANTE, C. (eds) 2004*a*. *Il nubifragio dell'ottobre 1954 a Vietri sul Mare, Costa di Amalfi, Salerno*. Pubblicazione Gruppo Nazionale per la Difesa dalle Catastrofi Idrogeologiche, **2870**.

ESPOSITO, E., PORFIDO, S., VIOLANTE, C., BISCARINI, C., ALAIA, F. & ESPOSITO, G. 2004*b*. Water events and historical flood recurrences in the Vietri sul Mare coastal area (Costiera Amalfitana, southern Italy). *In*: RODDA, G. & UBERTINI, L. (eds) *The Basis of Civilization—Water Science?* International Association of Hydrological Sciences Publication, **286**, 95–106.

FATELA, F. & TABORDA, R. 2002. Confidence limits of species proportions in microfossils assemblages. *Marine Micropaleontology*, **45**, 169–174.

FERNÁNDEZ-SALAS, L. M., LOBO, F. J., HERNÁNDEZ-MOLINA, F. J., SOMOZA, L., RODERO, J., DÍAZ DEL RÍO, V. & MALDONADO, A. 2003. High-resolution architecture of late Holocene highstand prodeltaic deposits from southern Spain: the imprint of high-frequency climatic and relative sea-level changes. *Continental Shelf Research*, **23**, 1037–1054.

FERRANTI, L., OLDOW, J. S. & SACCHI, M. 1996. Pre-Quaternary orogen-parallel extension in the Southern Apennine belt, Italy. *Tectonophysics*, **260**, 325–347.

FIGLIOLO, B. & MARTURANO, A. 1994. The eruptions of Vesuvius from the 7th to the 12th centuries. *In*: MORELLO, N. (ed.) *Volcanoes and History*. Brigati, Genova, 133–156.

FOLK, R. L. & WARD, W. C. 1957. Brazos river bar: a study in the significance of grain-size parameters. *Journal of Sedimentary Petrology*, **27**, 3–26.

FREY, R. W. & SEILACHER, A. 1980. Uniformity in marine invertebrate ichnology. *Lethaia*, **13**, 183–207.

GILBERT, R. 1975. Sedimentation in Lillooet Lake, British Columbia. *Canandian Journal of Earth Sciences*, **12**, 1697–1711.

GINGRAS, M. K., PEMBERTON, S. G. & SAUNDERS, T. 2001. Bathymetry, sediment texture and substrate cohesiveness; their impact on modern *Glossifungites* trace assemblages at Willapa Bay, Washington. *Palaeogeography, Palaeoclimatology, Palaeoecology*, **169**, 1–21.

GRÉMARE, A., SARDA, R. *ET AL.* 1998. On the dramatic increase of *Ditrupa arietina* O. F. Müller (Annelida: Polychaeta) along both the French and Spanish Catalan coasts. *Estuarine and Coastal Shelf Science*, **47**, 447–457.

HASIOTIS, T., CHARALAMPAKIS, M., STEFATOS, A., PAPATHEODOROU, G. & FERENTINOS, G. 2006. Fan-delta development and processes offshore a seasonal river in a seismically active region, NW Gulf of Corinth. *Geo-Marine Letters*, **26**, 199–211.

HOEFS, J. 1987. *Stable Isotope Geochemistry*. 3rd edn. Springer, Berlin.

HUNT, D. & TUCKER, M. E. 1992. Stranded parasequences and the forced regressive wedge systems tract: deposition during base-level fall. *Sedimentary Geology*, **81**, 1–9.

INSINGA, D. 2003. *Tefrostratigrafia dei depositi tardoquaternari della fascia costiera campana*. PhD thesis, University of Naples Federico II.

INSINGA, D., MOLISSO, F., LUBRITTO, C., SACCHI, M., PASSARIELLO, I. & MORRA, V. 2008. The proximal marine record of Somma–Vesuvius volcanic activity in the Naples and Salerno bays (eastern Tyrrhenian Sea) during the last 3 kyrs. *Journal of Volcanology and Geothermal Research*, **177**, 170–186.

IORIO, M., SAGNOTTI, L. *ET AL.* 2004. High-resolution petrophysical and paleomagnetic study of late-Holocene shelf sediments, Salerno Gulf, Tyrrhenian Sea. *Holocene*, **14**, 433–442.

JORISSEN, F. J. 1987. The distribution of benthic foraminifera in the Adriatic Sea. *Marine Micropalaeontology*, **12**, 21–48.

JORISSEN, F. J. 1988. *Benthic foraminifera from the Adriatic Sea; principles of phenotypic variation*. Utrecht Micropaleontology Bulletin, **37**.

JORISSEN, F. J. & WITTLING, I. 1999. Ecological evidence from live–dead comparison of benthic foraminiferal faunas off Cap Blanc (Northwest Africa). *Paleogeography, Paleoclimatology, Paleoecology*, **149**, 151–170.

KALLEL, N., PATERNE, M., LABEYRIE, L., DUPLESSY, J. C. & ARNOLD, M. 1997. Temperature and salinity records of the Tyrrhenian Sea during the last 18000 years. *Paleogeography, Paleoclimatology, Palaeoecology*, **135**, 97–108.

LEE, H. J., SYVITSKY, J. P. M., PARKER, G., ORANGE, D., LOCAT, J., HUTTON, J. H. W. & IMRAN, J. 2002. Distinguishing sediment waves from slope failure deposits: field examples, including the 'Humboldt Slide' and modelling results. *Marine Geology*, **192**, 79–104.

LINDSAY, J. F., PRIOR, D. B. & COLEMAN, J. M. 1984. Distributary mouth bar development and role of submarine landslides in delta growth, South Pass, Mississippi Delta. *AAPG Bulletin*, **68**, 1732–1743.

LINKE, P. & LUTZE, G. F. 1993. Microhabitat preferences of benthic foraminifera—a static concept or a dynamic adaptation to optimize food acquisition? *Marine Micropaleontology*, **20**, 215–234.

LIQUETE, C., ARNAU, P., CANALS, M. & COLAS, S. 2005. Mediterranean river systems of Andalusia, southern Spain, and associated deltas: a source to sink approach. *Marine Geology*, **222–223**, 471–495.

LIRER, L., PESCATORE, T. S., BOOTH, B. & WALKER, G. P. L. 1973. Two Plinian pumice-fall deposits from Somma–Vesuvius, Italy. *Geological Society of America Bulletin*, **84**, 759–772.

LIRER, L., VINCI, A., ALBERICO, I., GIFUNI, T., BELLUCCI, F., PETROSINO, P. & TINTERRI, R. 2001. Occurrence of inter-eruption debris-flow and hyperconcentrated flood-flow deposits on Vesuvio volcano, Italy. *Sedimentary Geology*, **139**, 151–167.

LIRER, L., CHIROSCA, M. C., MUNNO, R., PETROSINO, P. & GRIMALDI, M. (eds) 2005. *Il Vesuvio ieri, oggi, domani*. Regione Campania, Naples.

LOBO, F. J., FERNÁNDEZ-SALAS, L. M., MORENO, I., SANZ, J. L. & MALDONADO, A. 2006. The sea-floor morphology of a Mediterranean shelf fed by small rivers, northern Alboran Sea margin. *Continental Shelf Research*, **26**, 2607–2628.

LOEBLICH, A. R. & TAPPAN, H. 1988. *Foraminiferal Genera and their Classification*. Van Nostrand Reinhold, New York.

MACEACHERN, J. A., RAYCHAUDHURI, I. & PEMBERTON , S. G. 1992. Stratigraphic applications of the *Glossifungites* ichnofacies: Delineating discontinuities in the rock record. *In*: PEMBERTON, S. G. (ed.) *Applications of Ichnology to Petroleum Exploration, a Core Workshop*. Society of Economic Paleontologists and Mineralogists, Core Workshops, **17**, 169–198.

MAJOR, J. J., PIERSON, T. C., DIEARTH, R. L. & COSTA, J. E. 2000. Sediment yield following severe volcanic disturbance—A two decade perspective from Mount St. Helens. *Geology*, **28**, 819–822.

MALINVERNO, A. & RYAN, W. B. F. 1986. Extension in the Tyrrhenian Sea and shortening in the Apennines as result of arc migration driven by sinking of the lithosphere. *Tectonics*, **5**, 227–245.

MARIANI, M. & PRATO, R. 1988. I bacini neogenici costieri del margine tirrenico: approccio sismostratigrafico. *Memorie della Società Geologica Italiana*, **41**, 519–531.

MCCONNICO, T. S. & BASSETT, K. N. 2007. Gravelly Gilbert-type fan-delta on the Conway Coast, New Zealand: Foreset depositional processes and clast imbrications. *Sedimentary Geology*, **198**, 147–166.

MILIA, A. & TORRENTE, M. M. 1999. Tectonics and stratigraphic architecture of a peri-Tyrrhenian half-graben (Bay of Naples, Italy). *Tectonophysics*, **315**, 297–314.

MILIA, A., MIRABILE, L., TORRENTE, M. M. & DVORAK, J. J. 1998. Volcanism offshore of Vesuvius volcano in Naples Bay. *Bulletin of Volcanology*, **59**, 404–413.

MILIA, A., MOLISSO, F., RASPINI, A., SACCHI, M. & TORRENTE, M. M. 2008. Syn-eruptive features and sedimentary processes associated with pyroclastic currents entering the sea: the AD 79 eruption of Vesuvius,

Naples Bay, Italy. *Journal of the Geological Society, London*, **165**, 839–848.

MILLER, C. D. 1989. *Potential hazards from future volcanic eruptions in California*. US Geological Survey Bulletin, **1847**.

MILLER, C. D., MULLINEAUX, D. R. & CRANDELL, D. R. 1982. *Potential hazards from future volcanic eruptions in the Long Valley–Mono Lake area, east–central California and southwest Nevada: a preliminary assessment*. US Geological Survey Circular, **877**.

MOSHER, D. C. & SIMPKIN, P. G. 1999. Status and trends of marine high-resolution seismic reflection profiling: data acquisition. *Geoscience Canada*, **26**, 174–187.

MULDER, T. & SYVITSKI, J. P. M. 1995. Turbidity currents generated at river mouths during exceptional discharges to the world oceans. *Journal of Geology*, **103**, 285–299.

MULDER, T., SYVITSKI, J. P. M., MIGEON, S., FAUGÈRES, J.-C. & SAVOYE, B. 2003. Marine hyperpycnal flows: initiation, behavior and related deposits. A review. *Marine and Petroleum Geology*, **20**, 861–882.

MUNNO, R. & PETROSINO, P. 2004. New constraints on the occurrence of Y-3 upper Pleistocene tephra marker layer in the Tyrrhenian Sea. *Quaternario*, **17**, 11–20.

MURRAY, J. W. 1973. *Distribution and Ecology of Living Benthic Foraminiferids*. Cran, Russak, New York.

MURRAY, J. W. 1976. Comparative studies of living and dead benthic foraminiferal distribution. *In*: HEDLEY, R. H. & ADAMS, C. G. (eds) *Foraminifera, 2*. Academic Press, London, 45–109.

MURRAY, J. W. 1991. Ecology and distribution of benthic foraminifera. *In*: LEE, J. J. & ANDERSON, O. R. (eds) *Biology of Foraminifera*. Academic Press, New York, 221–253.

MUTTI, E., TINTERRI, R., BENEVELLI, G., DI BIASE, D. & CAPANNA, G. 2003. Deltaic, mixed and turbidite sedimentation of ancient foreland basins. *Marine and Petroleum Geology*, **20**, 733–755.

NAVA-SANCHEZ, E. H., GORSLINE, D. S., CRUZ-OROZCE, R. & GODINEZ-ORTA, L. 1999. The El Coyote fan-delta: a wave-dominated example from the Gulf of California, Mexico. *Quaternary International*, **56**, 129–140.

NEMEC, W. 1990*a*. Deltas—remarks on terminology and classification. *In*: COLELLA, A. & PRIOR, D. B. (eds) *Coarse-Grained Deltas*. International Association of Sedimentologists, Special Publications, **10**, 3–12.

NEMEC, W. 1990*b*. Aspects of sediment movement on steep delta slopes. *In*: COLELLA, A. & PRIOR, D. B. (eds) *Coarse-Grained Deltas*. International Association of Sedimentologists, Special Publications, **10**, 29–73.

NEMEC, W. 1995. The dynamics of deltaic suspension plumes. *In*: OTI, M. N. & POSTMA, G. (eds) *Geology of Deltas*. Balkema, Rotterdam, 31–93.

NEMEC, W. & STEEL, R. J. 1988. What is a fan-delta and how do we recognize it? *In*: NEMEC, W. & STEEL, R. J. (eds) *Fan-Deltas: Sedimentology and Tectonic Settings*. Blackie, Glasgow, 3–13.

OLDOW, J. S., D'ARGENIO, B., FERRANTI, L., PAPPONE, G., MARSELLA, E. & SACCHI, M. 1993. Large-scale longitudinal extension in the southern Apennines contractional belt, Italy. *Geology*, **21**, 1123–1126.

PAPPALARDO, L., CIVETTA, L. *ET AL*. 1999. Chemical and Sr-isotopical evolution of the Phlegrean magmatic system before the Campanian Ignimbrite and the Neapolitan Yellow Tuff eruptions. *Journal of Volcanology and Geothermal Research*, **91**, 141–166.

PARSONS, J. D., BUSH, J. W. M. & SYVITSKI, J. P. M. 2001. Hyperpycnal plume formation from riverine outflows with small sediment concentrations. *Sedimentology*, **48**, 465–478.

PASSARIELLO, I., MARZAIOLI, F. *ET AL*. 2007. Radiocarbon sample preparation at the CIRCE AMS laboratory in Caserta, Italy. *Radiocarbon*, **49**, 225–232.

PATACCA, E. & SCANDONE, P. 2007. Geology of the Southern Apennines. *Bollettino della Società Geologica Italiana, Special Issue*, **7**, 75–119.

PATERNE, M., GUICHARD, F. & LABEYRIE, J. 1988. Explosive activity of the south Italian volcanoes during the past 80 000 years as determined by marine tephrochronology. *Journal of Volcanology and Geothermal Research*, **34**, 153–172.

PEMBERTON, S. G. & FREY, R. W. 1985. The *Glossifungites* ichnofacies: modern examples from the Georgia Coasts, U.S.A. *In*: CURRAN, H. A. (ed.) *Biogenic Structures: their Use in Interpreting Depositional Environments*. Society of Economic Paleontologists and Mineralogists, Special Publications, **35**, 237–259.

PLINK-BJORKLUND, P. & STEEL, R. J. 2004. Initiation of turbidity currents: outcrop evidence for Eocene hyperpycnal-flow turbidites. *Sedimentary Geology*, **165**, 29–52.

PORFIDO, S., ESPOSITO, E., ALAIA, F., MOLISSO, F. & SACCHI, M. 2009. The use of documentary sources for reconstructing flood chronologies on the Amalfi rocky coast (southern Italy). *In*: VIOLANTE, C. (ed.) *Geohazard in Rocky Coastal Areas*. Geological Society, London, Special Publications, **322**, 173–187.

POSAMENTIER, H. W. & ALLEN, G. P. 1993. Variability of the sequence stratigraphic model: effects of local basin factors. *Sedimentary Geology*, **86**, 91–109.

POSTMA, G. 1990. Depositional architecture and facies of river and fan-deltas: a synthesis. *In*: COLELLA, A. & PRIOR, D. B. (eds) *Coarse-Grained Deltas*. International Association of Sedimentologists Special Publications, **10**, 13–27.

POSTMA, G. 1995. Sea-level-related architectural trends in coarse-grained delta complexes. *Sedimentary Geology*, **9**, 3–12.

POSTMA, G., BABIC, L., ZUPANIC, J. & ROE, S. L. 1988. Delta-front failure and associated bottomset deformation in a marine, gravelly Gilbert style fan-Delta. *In*: NEMEC, W. & STEEL, R. J. (eds) *Fan-Deltas—Sedimentology and Tectonic Settings*. Blackie, Glasgow, 91–102.

PRIOR, D. B. & BORNHOLD, B. D. 1990. The underwater development of Holocene fan-deltas. *In*: COLELLA, A. & PRIOR, D. B. (eds) *Coarse-Grained Deltas*. International Association of Sedimentologists, Special Publications, **10**, 75–90.

PRIOR, D. B., BORNHOLD, B. D., WISEMAN, W. J., JR. & LOWE, D. R. 1987. Turbidity current activity in a British Columbia fjord. *Science*, **237**, 1330–1333.

READING, H. G. 1996. *Sedimentary Environments: Processes, Facies and Stratigraphy*. Blackwell Science, Oxford.

REINECK, H. E. & SINGH, I. B. 1975. *Depositional Sedimentary Environments*. Springer, New York.

ROLANDI, G., BARRELLA, A. M. & BORRELLI, A. 1993. The 1631 eruption of Vesuvius. *Journal of Volcanology and Geothermal Research*, **58**, 183–201.

ROLANDI, G., PETROSINO, P. & MCGEEHIN, J. 1998. The inter-Plinian activity at Somma–Vesuvius in the last 3500 years. *Journal of Volcanology and Geothermal Research*, **82**, 19–52.

ROLANDI, G., MUNNO, R. & POSTIGLIONE, I. 2004. The A.D. 472 eruption of the Somma volcano. *Journal of Volcanology and Geothermal Research*, **129**, 291–319.

ROSI, M. & SANTACROCE, R. 1983. The A.D. 472 'Pollena' eruption: volcanological and petrological data for this poorly-known, Plinian-type event at Vesuvius. *Journal of Volcanology and Geothermal Research*, **17**, 249–271.

ROSI, M. & SBRANA, A. (eds) 1987. *Phlegrean Fields*. Quaderni de 'La Ricerca Scientifica', **114**.

ROSI, M., PRINCIPE, C. & VECCI, R. 1993. The 1631 Vesuvius eruption. A reconstruction based on historical and stratigraphic data. *Journal of Volcanology and Geothermal Research*, **58**, 151–182.

SACCHI, M., INFUSO, S. & MARSELLA, E. 1994. Late Pliocene–Early Pleistocene compressional tectonics in offshore Campania (eastern Tyrrhenian Sea). *Bollettino di Geofisica Teorica ed Applicata*, **36**, 141–144.

SACCHI, M., CONFORTI, A., MILIA, A., MOLISSO, F. & VIOLANTE, C. 2004. Il Golfo di Salerno. *In*: ESPOSITO, E., PORFIDO, S. & VIOLANTE, C. (eds) *Il nubifragio dell'Ottobre 1954 a Vietri sul Mare, costa di Amalfi, Salerno. Scenario ed effetti di una piena fluviale catastrofica in un'area di costa rocciosa.* Pubblicazione Gruppo Nazionale per la Difesa dalle Catastrofi Idrogeologiche, **2870**, 45–57.

SACCHI, M., INSINGA, D., MILIA, A., MOLISSO, F., RASPINI, A., TORRENTE, M. M. & CONFORTI, A. 2005. Stratigraphic signature of the Vesuvius 79 AD event off the Sarno prodelta system, Naples Bay. *Marine Geology*, **222–223**, 443–469.

SANTACROCE, R. (ed.) 1987. *Somma–Vesuvius. Quaderni de 'La Ricerca Scientifica'.*

SANTACROCE, R. & SBRANA, A. 2003. *Carta Geologica del Vesuvio.* Progetto CARG-Carta Geologica d'Italia, Servizio Geologico d'Italia. SELCA, Firenze.

SANTACROCE, R., SULPIZIO, R. *ET AL.* 2008. Age and whole rocks–glass composition of proximal pyroclastics from the major explosive eruptions of Vesuvius: a review as a tool for distal tephrostratigraphy. *Journal of Volcanology and Geothermal Research*, **177**, 1–18.

SCHMIEDL, G., DE BOVÉE, F., BUSCAIL, R., CHARRIÈRE, B., HEMLEBEN, C., MEDERNACH, L. & PICON, P. 2000. Trophic control of benthic foraminiferal abundance and microhabitat in the bathyal Gulf of Lions, western Mediterranean *Sea. Marine Micropaleontology*, **40**, 167–188.

SGARRELLA, F. & BARRA, D. 1984. Distribuzione dei foraminiferi bentonici nel Golfo di Salerno (Basso Tirreno, Italia). *Bollettino della Società dei Naturalisti Napoli*, **93**, 1–58.

SGARRELLA, F. & MONTCHARMONT-ZEI, M. 1993. Benthic foraminifers of the Gulf of Naples (Italy): systematics and autoecology. *Bollettino della Società Paleontologica Italiana*, **32**, 145–264.

SIANI, G., SULPIZIO, R., PATERNE, M. & SBRANA, A. 2004. Tephrostratigraphy study for the last 18,000 ^{14}C years in a deep-sea sediment sequence for the South *Adriatic. Quaternary Science Review*, **23**, 2485–2500.

SIGURDSSON, H., CAREY, S., CORNELL, W. & PESCATORE, T. 1985. The eruption of Vesuvius in AD 79. *National Geographic Research*, **1**, 332–387.

SIMPKIN, P. G. & DAVIS, A. 1993. For seismic profiling in very shallow water, a novel receiver. *Sea Technology*, **34**, 21–28.

SULPIZIO, R., MELE, D., DELLINO, P. & LA VOLPE, L. 2005. A complex sub-Plinian-type eruption from low-viscosity, phonolitic to tephri-phonolitic magma: the AD 472 (Pollena) eruption of Somma–Vesuvius, Italy. *Bulletin of Volcanology*, **67**, 743–767.

SULPIZIO, R., ZANCHETTA, G., DEMI, F., DI VITO, M., PARESCHI, M. T. & SANTACROCE, R. 2006. The Holocene syneruptive volcaniclastic debris flow in the Vesuvian area: geological data as a guide for hazard assessment. *In*: SIEBE, C., MACIAS, J. L. & AGUIRRE-DIAZ, G. J. (eds) *Neogene–Quaternary Continental Margin Volcanism: a Perspective from Mexico*. Geological Society of America, Special Papers, **402**, 203–221.

SULTAN, N., COCHONAT, P. *ET AL.* 2004. Triggering mechanisms of slope instability processes and sediment failures on continental margins: a geotechnical approach. *Marine Geology*, **213**, 291–321.

TRINCARDI, F. & SYVITSKI, J. P. M. 2005. Advances on our understanding of delta/prodelta environments: A focus on southern European margins. *Marine Geology*, **222–223**, 1–5.

TRINCARDI, F., CATTANEO, A., CORREGGIARI, A. & RIDENTE, D. 2004. Evidence of soft sediment deformation, fluid escape, sediment failure and regional weak layers within the late Quaternary mud deposits of the Adriatic Sea. *Marine Geology*, **213**, 91–119.

VEZZOLI, L. (ed.) 1988. *Island of Ischia*. Quaderni de 'La Ricerca Scientifica', **114**.

VIOLANTE, C. 2009. Rocky coast: geological constraints for hazard assessment. *In*: VIOLANTE, C. (ed.) *Geohazard in Rocky Coastal Areas*. Geological Society, London, Special Publications, **322**, 1–31.

VIOLANTE, C., BISCARINI, C., ESPOSITO, E., MOLISSO, F., PORFIDO, S. & SACCHI, M. 2009. The consequences of hydrologic events on steep coastal watersheds: the Costa d'Amalfi, eastern Tyrrhenian sea. *In*: LIEBSHER, H. J. *ET AL.* (eds) *The Role of Hydrology in Water Resource Management. Capri Italy, 2008.* IAHS Publication, **327**, 102–113.

VOGEL, J. S., SOUTHON, J. R., NELSON, D. E. & BROWN, T. A. 1984. Performance of catalytically condensed carbon for use in accelerator mass spectrometry. *Nuclear Instruments and Methods*, **B5**, 289–293.

WENINGER, B. & JÖRIS, O. 2007. A ^{14}C age calibration curve for the last 60 ka: the Greenland-Hulu U/Th timescale and its impact on understanding the Middle to Upper Paleolithic transition in Western Eurasia. *Journal of Human Evolution*, **55**, 772–781.

WENINGER, B., JÖRIS, O. & DANZEGLOCKE, U. 2008. CalPal-2007. Cologne Radiocarbon Calibration & Palaeoclimate Research Package. World Wide Web Address: http://www.calpal.de/

WESCOTT, W. A. & ETHRIDGE, F. G. 1990. Fan-deltas—Alluvial fans in coastal settings. *In*: RACHOCKI, A. H. & CHURCH, M. (eds) *Alluvial Fans: a Field Approach*. Wiley, New York, 195–211.

WULF, S., KRAML, M., BRAUER, A., KELLER, J. & NEGENDANK, J. F. W. 2008. Tephrochronology of the 100 ka lacustrine sediment record of Lago Grande di Monticchio (southern Italy). *Journal of Volcanology and Geothermal Research*, **177**, 118–132.

WULF, S., KRAML, M., BRAUER, A., KELLER, J. & NEGENDANK, J. F. W. 2004. Tephrochronology of the 100 ka lacustrine sediment record of Lago Grande di Monticchio (southern Italy). *Quaternary International*, **122**, 7–30.

Catastrophic landslides off Ischia volcanic island (Italy) during prehistory

G. DE ALTERIIS[1,2]* & C. VIOLANTE[1]

[1]*IAMC-CNR Institute for Coastal Marine Environments, National Research Council, Calata Porta di Massa, 80133 Napoli, Italy*

[2]*GeoLab-Marine Surveys srl, Via Monteruscello 75, 80078, Pozzuoli, Napoli, Italy*

**Corresponding author (e-mail: giovanni.dealteriis@geolab.it)*

Abstract: Generally the evaluation of the geological hazards from active volcanoes chiefly concerns the prediction of eruptions whereas less attention is generally paid to other volcanic-related phenomena, such as avalanching–landsliding. This is the case for Ischia (Italy), an active volcanic complex, whose collapse behaviour is only now being evaluated and recognized following extensive marine geophysical and geological investigations. The island of Ischia represents the emerged section of a larger east–west-trending volcanic ridge. The central sector of the island, Mt. Epomeo, has risen to at least 800 m above sea level (a.s.l.) in the past *c.* 30 ka, at an average rate of 20 mm a^{-1}. The major consequence of such volcano-tectonic uplift includes either sudden collapses, with attendant debris avalanches, or other mass movements in the form of mud-debris flows, debris slides and rock-falls, all radiating out from Mt. Epomeo and most of them entering the sea. During prehistoric times the island of Ischia underwent major catastrophic collapses resulting in debris avalanche deposits of >1 km^3 to <0.5 km^3 that have been recognized offshore both NW and south of the island. This study provides possible scenarios for the emplacement of these deposits, with particular reference to the resulting landslide-related tsunami hazard.

Ischia is a densely populated volcanic island bordering the Bay of Naples, Italy, to the west (Figs 1 and 2). The island is inhabited by some 45 000 people, and the population increases several-fold during the summer as a result of tourism.

Ischia's natural history includes a broad range of volcano-related phenomena such as eruptions, shallow earthquakes and landslides, whose influence on human activity is proven since Neolithic (3000–4000 years BP), when the island was already inhabited by Italic populations. Then, after a long unwitnessed interval, the island was settled by Greeks around the seventh century BC and became the first Greek colony in southern Italy under the name of Pithecussai (Buchner 1986). The first traces of the Roman civilization date back to the fourth century BC and partly overlap with the Greek former colonization. According to archaeologists the Roman village of Aenaria suddenly disappeared around AD 130–150, probably after a natural disaster (i.e. an eruption or a volcanic earthquake; see Buchner 1986). Eruptions lasted until the Middle Ages, and the last occurred in AD 1302 when the Arso lava flow covered the eastern side of the island (Vezzoli 1988). The last destructive earthquake occurred in 1883 at Casamicciola village on the north side of the island (Cubellis & Luongo 1998). Volcano-tectonic ground uplift, coupled with intense hydrothermal weathering and seismic shaking, has strongly favoured recurrent landsliding over the subaerial flanks of Mt. Epomeo, the major relief of the island. These landslides, in the form of debris slides, debris flows or mud flows, and their effects on humans have been frequently reported in historical chronicles from the Middle Ages to the present (see Guadagno & Mele 1995; Mele & Del Prete 1998).

Ischia is one of the seven Italian volcanoes considered 'active' by the national scientific community, the others being Campi Flegrei and Vesuvius in the Neapolitan region; the islands of Stromboli, Volcano and Pantelleria; and Etna in Sicily. Provided an average recurrence interval can be determined (based on its historical–stratigraphical record) a volcano may be considered active if the recurrence period is greater than the time elapsed since the last eruption. This definition suggests the presence of a magma chamber at relatively shallow depths, with the potential hazard of future eruptions.

After the discovery of large underwater hummocky fields in deep water to the south of Ischia (Chiocci *et al.* 1998) the island offshore has been the subject of a wide range of marine geophysical and geological surveys. The major outcome was the recognition of the 'Ischia debris avalanche', the first unequivocal example of a mass-transport deposit related to a prehistoric collapse so far recognized in the Mediterranean (Chiocci & de Alteriis 2006).

From: VIOLANTE, C. (ed.) *Geohazard in Rocky Coastal Areas*. The Geological Society, London, Special Publications, **322**, 73–104. DOI: 10.1144/SP322.3

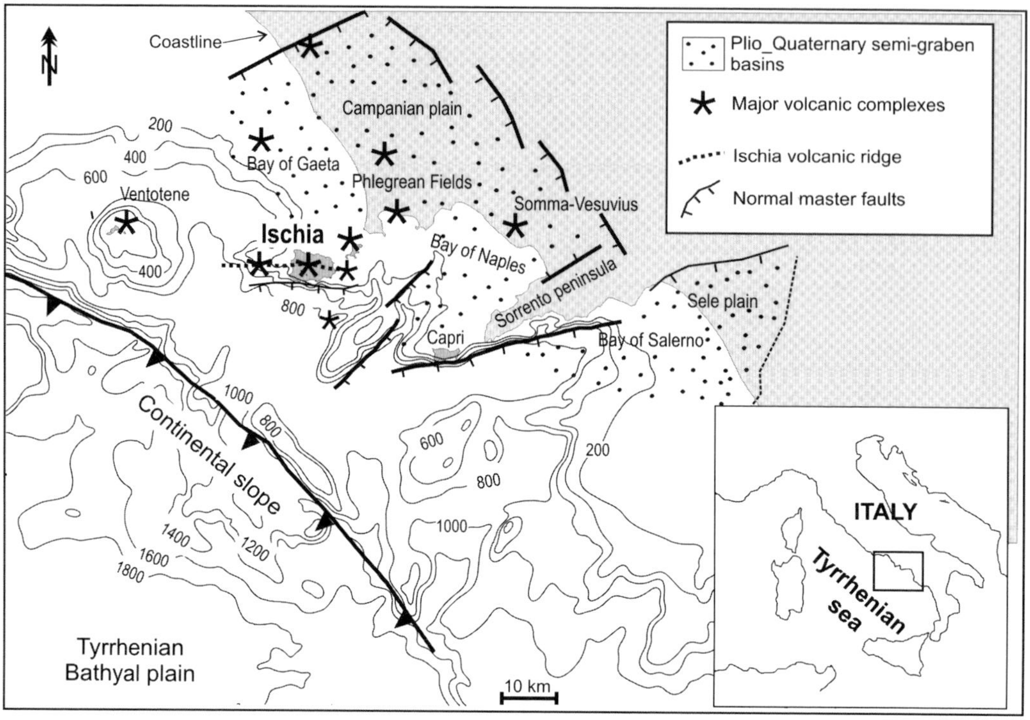

Fig. 1. Location of Ischia and the Bay of Naples on the eastern Tyrrhenian extensional margin, Italy. In this region, volcanism followed east–west and ENE–WSW lineaments, and Ischia volcano lies at their junction. Fine contour lines are bathymetry from GEBCO Digital Atlas; contour interval is 200 m.

At the same time, previously unreported hummocky terrains were detected at shallow depths offshore to the north and west of the island (Violante *et al.* 2004).

These studies have revealed the importance of gravity-driven processes leading to the development of extremely rapid (catastrophic) landslides, such as debris avalanches and debris flows, entirely or partially underwater. The studies also established that such deposits are all related to the volcano-tectonic uplift of Mt. Epomeo, and that they are one or more orders of magnitude greater than the on-land landslides mentioned above. Such behaviour makes this volcanic complex, together with the island of Stromboli, among the most likely potential sources for landslide-related tsunami in the Tyrrhenian Sea.

The aim of this paper is to provide new geophysical and geological data on Ischia submarine mass movements and to discuss their origin and behaviour. In addition, an attempt is made to assess the geological hazard affecting the island and the neighbouring coastal zone.

Geological setting

Ischia, together with Procida, Campi Flegrei and Somma–Vesuvius, is part of the volcanic complex encircling the Bay of Naples that originates from the crustal extension and associated magmatism that occurred in this sector of the eastern Tyrrhenian margin during the Pleistocene (Fig. 1). Crustal extension in the Bay of Naples resulted in a half-graben basin confined by a NE–SW-trending master fault roughly perpendicular to the regional trend of the margin (Milia & Torrente 1999). Magnetic data also indicate that magma ascent and attendant volcano alignments have been primarily favoured by NE–SW or east–west faults rather than by other lineaments (Bruno *et al.* 2002).

The known volcanic record of the island dates back to *c.* 150 ka, which is the age of the older products cropping out in the southeastern and southwestern corners of the island. During this period roughly a hundred eruption events have been recognized, almost all related to the ejection of small volumes of alkali basalt to trachytic magma, leading to the emplacement of pyroclastic deposits, lava domes and, in a few cases, large ignimbrites (Fig. 3; Rittmann 1930; Gillot *et al.* 1982; Vezzoli 1988). The highest relief of the island, Mt. Epomeo, which is 787 m high (Fig. 4), chiefly consists of an ignimbrite, known in the literature as the Mt. Epomeo Green Tuff (MEGT), which

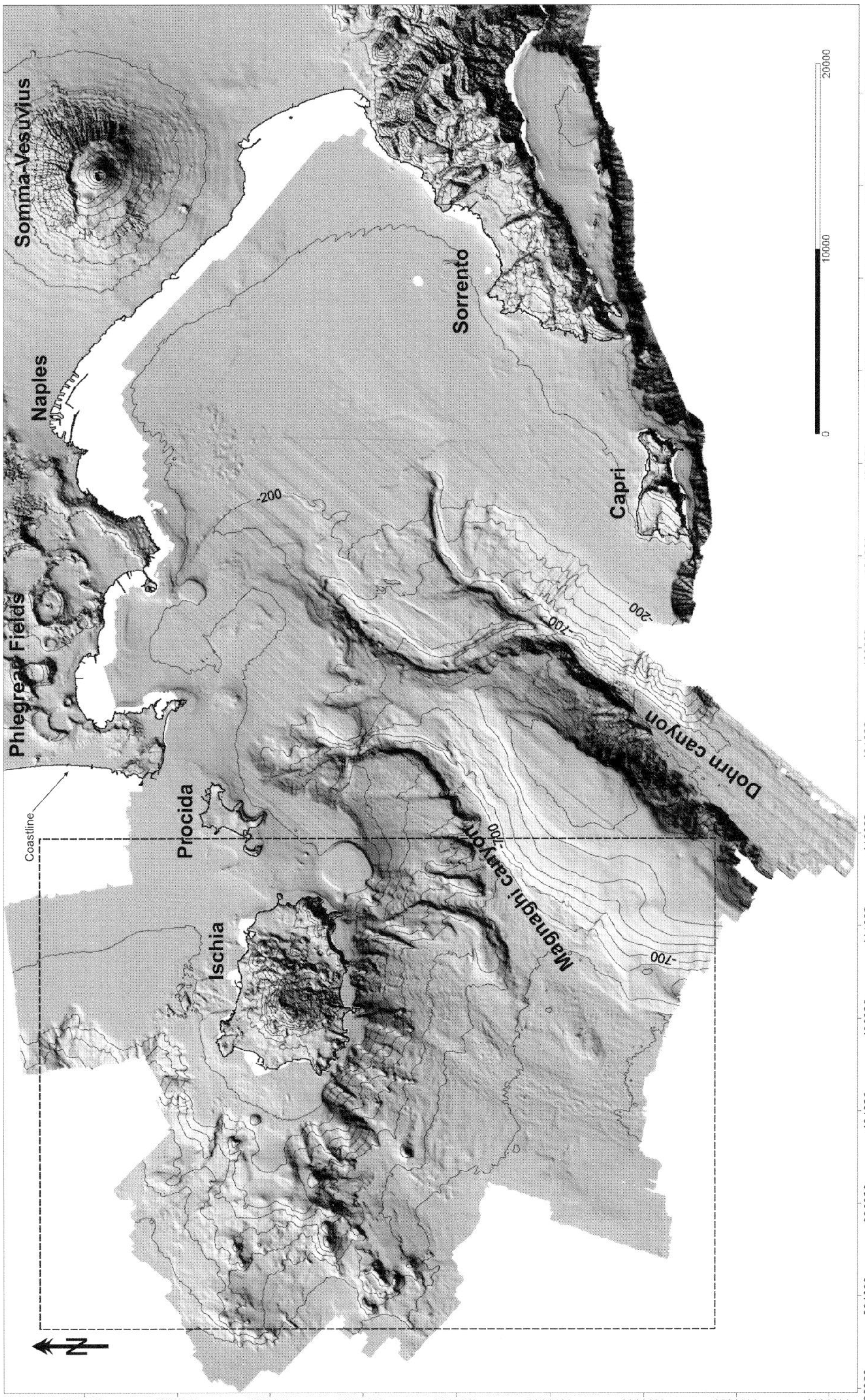

Fig. 2. Digital elevation model (DEM) of the Bay of Naples and its surrounding region rendered as a shaded relief map (modified after D'Argenio *et al.* 2004). It should be noted that Ischia represents the emerged part of a larger volcanic ridge extending to the west. The Magnaghi canyon is separated from the Dohrn canyon by the Banco di Fuori morphostructural high, which corresponds to a Mesozoic monocline. Water depth contour is in metres from sea level. Reproduced with permission of APAT-Dipartimento Difesa del Suolo, Rome, Italy.

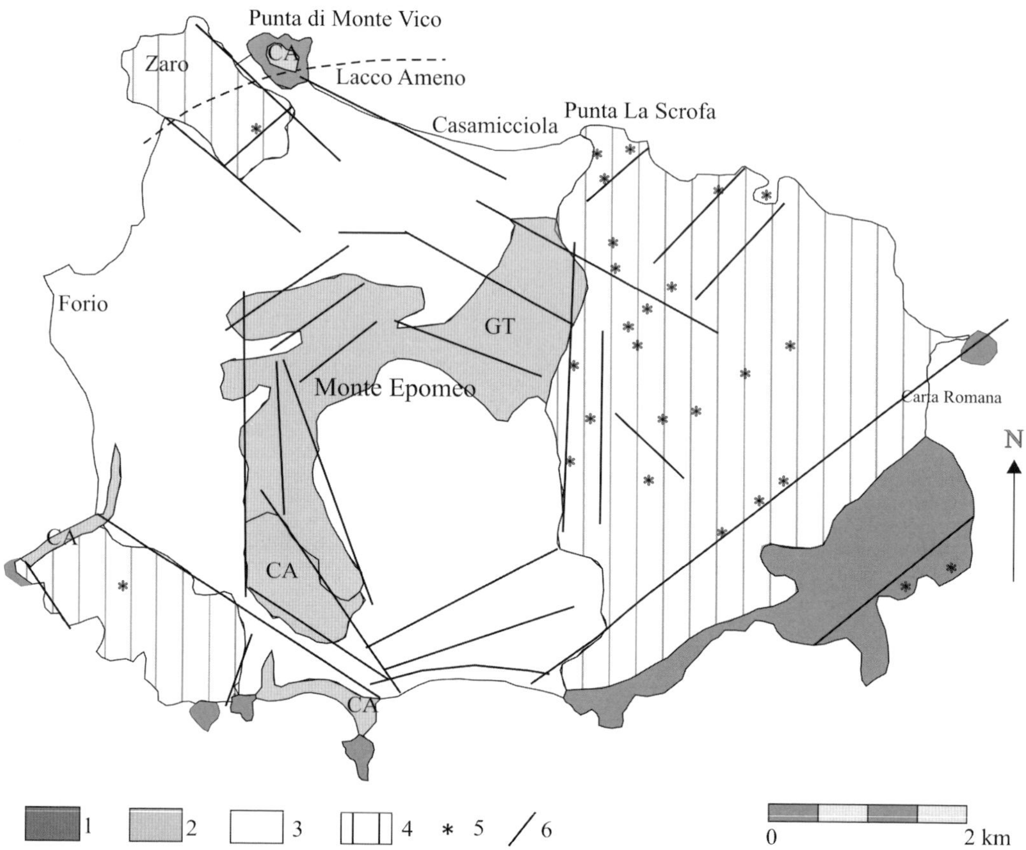

Fig. 3. Simplified geological map of Ischia. 1, volcanic units and lava domes preceding Monte Epomeo Green Tuff (147–75 ka); 2, Monte Epomeo Green Tuff (GT) and Colle Ietto formation (55 ka); Citara tuffs (CA); 3, landslide deposits; 4, Recent lava flows and pyroclastic deposits (<20 ka); 5, Holocene eruptive vents; 6, faults (modified from Gillot *et al.* 1982; Tibaldi & Vezzoli 1998).

was emplaced *c.* 55 ka. The MEGT is the most widespread pyroclastic deposit on the island and its welded facies is thickest around Mt. Epomeo. Most researchers agree that there was a vertical collapse associated with the MEGT eruption that led the central sector of the island being invaded by the sea, and then uplifted as a caldera resurgent block (Orsi *et al.* 1991; Tibaldi & Vezzoli 1998) whereas in former interpretations a simple volcano-tectonic horst model had been proposed (Rittmann 1930). However, the geological evidence, including drag folding of a younger formation, 43–33 $\pm$ 5 ka old, in contact with the MEGT (Gillot *et al.* 1982) and bio-stratigraphical data (Barra *et al.* 1992), proves that the block has risen in the past 33 ka by at least 800 m. This also implies that uplift might have started before 33 ka. However, adopting a 35 ka time span and a total displacement of 700 m, one obtains an average, conservative rate of *c.* 20 mm a^{-1} (Fig. 5).

Despite uncertainties about the onset and duration of such uplift, geological and archaeological evidence (Alessio *et al.* 1996; Buchner *et al.* 1996) indicates that it has lasted until the Holocene. However, data collected in the first ground levelling experiments in 1892–1912, during the period 1980–1987, and in recent studies, most by Differential Global Positioning System (DGPS) surveys (1991–2003), indicate that neither Mt. Epomeo nor other sectors of the island are currently rising (Fig. 6) but instead long-term subsidence (3 mm a^{-1}) as a result of slow landsliding is continuing (Manzo *et al.* 2006).

Data and methods

The dataset used in this study includes partially published swath bathymetric data, acoustic imagery from side-scan sonar records, and some high-resolution seismic and sub-bottom acoustic profiles.

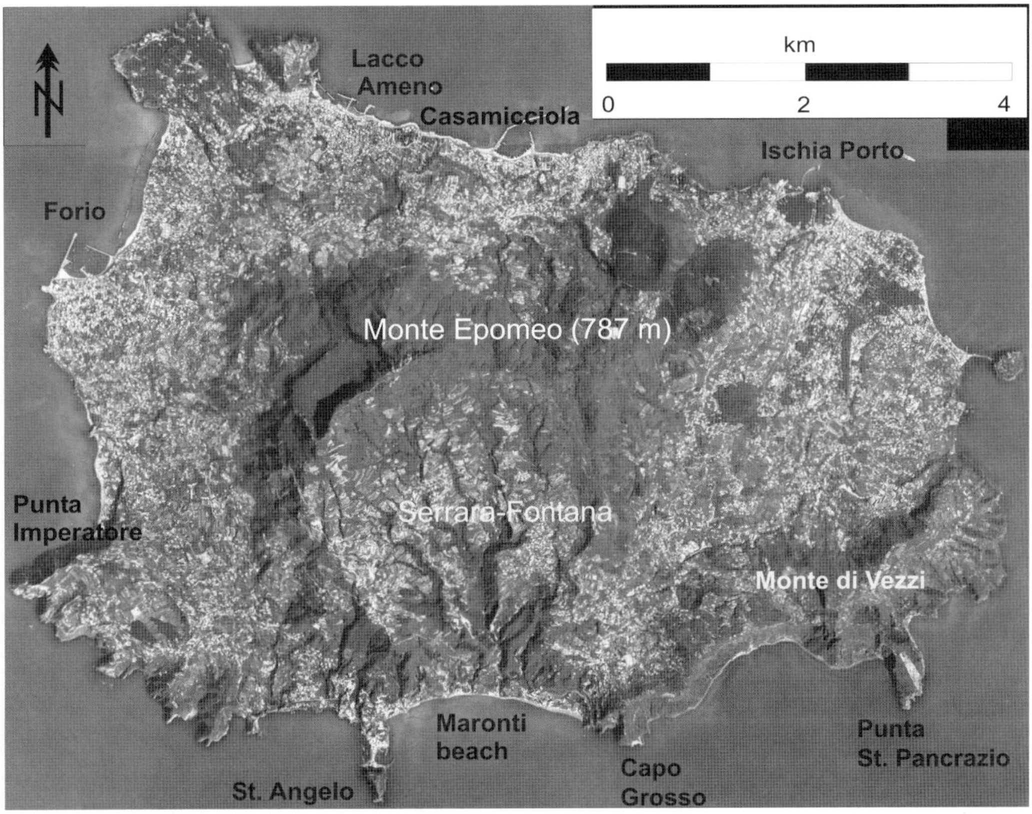

Fig. 4. Satellite image of Ischia with indication of the main localities mentioned in the text.

Data from direct observations (remotely operated vehicles and scuba-diving) and sea-bottom sampling (dredge hauls, gravity and vibro-cores) have also been taken into account.

The bathymetric data have been collected in several multibeam sonar surveys over a wide depth range, with different spatial resolutions, in the period 2001–2003. The merging of these surveys with the island topography resulted in a 20 m × 20 m digital elevation model (DEM) (Fig. 7; de Alteriis *et al.* 2006). This DEM allows us to recognize large-scale features and sea-floor fabric with a spatial resolution of a few tens of metres, which is sufficient to detect most of the blocky fields described below. Furthermore, a higher resolution DEM with a 5 m × 5 m cell size extracted from the field dataset permits us to recognize metre-scale sea-floor fabric such as sedimentary structures and minor blocks in the northwestern offshore areas (Fig. 8).

Side-scan sonar data were acquired with a 100 kHz Edgetech DF1000 tow fish and processed through the Isis-Delph Map^{TM} software package for navigational errors and slant-range correction. Processing and mosaicing of the sonar records resulted in backscatter maps with sub-metre resolution. High-resolution side-scan sonar coverage was almost complete over the northwestern submerged flank whereas the southern slope was studied using a wide range, deep-towed (TOBI) vehicle (Chiocci *et al.* 1998; Chiocci & de Alteriis 2006).

Seismic single-channel data include some digital and analogue records both acquired with a 1–4 kJ sparker source. Digital records were processed in the 200–1500 Hz frequency range after de-bias, filtering, time variant gain, automatic gain control and muting. The vertical and spatial (horizontal) resolution allowed by this acquisition system was 1 m and 2–3 m, respectively.

Results

Morpho-structural overview

The island of Ischia is the subaerial section of a large volcanic ridge characterized by east–west and ENE–WSW structural patterns. The ridge constrains southwards the wide shelf area of the Bay of

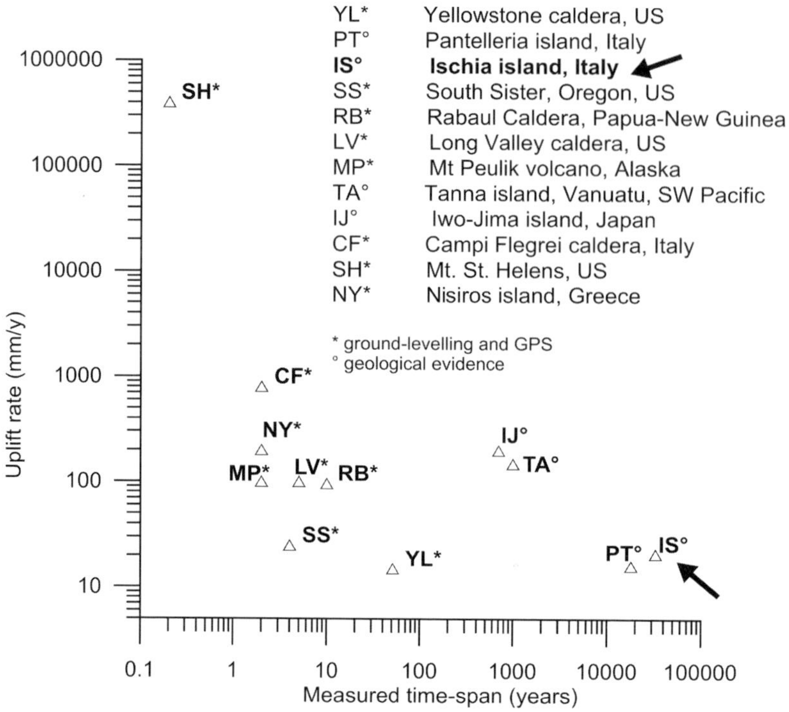

Fig. 5. Uplift rates in volcanic areas taken from the literature compared with measured time span. It should be noted that the lowest uplift rates, as in the case of Ischia (arrow) and Pantelleria, are those averaged over $10^3 - 10^4$ year time spans. Thus the Ischia volcano-tectonic uplift may have attained higher rates in the past several thousand years. YL, Yellowstone caldera, wyoming, USA (Dzurisin *et al.* 1999); PT, Pantelleria island, Italy (Mahood & Hildreth 1983); IS, Ischia island, Italy (Gillot *et al.* 1982); SS, South Sister volcano, Oregon, USA, (www.usgs.gov); RB, Rabaul caldera, Papua New Guinea (McKee *et al.* 1989); LV, Long Valley caldera, California, USA (Hill 1984); MP, Mt. Peulik volcano, Alaska, (www.usgs.gov); TA, Tanna island, Vanuatu, SW Pacific (Chen *et al.* 1995); IJ, Iwo-Jima, Japan (Kaizuka, 1992); CF, Campi Flegrei caldera, Italy (Berrino *et al.* 1984); SH, Mount St. Helens volcano, Washington, USA (www.usgs.gov); NY, Nisiros island, Greece (www.gein.noa.gr).

Gaeta and forms part of the steep continental slope bordering the Bay of Naples to the north. In the Gaeta embayment an overfilled sedimentary regime has been favoured by high river input and progradation during the Pleistocene whereas the Naples embayment has partly undergone an erosional regime along the Magnaghi and Dohrn canyons (Figs 2 and 7). This resulted in sediment damming and burial of the volcanic basement to the north and a very narrow or absent shelf to the south of the island. Such asymmetry is illustrated by a north–south topographic–bathymetric profile across the volcanic edifice (Fig. 9a); the steep northern subaerial slope and an equally steep underwater southern slope should be noted. At the same time an east–west section (Fig. 9b) shows oversteepening of the western Mt. Epomeo flank, as opposed to a gentler eastern flank resulting from the stack of recent volcanic products and monogenic vents. Thus, only the eastern side of the volcanic edifice may be considered entirely buttressed.

The shelf area, which is 3–4 km wide NW and west of the island, is extensively incised by canyon heads or erosional 'chutes', some with typical trapezoidal shape. The shelf width is reduced to less than 1 km on the southern flank. Remnants of the depositional shelf break located at depths of 140–180 m can be recognized between canyon heads. A shallower terrace rim, 80 m deep, can be followed along the northern and the south–SW shelf sectors. A third, shallowest terrace level, corresponding to the present-day coastal wedge although discontinuous, encircles the island along the 15–20 m isobaths, except for its western side, where a deeper rim develops at depths of 30–35 m. Locally some terrace rims are affected by incipient slope failure (Fig. 10).

Onshore, the central sector of the island is dominated by the Mt. Epomeo block, which has a polygonal shape and is bordered to the west, NW and north by fault-controlled scarps or steep slopes (dipping 40–50° to near vertical). Not all

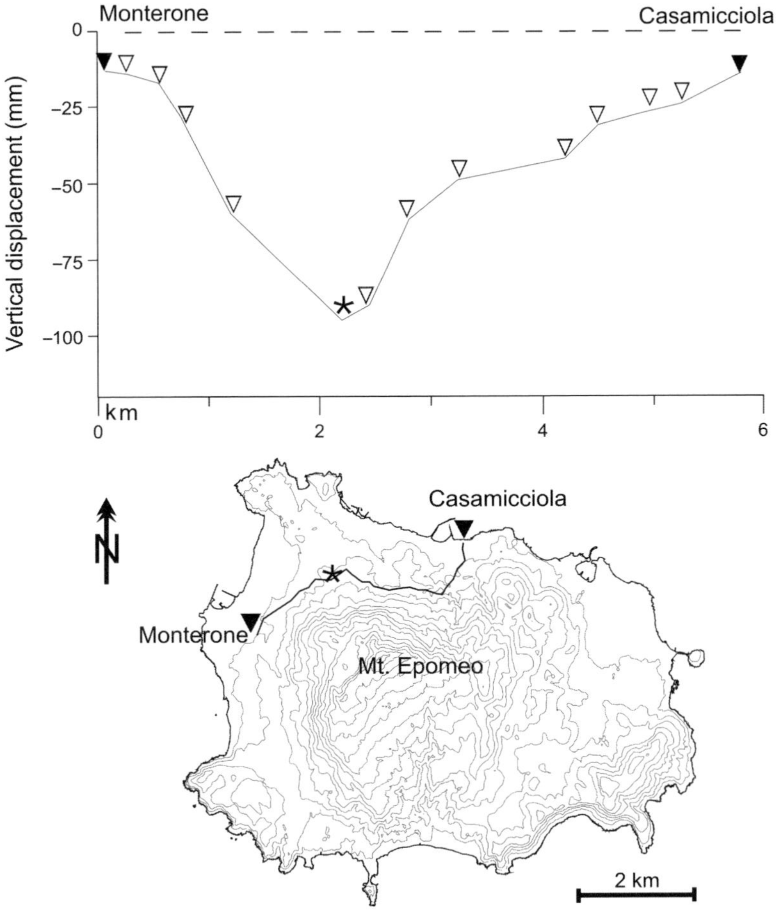

Fig. 6. Plot of the ground deformation resulting from DGPS-based monitoring survey along the 'Borbonica line' along northwestern flank of Mount Epomeo measured in the period 1987–2003 (simplified after Manzo *et al.* 2006). The overall subsidence and the −100 mm peak located at the northwestern toe of Mt. Epomeo should be noted. The dashed line is the 1987 reference datum. ▼ symbol, Casamicciola and Monterone benchmarks.

researchers agree about the exact geometry of these faults and whether they can be considered as inward or outward dipping, as this has implications for the stress regime that produced the uplift (Orsi *et al.* 1991; Tibaldi & Vezzoli 2000).

Conversely, the Mt. Epomeo southern flank has a gentler slope with an overall dip of 12–15° towards the SSE. This feature has led some researchers to invoke a SSE-dipping monocline (Acocella & Funiciello 1999) producing a structural rather than morphological slope.

Three main areas of hummocky topography, typically indicative of debris avalanche (hereafter DA) deposits radiate out from Mt. Epomeo and spread over the southern continental slope, and over the western and northern shelf areas (Figs 7 and 11). For simplicity they have been named the southern debris avalanche (SDA), western debris deposit (WDD) and northern debris deposit (NDD), respectively. The SDA has already been recognized as a DA (Chiocci & de Alteriis, 2006), and the interpretation of the WDD and NDD as debris avalanche deposits will be discussed below.

Offshore evidence of flank instability

The southern debris avalanche (SDA)

The southern undersea section of the island consists of a very steep upper slope, which shows a 4 km wide trapezoidal scar bordered to the west by a very steep north–south sidewall aligned with the S. Angelo lava dome on land. Upslope, off the Maronti shoreline between the S. Angelo promontory and

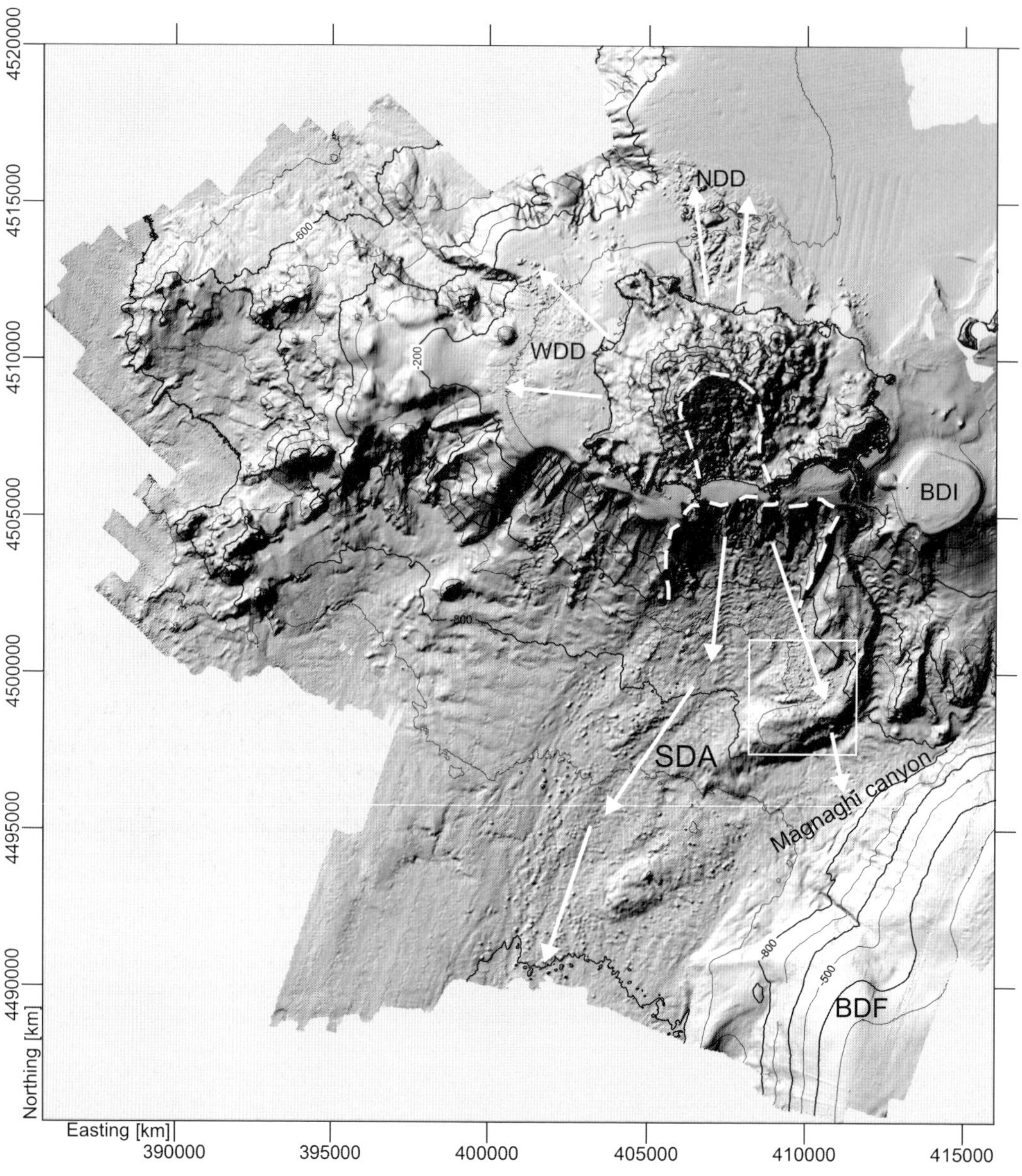

Fig. 7. Combined shaded relief and contour map of Ischia and surrounding marine areas based on a 20 m cell size DEM derived from swath bathymetric surveys and topographic data. Contours interval is 100 m, with bold contours every 200 m. White box shows the location of Figure 12. BDF, Banco di Fuori high; BDI, Banco d'Ischia undersea volcano. SDA, WDD and NDD are the landslides investigated in this study. White arrows indicate main runout axes. Dashed lines are inferred failure scars. Metric coordinates, UTM projection, Zone 33; datum: WGS84.

Capo Grosso, a subhorizontal terrace with its rim located at −20 m extends for 800–1000 m from the shoreline (Fig. 7).

The SDA, first observed by Chiocci *et al.* (1998) and later investigated by Chiocci & de Alteriis (2006), appears on swath bathymetry as an elongated field of blocks covering an area of about 150 km^2 extending from the toe of the submarine slope at −550 m down to −1100 m, reaching a distance of 40–45 km from the source area. Along the runout axis, slopes range from 5–6° to less than 1–2° in the distal areas. The main tongue of the deposit is constrained in a relatively narrow chute that curves westward as it approaches the Capo Grosso

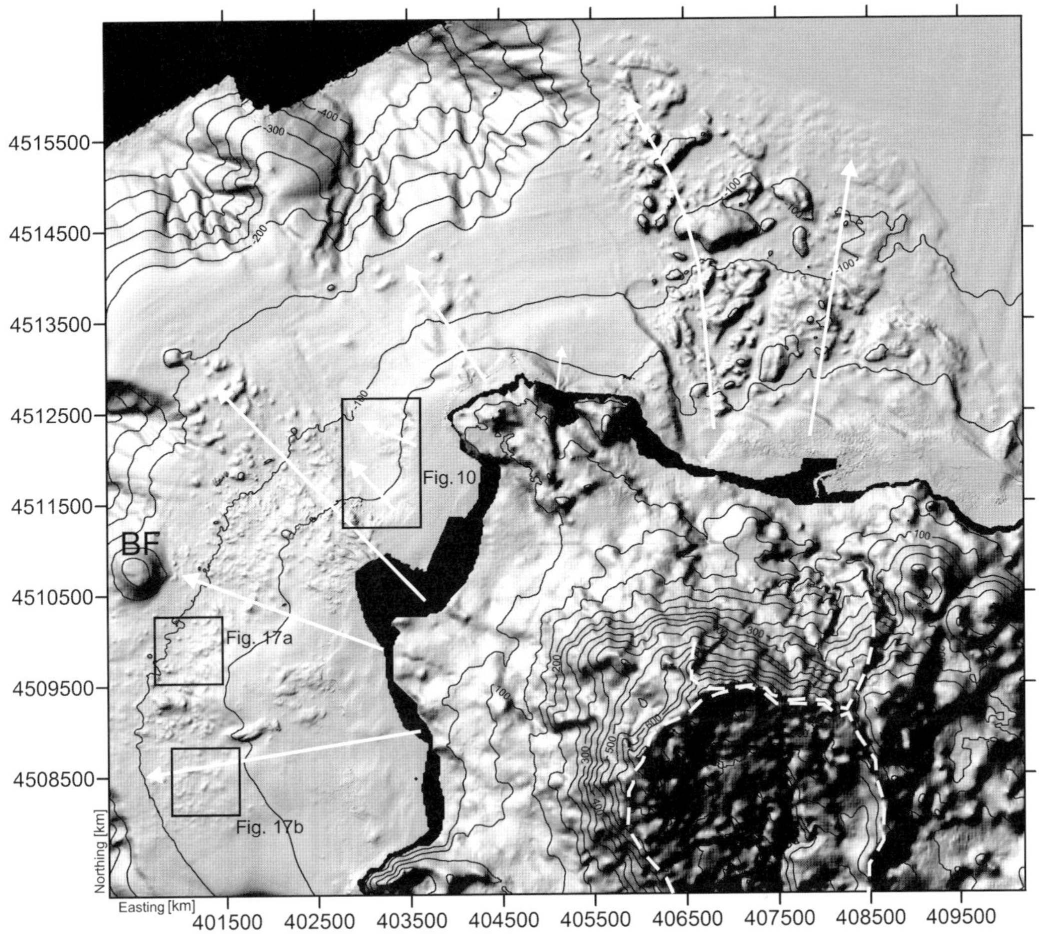

Fig. 8. Combined shaded relief and contour map of northwestern flank of Ischia based on a 5 m cell size DEM derived from swath bathymetric surveys and land topography. Contours interval is 50 m. BF, Banco di Forio submerged volcano. White arrows indicate main runout axes; dashed lines indicate inferred failure scars. Coordinates as in Figure 7.

bank, a volcanic basement outcrop at 500 m depth (Fig. 11). It forms elongated lobes from −800 to −1000 m with heights of 15–20 m above the surrounding sea floor. Traces of erosion, especially over the western levee of the DA, appear on sub-bottom profiles as well as on detailed bathymetry. A secondary debris tongue, which possibly originated from an eastern scar, travelled downslope towards the SSE, climbing a 70 m submarine high before falling into the thalweg of the Magnaghi canyon (Fig. 12). This tongue seems to have smaller and better sorted blocks compared with the main tongue, suggesting a different source area. At 40 km from the source areas some outrunner blocks are located on flat surfaces and appear to have travelled much further than the DA itself.

Side-scan sonograms allow us to detect with a higher accuracy than multibeam data the main block sizes, abundance and post-avalanche sea-floor fabric. Blocks are on average 20–30 m across and rise 10–15 m above the surrounding sea floor, with some boulders attaining a size of 150 m × 100 m and a height of 30–35 m (Fig. 13). Post-avalanche sea-bottom features include incipient creep with concave failure patterns, comet marks and abundant minor slope instabilities, mostly in the form of debris chutes along the upper slope (see Figs 7 and 11).

The SDA deposits have been repeatedly sampled, by almost 20 gravity cores and some dredge hauls (Chiocci & de Alteriis 2006). These have recovered both a ‘blocky’ facies and a ‘mixed’ or ‘matrix’ facies composed of slightly coherent sandy and clayey deposits, including extremely heterogeneous volcano-sedimentary clasts. Furthermore, a ‘debris-flow’ facies (hereafter DF) composed of a mud-supported deposit incorporating

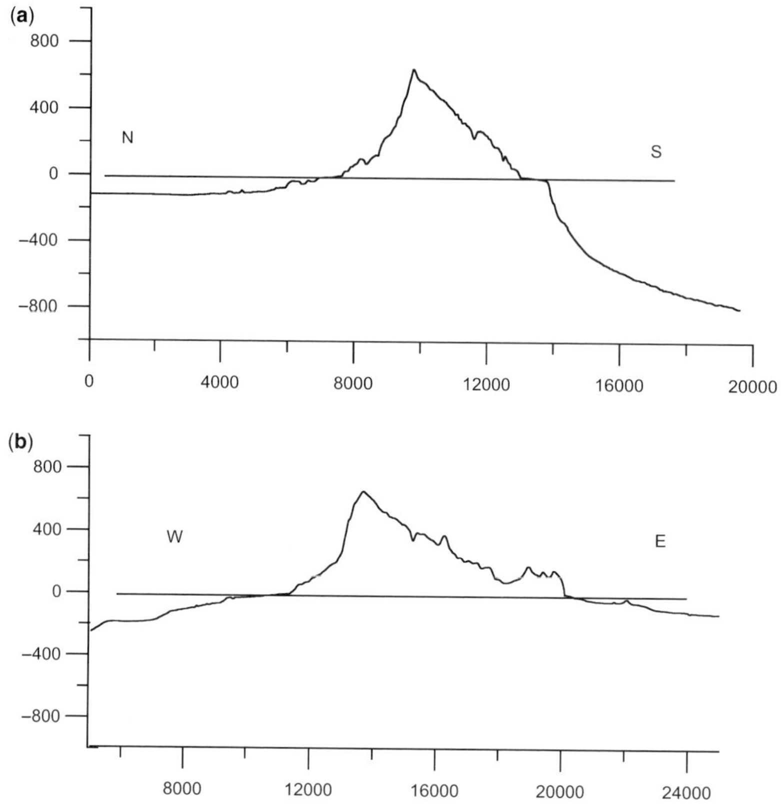

Fig. 9. Elevation profiles of Ischia and its offshore areas: (**a**) north–south; (**b**) east–west. Note worthy features are the strong asymmetry across the north–south axis, the Maronti shallow-water depositional terrace and the upper slope. Vertical exaggeration is *c*. 15:1.

centimetre-size clasts has been widely recognized in almost all cores.

The SDA acoustic facies on chirp sub-bottom (3–5 kHz) profiles is typically characterized by dense diffraction hyperbolas (Fig. 14), similar to those of many other DAs elsewhere (see review by Canals *et al.* 2004), and only in a few instances can the post-avalanche hemipelagic drape be detected by geophysics.

In a few cases the base of the DA has been detected through sparker-source seismic profiles at mid-distal ranges in areas of lower block density (Fig. 15), and possibly whenever a DA evolves into a debris-flow deposit. In these sections the DA thickness ranges from 30–50 ms two-way travel time (twt) (18–30 m) at medium ranges (10–12 km from source) to some 10–15 ms twt (6–10 m) at distal ranges.

The western debris deposit (WDD)

This deposit spreads west of the island over an area of *c*. 11 km^2 with hummocky topography extending from about −30 m to the shelf break at −150 m (Fig. 16). Along the dispersion axis, from the toe of Mt. Epomeo to the coastline and out to sea, the mean terrain slope ranges from 4–5° in proximal areas to less than 2° at −140 m. The nearshore sector (a few metres deep) is characterized by block populations related to the most recent landslides sourced from Mt. Epomeo during historical times (see the next section). In contrast to what is observed to the south, here the overall radial pattern suggests spreading of the avalanche deposits in an overall fan shape. Further seaward, from −40 to −60 m, the distribution of hummocks is not uniform, with most blocks buried under organogenic sediments (Fig. 17). Most of the hummocky facies is represented by circular mounds resulting from blocks embedded in matrix and draped by the most recent sediments. Single mounds range in size from a few metres to some 30 m across and are a few metres to some 20 m above the surrounding sea floor. The size of blocks seems to increase towards the NW sector of the deposit. Some previous morphological highs in the form of elongated ridges

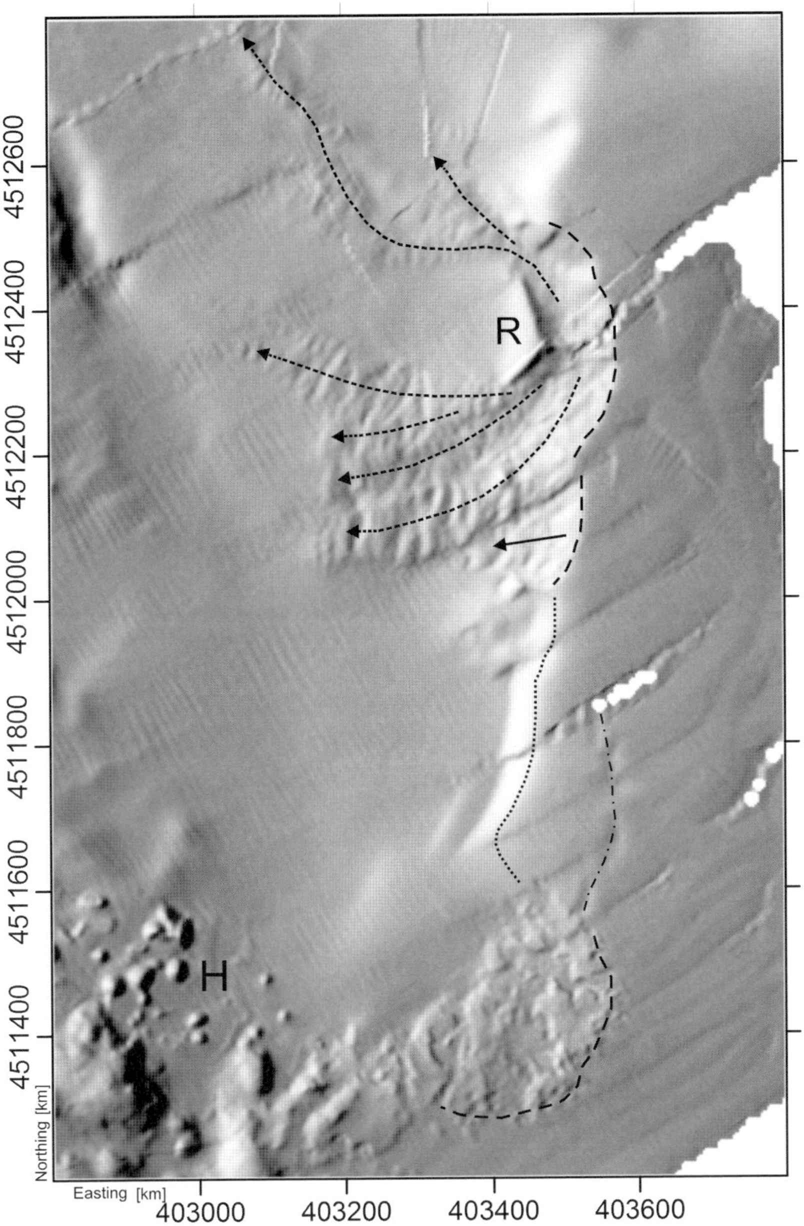

Fig. 10. Slope failure at a terrace rim (−30 m) offshore Forio village. R, toppled block; H, hummocks; dashed line, failure scarp; dotted line, terrace rim; dash–dot line, incipient failure scarp; arrow, sediment flow. Metric coordinates, UTM projection, Zone 33; datum: WGS84; location shown in Figure 8.

(possibly dykes) or tuff edifices have been encircled by blocks. On its southern side the debris avalanche deposit overlies and truncates the terraced surface located at 80 m below sea floor (bsf) (Fig. 16).

Cores sampled in the blocky area have retrieved tuffs similar to the MEGT and tuffites, and some matrix facies overlain by a sedimentary drape 10–150 cm thick, consisting of silt and clay or organogenic and volcanoclastic sands. The 'matrix' facies is similar to that occurring in the SDA but it seems to be more widespread.

Some sparker seismic profiles over the WDD show a typical chaotic facies that overlies a lower seismic unit, consisting of subhorizontal parallel

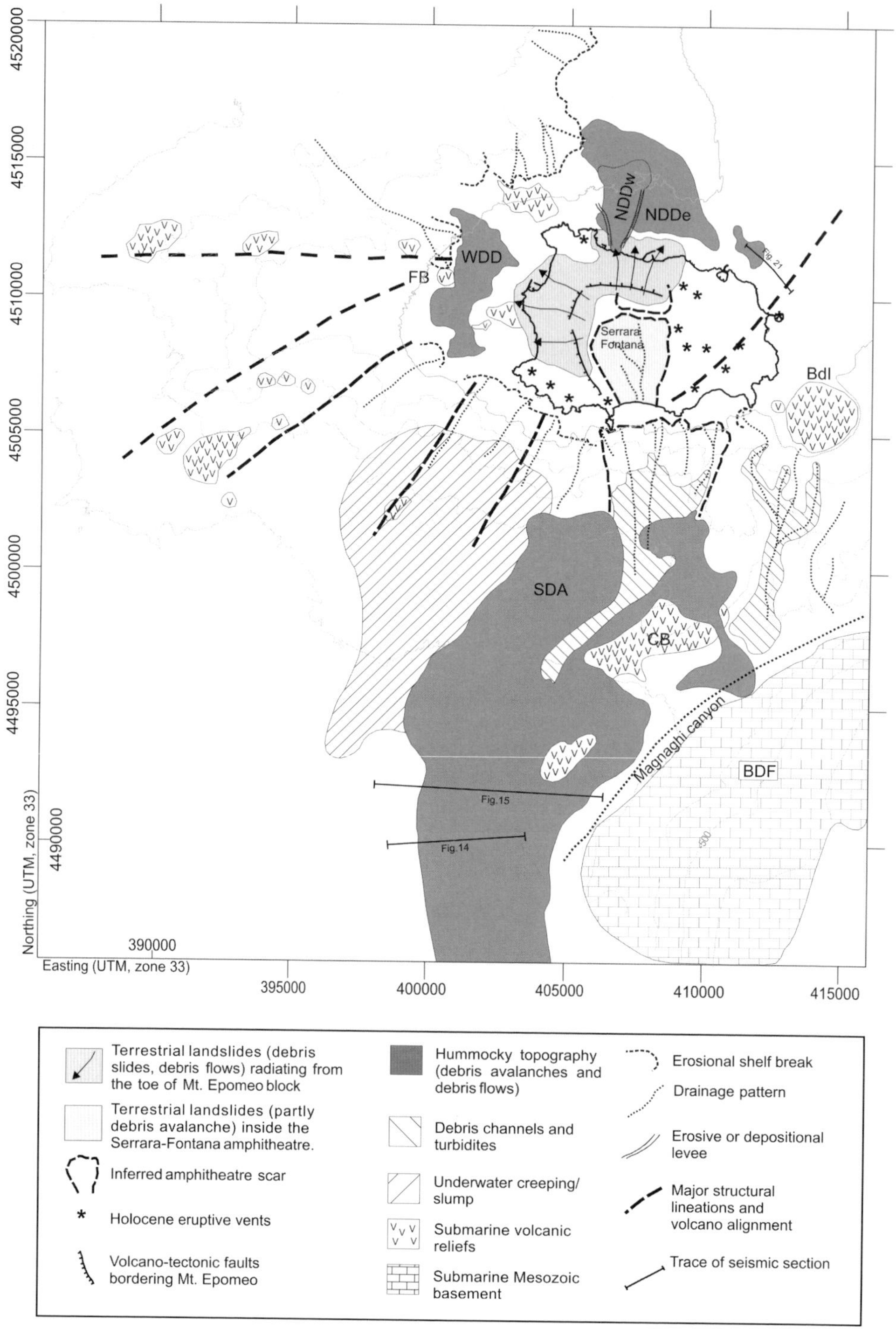

Fig. 11. Interpretative geomorphological map of Ischia volcanic island and its offshore region. The map shows the area surveyed in Figure 7. SDA, southern debris avalanche; WDD, western debris deposit; NDD_E, northern debris deposit; NDD_W, northern debris deposit (latest event); BDF, Banco di Fuori; BdI, Banco d'Ischia; FB, Banco di Forio; CB, Banco di Capogrosso. Coordinates as in previous figures. Thin lines are bathymetry contours.

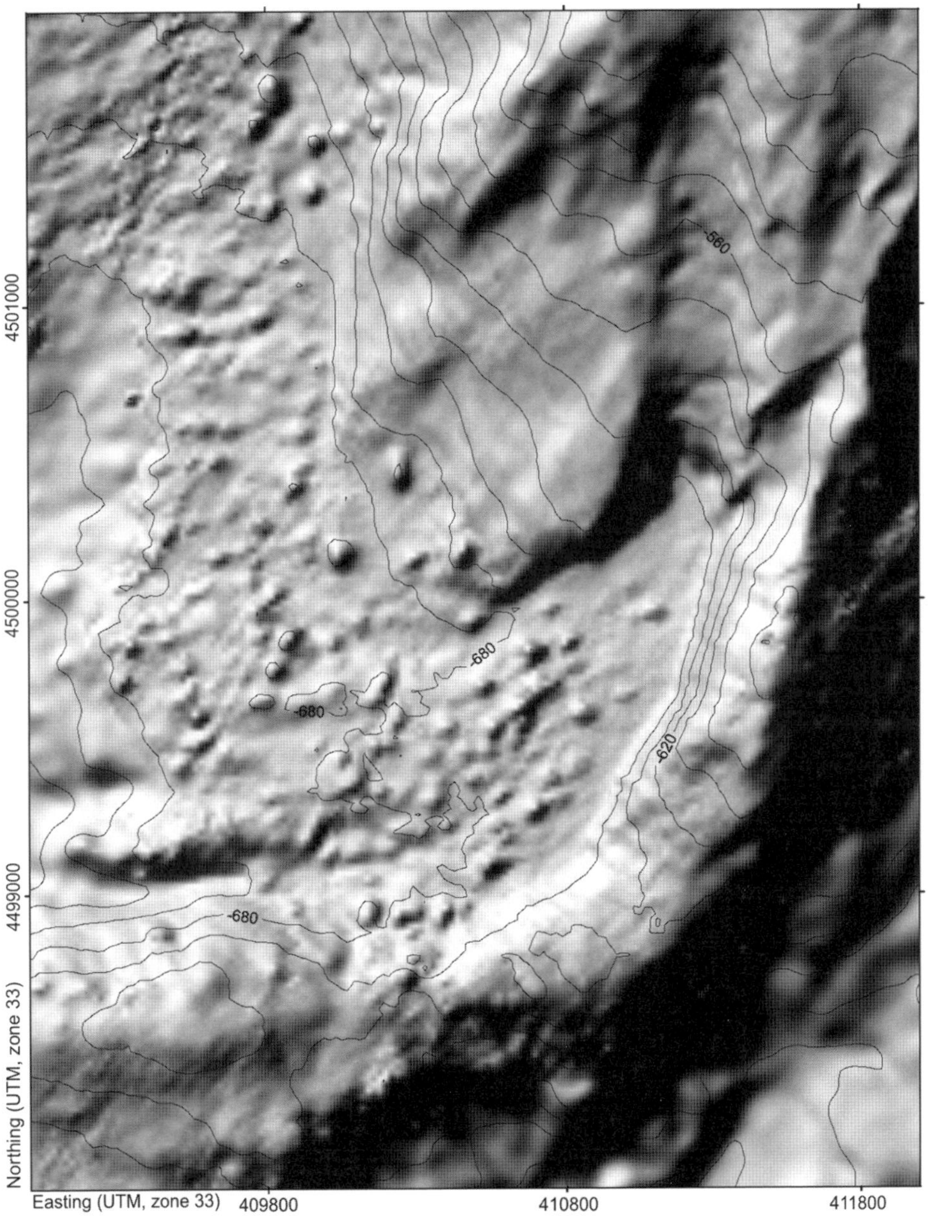

Fig. 12. Shaded relief map based on a 10 m cell size DEM showing boulders of different sizes extending up a 70 m high submerged relief offshore southern Ischia. Coordinates as in previous figures. Location shown in Figure 7.

reflectors laterally evolving into clinoforms. The WDD lies on planar-parallel reflectors probably corresponding to the distal section of a lowstand prograding wedge (Fig. 18). Its overall acoustic homogeneity indicates no significant hiatuses during emplacement. Therefore the WDD either resulted from a single major episode or from multiple landslides that occurred in a very short time span.

The avalanche deposit is thicker in its central section, attaining 40–50 ms twt (35 m) at about 1.5 km from the coastline, and creates, on the whole, a positive topography in the form of a fan-shaped apron. The axis of this fan is centred on

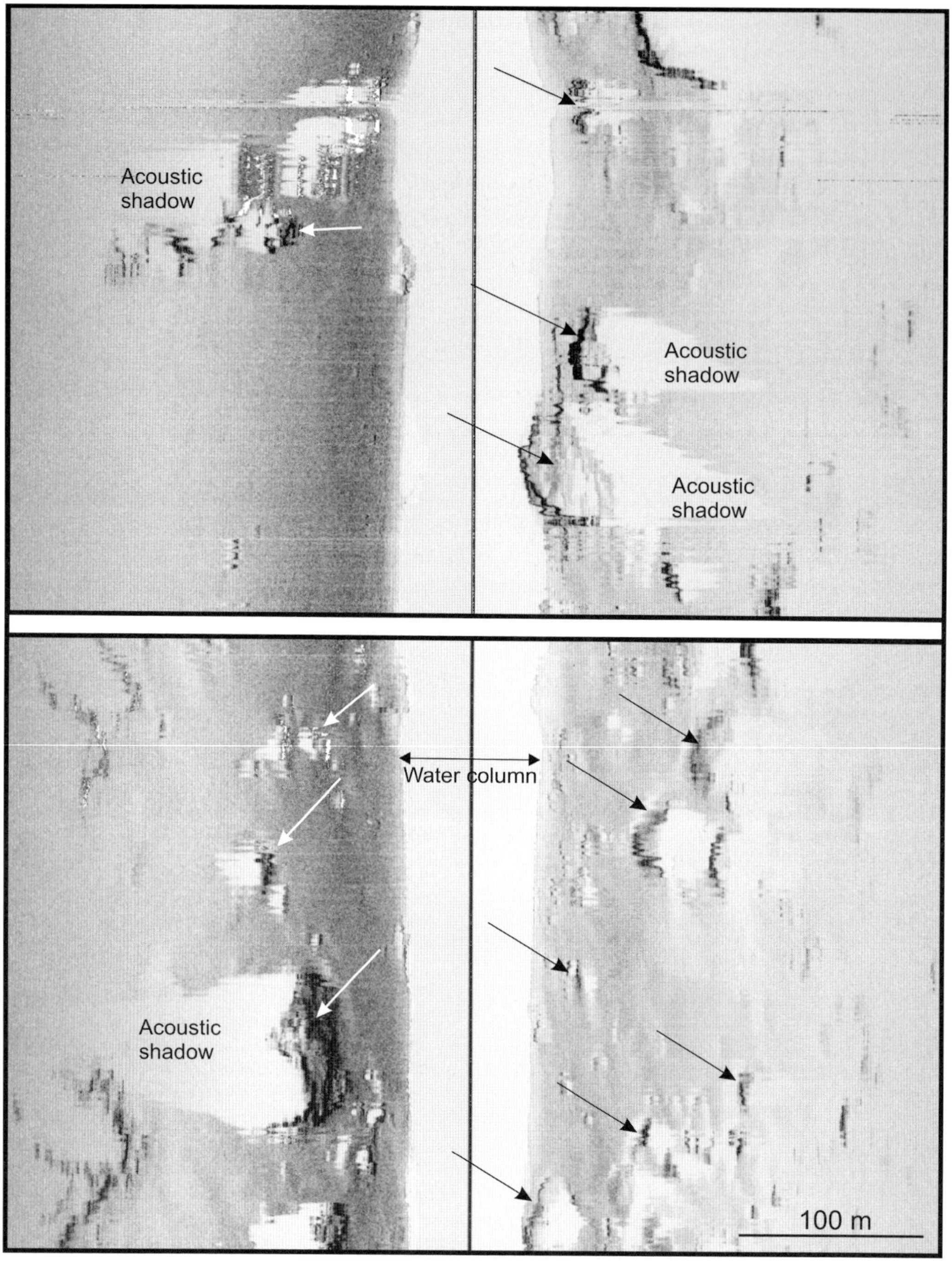

Fig. 13. Unprocessed side-scan sonar image showing blocks of different sizes (arrows) characterizing the central areas of the southern debris avalanche (SDA) at about −800 m.

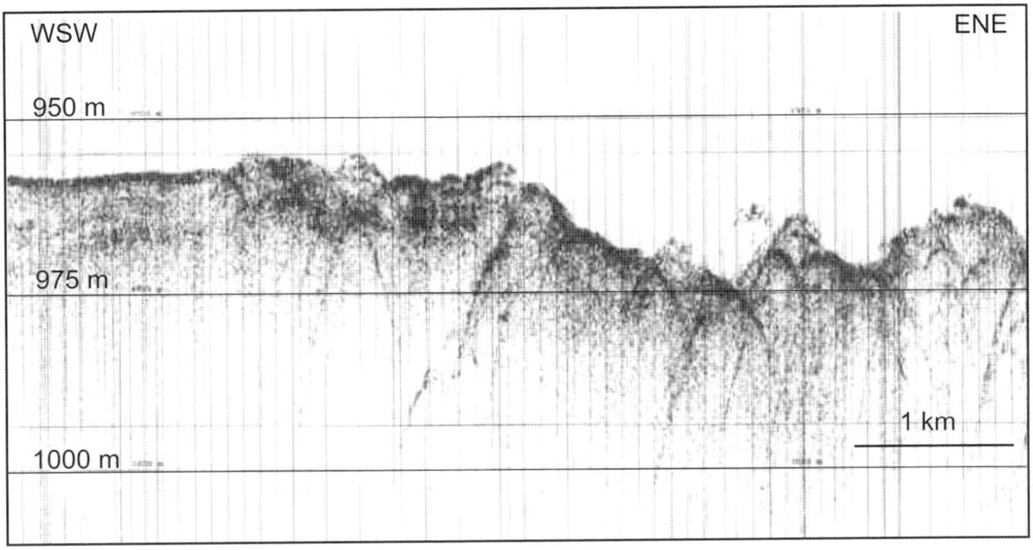

Fig. 14. 'Chirp' (2–7 kHz) sub-bottom profile crossing SDA and showing transition from flat sea-floor morphology to hummocky topography with typical dense diffraction hyperbolas. Location shown in Figure 11.

Forio village with a WNW trend. The WDD thickness decreases to some 25 ms twt (15 m) after less than 1 km from the depocentre along its axis. Lateral thinning also occurs over relatively short distances (600–800 m). Estimates of the volume of the landslide deposit based on seismic reflection profiling and swath bathymetric data range from 80×10^6 m^3 to 150×10^6 m^3.

The northern debris deposits (NDD_E and NDD_W)

In contrast to the WDD and SDA, these deposits are characterized by an overall major topographic relief and a particularly uneven morphology (see Fig. 8). They extend off the Lacco Ameno and Casamicciola villages for a total length of 12 km and have a variable width ranging from 5 to 8 km; blocks are found from a few metres to 140 m bsf. The overall shape of the hummocky terrain is that of a tongue slightly verging to the west and downslope to the shelf edge. More than in the WDD, the blocks are extremely variable in size, from a few metres to 100 m across (Fig. 19). With respect to the flow pattern, longitudinal (or slightly oblique) and latitudinal (*sensu* Masson *et al.* 1998) ridges characterize the entire deposit. The largest boulder is a latitudinal block, located 2.7 km from the coastline, which is 400 m × 900 m in plan-view and rises from −110 to −50 m. This implies a volume of at least $(15–20) \times 10^6$ m^3.

The most evident morphological feature off northern Ischia is a fan-shaped proximal area lying in a 15–20 m depression and constrained between a western, curved levee and an eastern rectilinear levee (see Figs 8 and 16). Both levees, but especially the western one, show erosional and depositional features and rise 15–25 m above the surrounding sea floor.

This western levee-constrained deposit, hereafter the NDD_W, clearly post-dates the emplacement of a basal landslide deposit located eastward, hereafter the NDD_E. This latter is post-dated by the present-day wave-cut marine terrace with the rim at −20 m and by the historical volcanic products of the eastern island sector (1–2 ka; i.e. of Roman age). In contrast, the NDD_W truncates both the depositional terrace with the rim at −80 m and the −20 m present-day terrace. These morphological correlations also seem to be confirmed by sub-bottom chirp profiles running parallel to the coastline. These high-resolution records do not acoustically penetrate the NDD_E and NDD_W but allow some insights into their stratigraphy. A sub-bottom profile crossing the NDD_W shows that the −80 m depositional terrace consists of prograding clinoforms interbedded with volcanic tephra. It also shows a very thin (<2 ms) acoustically transparent unit, which is correlative to the main landslide deposit. This thin layer overlies both the prograding clinoforms and a reflective basement (i.e. the Zaro lava flow, which crops out in the nearby coastal promontory; Fig. 20). Provided that the Zaro lava flow has an age of 6 ± 2 ka (according to published K–Ar data; Vezzoli 1988) or 9 ± 1 ka (according to recent Ar–Ar dating; H. Guillou, pers.

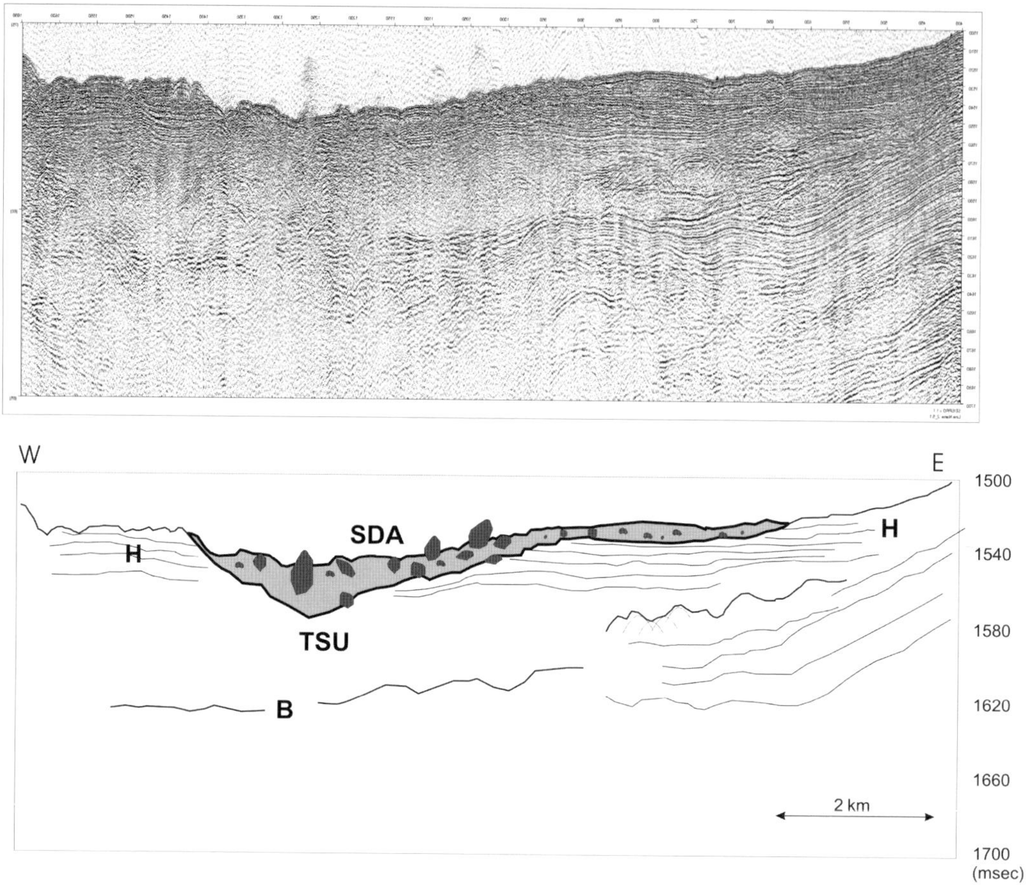

Fig. 15. Single-channel seismic profiles (Geo-Spark© multi-tip sparker source 800 J) across the SDA in the distal region of the avalanche constrained by the Banco di Fuori high to the east. Noteworthy features are the variable DA thickness and the excavation of a previous sea-floor depression, probably the former Magnaghi Canyon, characterized by transparent acoustic facies (TSU, transparent seismic unit). H, plane-parallel reflections (hemipelagic and possibly turbiditic sequences); B, possible volcanic or sedimentary basement. Two-way travel time (ms); location shown in Figure 11.

comm.) this puts a lower time constraint on the age of the NDD_W. Another, relatively transparent seismic unit less than 4 ms twt (3–4 m) thick, mostly occurring in the inter-block depressions, can be detected on top of both the NDD_E and NDD_W. At present, the lack of bottom samplings does not allow us to determine whether this sequence might be interpreted as a post-landslide hemipelagic lag or as a syndepositional muddy deposit (Fig. 20). Other minor landslide deposits, also characterized by hummocky topography, occur to the NE of the island just off the main island harbour (Fig. 21).

More penetrative analogue seismic profiles (4 kJ sparker source) located farther north allow us to detect the base of the NDD_E and NDD_W and to establish a tentative seismic stratigraphy. Although a distinction between the NDD_E and NDD_W can be only inferred, these records clearly show that they both lie over a lowstand prograding wedge, and partially excavate through a transgressive–highstand seismic sequence that appears to be Holocene in age. They also show an average thickness of 50–60 m (70 ms twt) at around 2 km from the coastline and 3 km from the toe of Mt. Epomeo, in an area where blocks rise some 30–40 from the sea floor (Figs 22 and 23).

Based on these data the incised fan-shaped valley developing offshore of Lacco Ameno village is likely to be related to the emplacement of a very recent landslide event that has chiefly eroded the basal deposit leaving a channel-like scar and an associated depositional levee to the west.

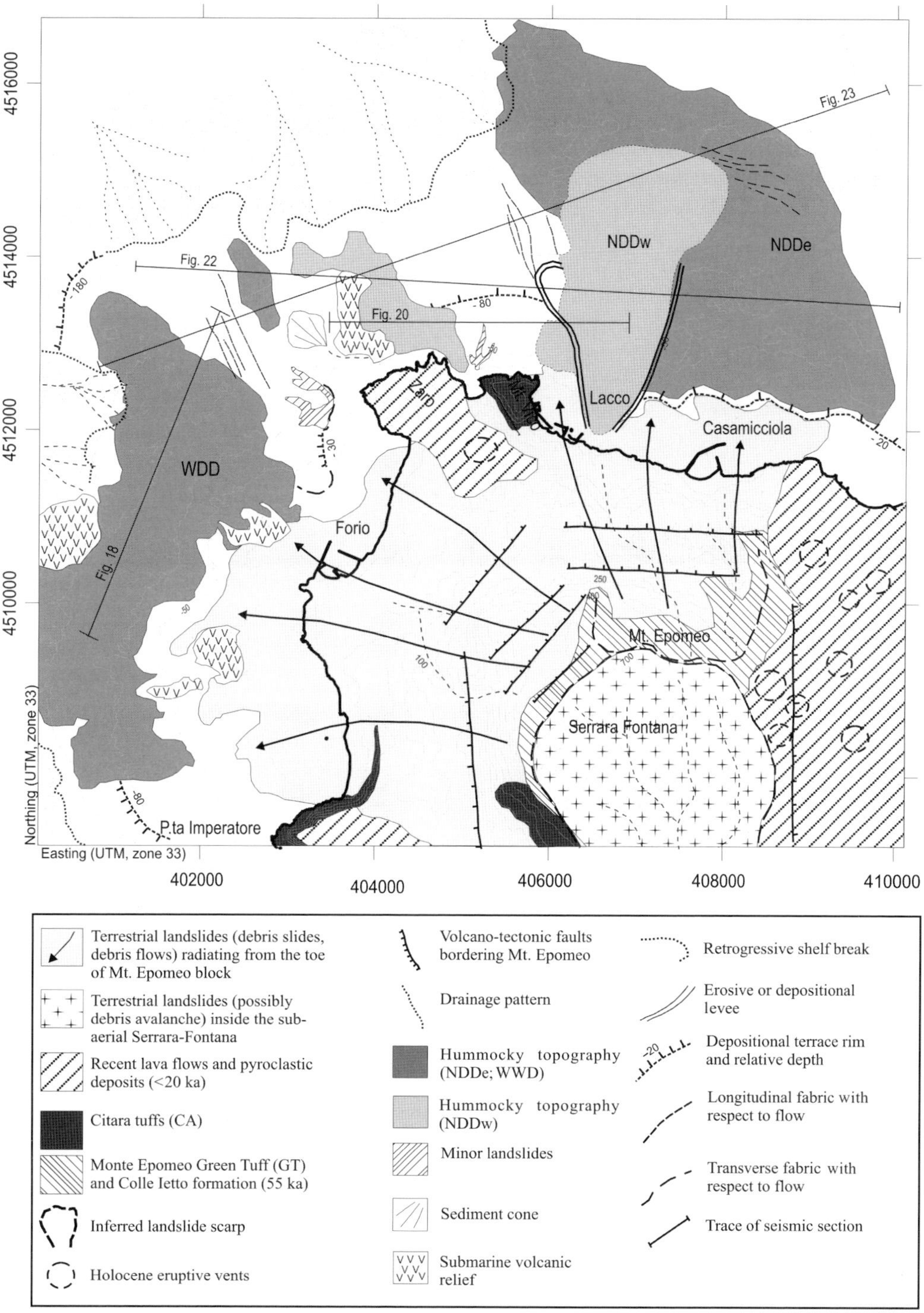

Fig. 16. Interpretative geomorphological map of the northwestern flank of Ischia volcanic island. The map shows the area surveyed in Figure 8. WDD, western debris deposit; NDD, northern debris deposit; NDD_W, northern debris deposit (latest event). Coordinates as in previous figures.

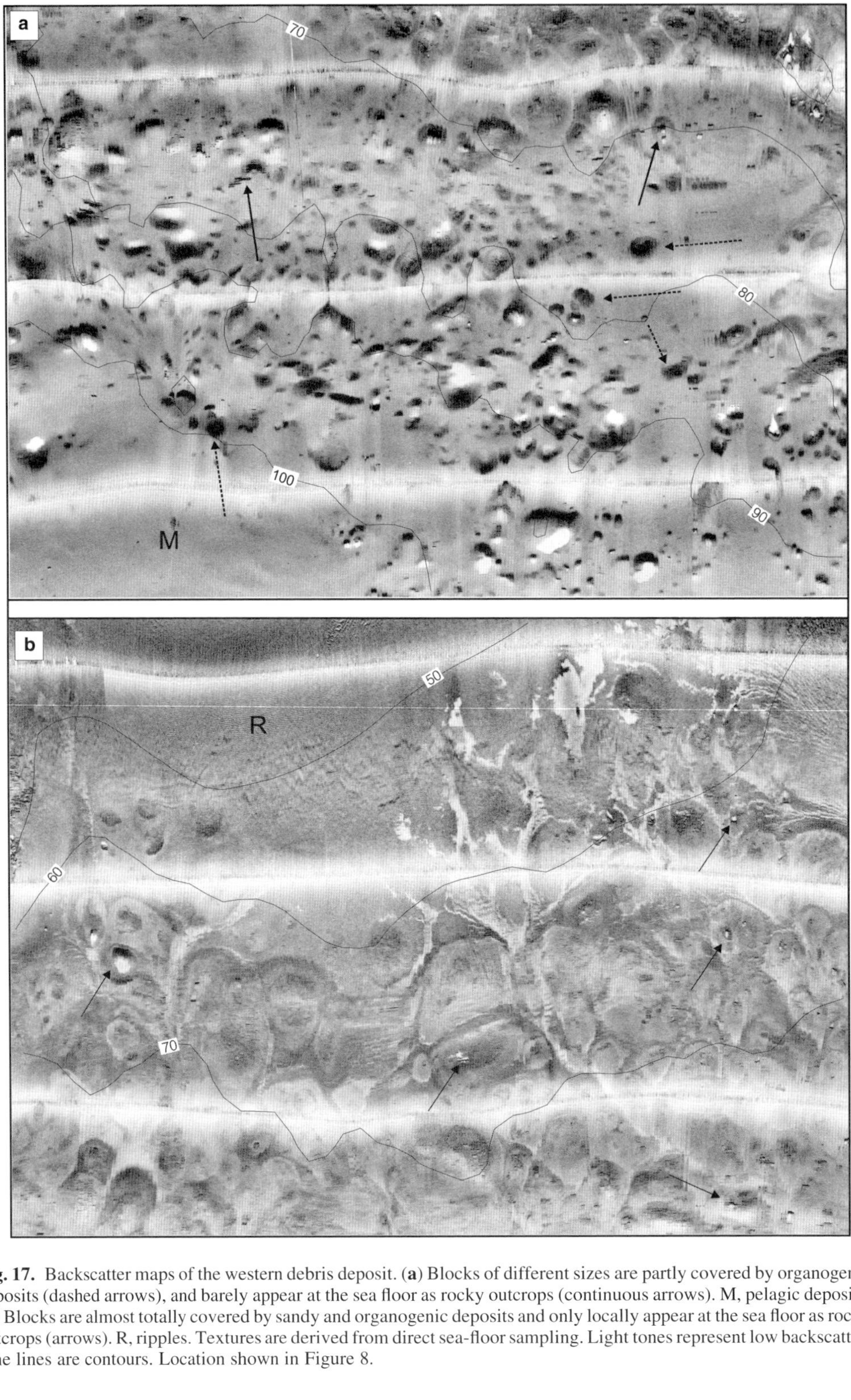

Fig. 17. Backscatter maps of the western debris deposit. (**a**) Blocks of different sizes are partly covered by organogenic deposits (dashed arrows), and barely appear at the sea floor as rocky outcrops (continuous arrows). M, pelagic deposits. (**b**) Blocks are almost totally covered by sandy and organogenic deposits and only locally appear at the sea floor as rocky outcrops (arrows). R, ripples. Textures are derived from direct sea-floor sampling. Light tones represent low backscatter. Fine lines are contours. Location shown in Figure 8.

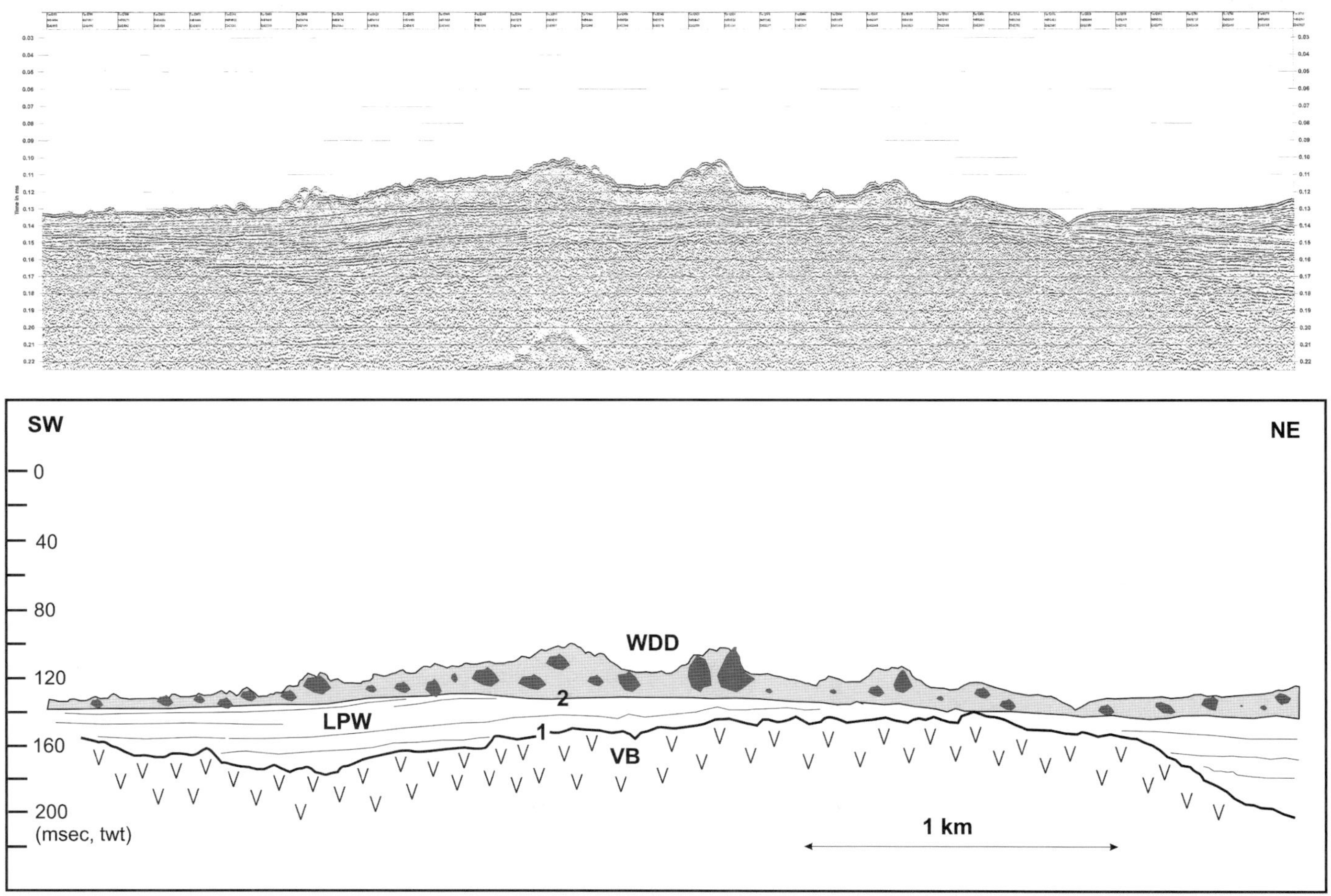

Fig. 18. Single-channel seismic profile (Line_11) crossing the WDD Geo-Spark© multi-tip sparker source (energy 400 J; frequency range 800–1200 Hz). Courtesy of GeoLab, Marine Surveys, Naples, Italy. Reflector 1 is the top of the volcanic basement (MEGT, Monte Epomeo Green Tuff?). Reflector 2 is an erosional unconformity located at the top of a Holocene marine–coastal prograding wedge and marking the base of the WDD debris avalanche. LPW, lowstand prograding wedge; VB, volcanic basement. Location shown in Figure 16.

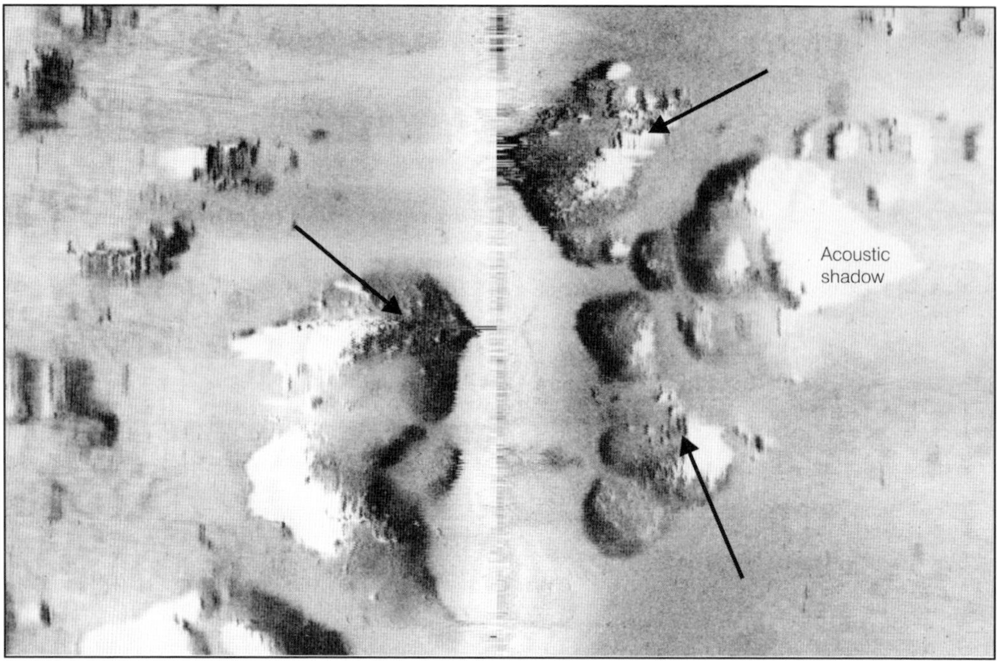

Fig. 19. Processed (corrected for slant range) side-scan sonar image showing blocks and boulders of different sizes partially covered by marine deposits. Rocky outcrops locally occur at top of large boulder (arrows). Light tone represents low backscatter. The sea-floor area surveyed is 320 m × 190 m.

Onshore landslides

The Mt. Epomeo block is encircled by a multitude of landslides, including debris slides, debris flows, mud flows and rockfalls. Apart from local rockfalls occurring along sea cliffs, all these slides correlate with the uplift of Mt. Epomeo. Based on detailed stratigraphy, a genetic relationship among uplift, volcanism and slope instability has been recently proposed for the past 5.5 ka (de Vita *et al.* 2006). Landslides pattern radiates out from Mt. Epomeo toward the northern, western and southern sectors of the island, and the only sectors that are not affected by landsliding are those buttressed by the most recent (10 ka) small volcanic edifices located in the eastern part of the island (see Figs 3, 11 and 16). All these mass movements have volumes of the order of 10^2–10^3 m^3; thus they are generally smaller than those found at sea. Along the western shore, near Forio village, the onshore landslide deposits overlap older debris-flow or mud-flow deposits outcropping along coastal cliffs and continuing undersea.

Analysis of the ground effects induced by historical earthquakes provides evidence that rock falls and debris flows are the most common types of slope failures. These phenomena are also controlled by strong weathering favoured by hydrothermal alteration and fractures affecting the northern and western faulted slopes of Mt. Epomeo (Mele & Del Prete 1998). This latter sector is characterized by continuous sliding phenomena (debris flows) mobilizing new masses of soil and old landslides. For instance, earthquake-induced landslides (debris flows and debris slides) occurred over a 600 year time span (1228–1883) in a narrow area located on the northern flank of Mt. Epomeo, with reactivation of the main landslide near Lacco Ameno village. To the west, in the plain of Forio village, larger debris flows repeatedly occurred during Greek, Roman and historical times (Rittmann & Gottini 1980; Buchner 1986; Guadagno & Mele 1995; Cubellis & Luongo 1998; Mele & Del Prete 1998). All the reported phenomena are summarized in Table 1, which relates landslide events to macroseismic parameters.

Discussion

Emplacement dynamics and relationship with subaerial or submarine failure scars

Underwater gravity-driven mass movements and their emplacement dynamics (see the review by

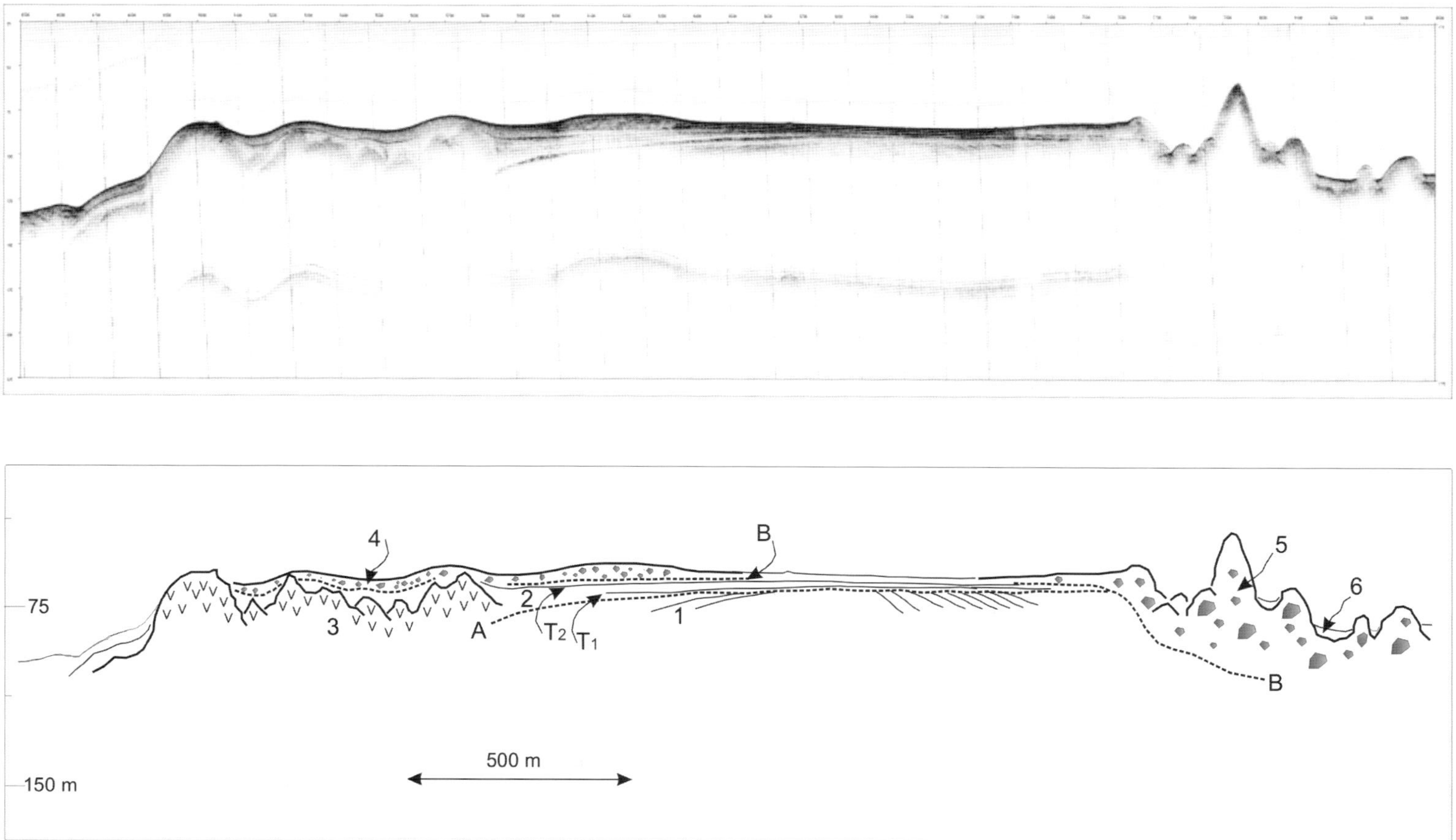

Fig. 20. 'Chirp' sub-bottom (2–7 kHz) profile (Line ISN_01) across the northern debris avalanche. 1, recent prograding coastal wedge (Holocene); 2, marine deposits; 3, lava flow deposits ('Zaro' formation); 4, NDD_W correlative deposits; 5, NDD_W debris avalanche deposits; 6, post-avalanche or syn-avalanche most recent muddy deposits; A, toplap surface on top of the coastal wedge; B, base of landslide deposits; T1 and T2, tephra layers. Location shown in Figure 16.

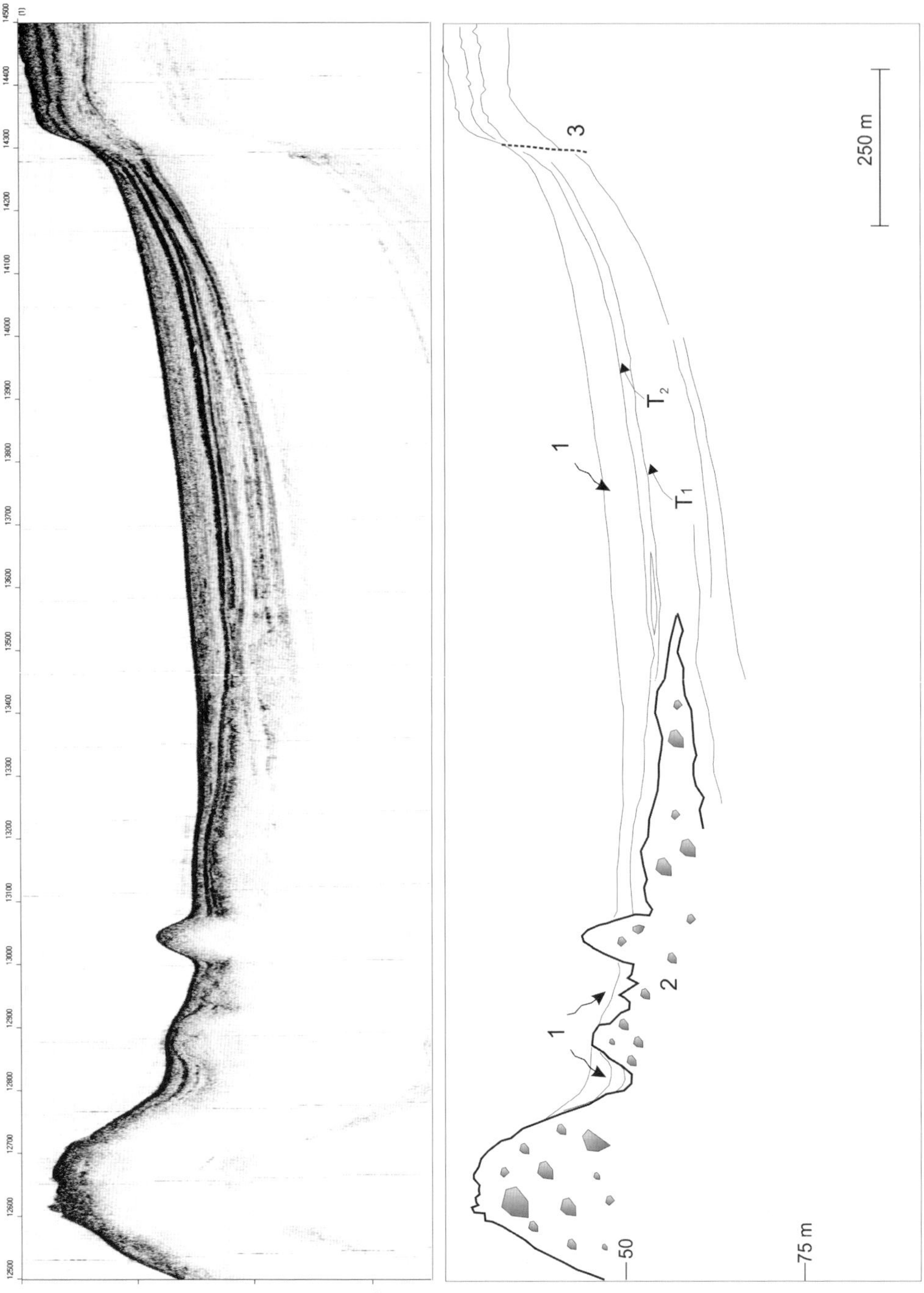

Fig. 21. 'Chirp' sub-bottom (2–7 kHz) profile (Line ISN_02) across minor landslide deposits offshore Ischia harbour NE of the island. 1, modern hemipelagic deposits; 2, landslide deposits; T1 and T2, tephra layers; 3, inferred fault. Location shown in Figure 11.

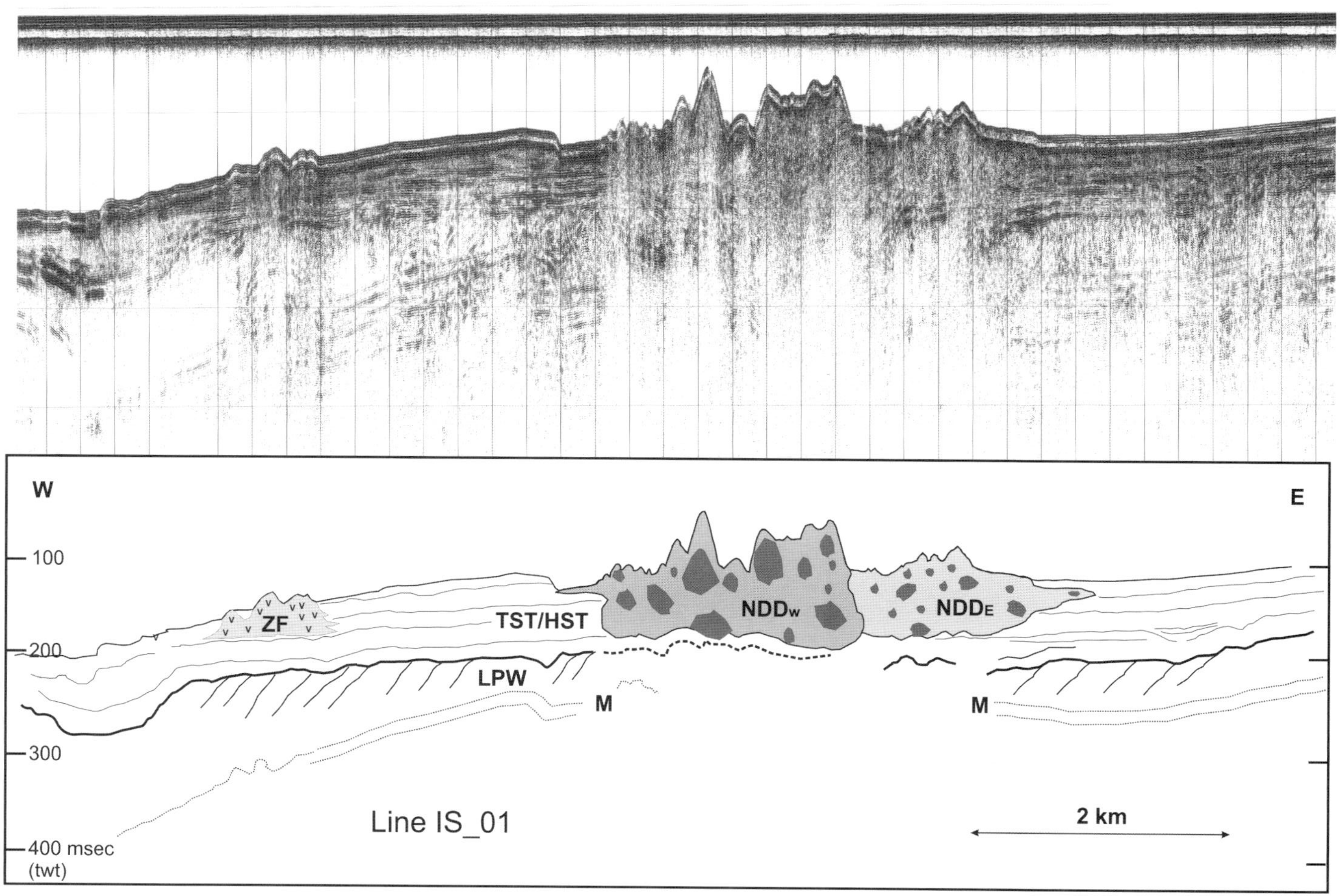

Fig. 22. IS_01 seismic line (sparker source 4 kJ) crossing the NDD. M, sea-bottom multiple; LPW, Late Quaternary lowstand prograding wedge; TST/HST, transgressive–highstand systems tract (Holocene?); ZF, Zaro lava formation. Vertical axis is two-way travel time (ms). It should be noted that the thickness of the avalanche here is of the order of 70 ms, corresponding to 50–60 m. Track location is shown in Figure 16.

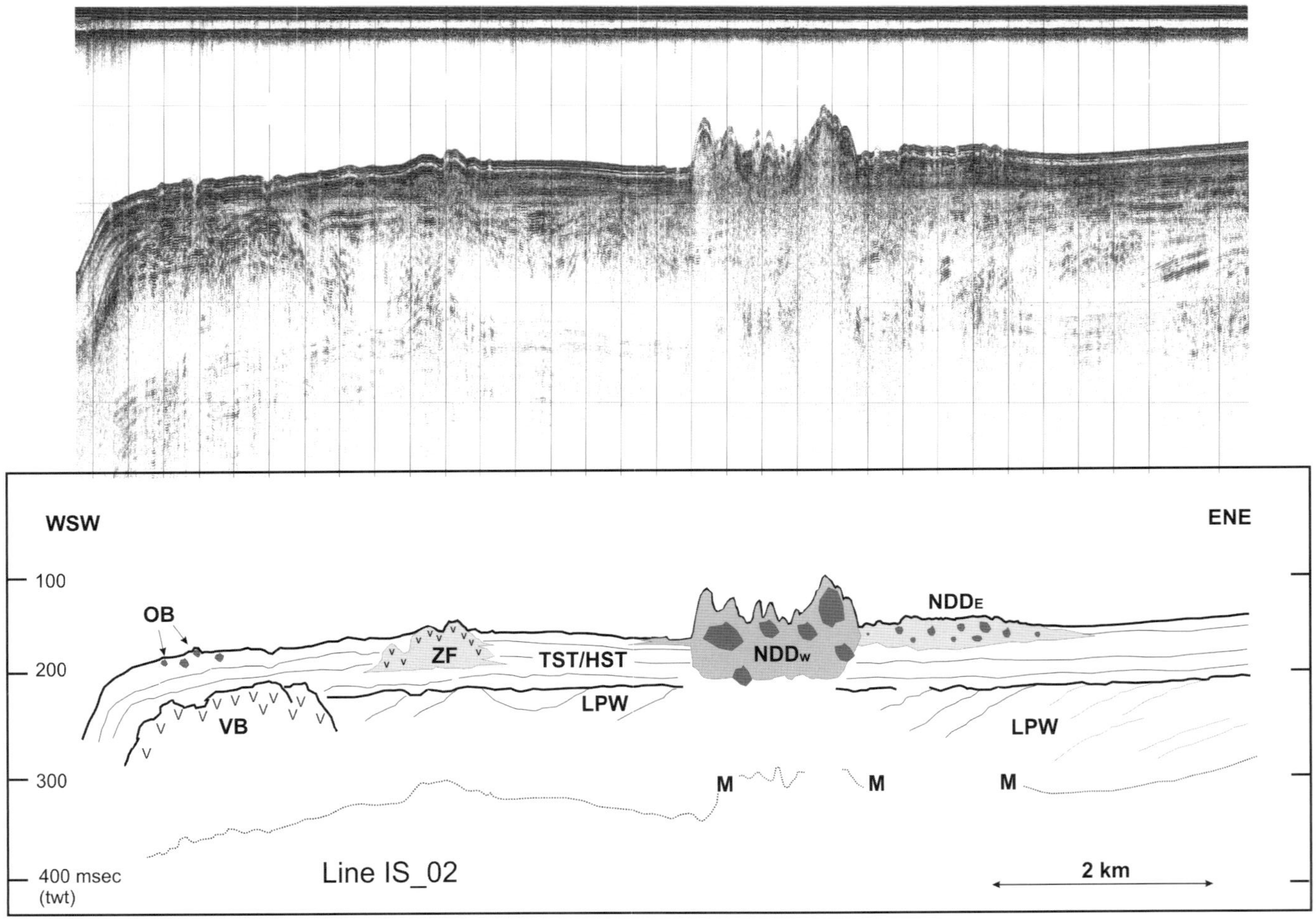

Fig. 23. IS_02 seismic line (sparker source 4 kJ). OB, outrunner blocks related to the WDD; VB, buried volcanic basement. Other abbreviations as in Figure 22. Track location is shown in Figure 16.

Table 1. *Onshore sliding phenomena triggered by earthquakes along the northern and western flanks of Mt. Epomeo, since the fourth century* BC

Date	Observed phenomena	Affected area	Macroseismic epicentre	Earthquake intensity (MCS)
470–460 BC	Volcano-tectonic events. Deep-seated gravitational slope deformation; the sliding phenomenon affected an area of about 3.5 km^2 lying between Mount Epomeo and Forio village	Western sector		
Second–third centuries AD	Debris slides, starting from the top of Mount Epomeo towards the Spiaggia di Citara–Cava dell'Isola	Western sector		
1228	Rock fall, starting from the steep slope of Mount Epomeo towards Casamicciola–Lacco Ameno. Debris avalanches	Northern sector	Casamicciola	IX–X
1767	Rock fall	Northern sector	Casamicciola	VII–VIII
1796	Rock fall, starting from the top of Mount Epomeo, on the northern flank	Northern sector		VIII
1828	Rock fall, starting from the top of Mount Epomeo. Debris avalanches, starting from the northern flank of Mount Epomeo towards Casamicciola and Lacco Ameno	Northern sector	Casamicciola	VIII–IX
1863	Rock fall, starting from the steep slope of Mount Epomeo	Northern sector	Casamicciola	VII
1881	Rock fall, starting from the top of Mount Epomeo. Debris slides, starting from the northern flank of Mount Epomeo towards Casamicciola and Lacco Ameno; the total area affected by sliding phenomena is about 3.5 km^2, including also the slides that occurred in 1828	Northern sector	Casamicciola	IX
1883	Rock fall, starting from the northern slope of Mount Epomeo towards Fango and on the slopes of Mount Nuovo and Mount Bastia. Debris slides, occurring on the northern flank of Mount Epomeo towards Lacco Ameno, and towards Mount Rotaro	Northern sector	Casamicciola	X

Sources: Violante *et al.* (2004) and references therein.

Mulder & Cochonat 1996) are still poorly understood despite the continuous improvement in marine geophysics. Investigation tools in the marine environment are mostly based on underwater acoustics (sonar imaging, seismic surveys, etc.) whereas direct observation and sampling are traditionally used for subaerial deposits. This difference in investigation tools is often one of the main sources of ambiguity when the same classification criteria for underwater and on-land mass movements are applied.

Of all the types of mass movements, debris avalanches (DAs) and debris flows (DFs) are the most widespread over both subaerial and submerged flanks of volcanic edifices. Although general agreement exists for the definition of DAs some ambiguities arise in the DF nomenclature, especially for those occurring undersea.

A DA is 'a flowing mixture of debris, rock and moisture that moves downslope under the effect of gravity' (Ui *et al.* 2000). DAs are widely recognized as extremely rapid and disintegrative landslides. The term 'extremely rapid' would imply a velocity of at least 5 m s^{-1} according to classical on-land landslide velocity classification, but is actually much greater (of the order of 50–100 m s^{-1}), as demonstrated by results from the few witnessed avalanching events (18 May 1980 at Mt. St. Helens volcano, USA; Huascaran, Nicaragua; Glicken 1998). The term 'disintegrative' implies an overall dissipation of shear strength inside the landslide body during collapse and transport that leads to its fragmentation into discrete blocks, ranging from centimetre-sized clasts to major boulders up to hundreds or thousands of metres across.

Almost all studies on the sedimentology and the texture of volcanic DAs relate to on-land cases. DAs incorporate a block facies and a mixed facies called the matrix. The block facies consists of centimetre- to kilometre-sized blocks of volcano-sedimentary rocks belonging to the failed edifice; the matrix may include clasts of micrometres to millimetres in size and groundmass (see Glicken 1991; Ui *et al.* 2000).

DFs are defined as water-saturated mixtures of debris and water having high sediment concentration. For volcanologists DFs coincide with the hyperconcentrated mixtures that characterize lahars. According to many workers (e.g. Vallance 2000) DAs differ from DFs in that they are not water-saturated and the load is entirely supported by particle-to-particle interactions. However, this distinction is not widely acknowledged. For instance, Canals *et al.* (2004) defined DFs as chiefly grain-supported and DAs as chiefly matrix-supported.

According to the sedimentological approach, DFs (both cohesive and granular flows) are characterized by a viscous laminar flow generally opposed to the turbulent Newtonian flow that should characterize turbidites (see the review by Shanmugan 1997). These different approaches indicate that some ambiguities remain as regards DF emplacement in the marine environment.

Ischia underwater blocky deposits described here show strong morphological and acoustical analogies with DAs found, at various scales of magnitude, off other volcanic islands (Lipman *et al.* 1988; Begét & Kienle 1992; Moore *et al.* 1994; Labazuy 1996; Urgeles *et al.* 1997, 1998; Funck & Schmincke 1998; Deplus *et al.* 2001; Satake & Kato 2001; Masson *et al.* 2002) or, in non-volcanic settings, at the toe of seismically active continental slopes (e.g. Normark *et al.* 2004). For almost all these analogues, the occurrence of a block field is generally considered diagnostic of very rapid and catastrophic landslide.

It is worth noting the case of sediment megablocks (up to 10 km across), which were first interpreted as rafted blocks in the so-called Canary Debris flow, NE Atlantic (Masson 1996) and later reinterpreted as *in situ* blocks surrounded by the debris flow itself (Masson *et al.* 1998). In the case of rafted blocks, the distance, or runout, travelled by these blocks may be significantly longer than that of many debris avalanches. In almost all these examples sediment blocks tend to maintain their internal coherency and move away as rafted blocks along a shallow, very low-angle ($<1.5°$) décollement surface parallel to the slope and probably coinciding with a localized high pore-water pressure layer.

As shown above, the major undersea mass movements reported in this paper (Table 2) are all characterized by a blocky appearance, and have volumes ranging from >1.5 km^3 (in the case of the SDA) to <0.5 km^3 (in the case of the WDD and NDD). The basic difference between the former and the latter resides in the degree of correlation between the failed masses and their submarine or submarine–subaerial failure scars. The SDA appears to have occurred as a single catastrophic event and is clearly related to a land–sea amphitheatre or an entirely underwater scar depending on the scenario adopted (Chiocci & de Alteriis 2006). Further evidence of its high kinetic energy is provided by the upslope climbing of a block sub-population that travelled SSE.

In contrast, the WDD does not correlate with a visible scar, and the NDD correlates only with a scar in the northern subaerial flanks of Mt. Epomeo (see Figs 11 and 16). Consequently, whereas the SDA may be unambiguously recognized as a debris avalanche related to a catastrophic flank collapse, some doubts may arise for the interpretation of the WDD and NDD. It is reasonable to assume that all the WDD and NDD displaced material at sea originated from the only available potential energy relief (i.e. the subaerial Mt. Epomeo flanks). Therefore the WDD and NDD emplacement may be explained by the following 'end-member' solutions or a combination of them.

(1) The WDD and NDD emplacement was not catastrophic. These deposits result from the stack of slow-moving multiple landslides occurring over a considerable time span (hundreds to thousands of years).

(2) WDD and NDD emplacement was catastrophic but their original scars have been mostly remodelled by erosion and/or faulting.

(3) WDD and NDD emplacement was catastrophic and most of the failed material originated from the Mt. Epomeo summit, implying its

Table 2. *Tentative chronology of landslide events and geomorphological features off Ischia*

Sequence	Event or process	Evidence	Inferred age
1	Deposition lowstand wedge	−190 m to −140 m shelf breaks	Around 18–20 ka
2	Depositional wedge	−80 m rimmed surfaces	?
3	Zaro lava flow entering the sea	Seismic stratigraphy	6 ka ± 2 ka (Vezzoli 1988); redated at 9 ka ± 1 ka (H. Guillou, pers. comm.)
4	Debris avalanche	SDA deposit	Holocene *sensu lato* (*s.l.*) (Chiocci & de Alteriis 2006)
5	Debris avalanche	WDD deposit	Holocene *s.l.*
6	Debris avalanche possibly remobilized	NDD_E deposit	Holocene *s.l.*
7	Coastal wedge	−20 m rim	Holocene *s.l.*
8	Debris avalanche or debris flow possibly remobilized	NDD_W deposit	Holocene *s.l.*–historical

All ages are years BP (before present).

self-decapitation at the moment of collapse (Fig. 24).

The first hypothesis can be discarded on the basis of the absence of multiple block patterns recognizable from sonar imagery. A single event is further confirmed by seismo-stratigraphic investigations carried out off the western shore, and no more than two events are detectable to the north.

The second and third scenarios seem more realistic considering: (1) the volcanic setting, (2) the existence of Mt. Epomeo uplift, and (3) the overall disintegrative appearance of the landslide deposits, with very poor block sorting, often with extremely irregular block shapes and, particularly in the case of the NDD, downslope distribution of major boulders suggesting a ‘shield’ effect during the

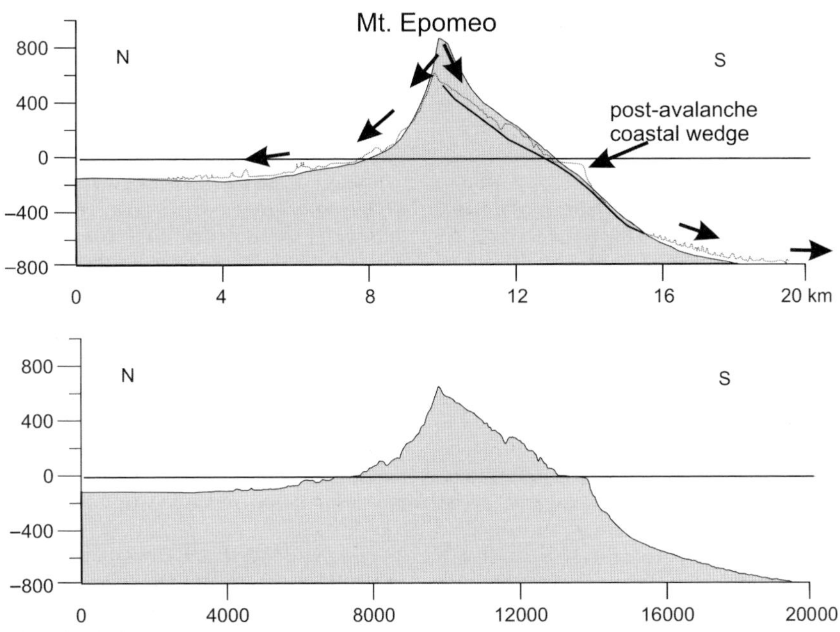

Fig. 24. Mt. Epomeo ‘self-decapitation’ model shown along a north–south profile and assuming an original height of >1000 m with most of the failed volume directed to the south.

avalanche. The absence of clear collapse scars on the northern and western slopes of Mt. Epomeo might be due to the high volcano-tectonic stress and the associate fracturing and hydrothermal activity (Guadagno & Mele 1995; Mele & Del Prete 1998), which have induced accelerated erosion and remodelling (Violante *et al.* 2004). However, some doubts regarding an entirely catastrophic failure dynamics still remain, as suggested by the local occurrence of sea-bottom flow-fabric structures (see Fig. 16) recalling pressure ridges (in the case of the NDD_E), and by the analogy with onshore landslides. On Ischia, slow-moving landslides along detachment surfaces (which may be shallow or deep rooted) have been proven in the NW sector and in the central southern sector of the island by levelling surveys (Manzo *et al.* 2006). The last (1883) earthquake triggered a slide that is still in motion at a rate of the order of centimetres a year.

Therefore, a combination of catastrophic and less catastrophic movements seems the more likely and realistic scenario, particularly for the northern avalanche. In this case, a first avalanche event (the NDD_E) was accompanied, after the emplacement, by slow sliding processes and finally followed by another avalanche or an erosional event (the NDD_W). Whether this latter further carried away blocks or chiefly eroded existing deposits cannot be determined at present. The NDD_W might have behaved as an erosional debris flow composed of a thick slurry of rocks and water. In analogy with some on-land examples (Capra *et al.* 2002) the NDD_w debris flow may have eroded the older deposits and incorporated additional water and wet sediment along its path, thus increasing in size.

Timing of the landslide events

At present only the age of emplacement of the SDA has been inferred with a certain degree of confidence in a time span ranging from 10 to 4 ka (Chiocci & de Alteriis 2006) whereas there are no data available for the timing of the NDD_E and NDD_W. However, the identification of submarine terraces along with their morphological and stratigraphic relationship with landslide deposits can be used for building up a first, tentative chronology (Table 2) and cross-correlation of debris avalanche events.

A clear distinction between glacio-eustatic fluctuations and ground deformation cannot be made at Ischia or in the surrounding areas, in common with other active volcanic areas characterized by uplift and subsidence. It is worth noting that the average ground deformation rate at Ischia is of the order of 20 mm a^{-1}, which is comparable with the Holocene post-glacial sea-level rise of the order of 10 mm a^{-1}. Thus some clearly recognizable marine terraced surfaces located between -140 m and -40 m might correlate to relative sea-level stillstands but their dating is problematic. The age of the depositional shelf rims located between -140 and -190 m, although possibly displaced by volcano-tectonism, can be assigned to 18 ka. At shallower depths another terrace rim, between -80 and -100 m, can be recognized along the western, northwestern and southern offshore areas and seems older than the WDD. No clear relationships are observable between the -80 m rim and the NDD_E. On the contrary, the NDD_E clearly lies below a -15 to -20 m terrace rim that should be very recent (the present-day coastal wedge). Finally, on acoustic profiles the NDD_W is younger than the Zaro formation, dated at 6 ± 2 ka, and truncates both the -80 m and the -20 m terraced rims (see also the interpretation in Fig. 20).

Predisposing or triggering factors and tsunamigenic potential

There is no doubt that the main cause of slope instability occurring both onshore and offshore Ischia is the volcano-tectonic uplift of Mt. Epomeo. Of all the recent to active volcanoes in southern Italy, Ischia and Stromboli are the only ones that have clearly undergone sector or flank collapse. This is confirmed by a simple comparison with the undersea slopes off the neighbouring Pontine volcanic islands. These have a high aspect ratio but do not show collapse scars or hummocky topography (F. L. Chiocci, personal communication).

Although inflation of volcanic edifices and consequent ground deformation is a relatively common response to volume or mass changes inside active magma reservoirs, an exhaustive dataset is not yet available in the literature. Uplift rates are normally measured by two different groups of methods, which either result from active monitoring (topographic, geodetic, DGPS levelling surveys) over very short time spans (of the order of 10^{-2}–10^{0} years), or result from geological evidence over longer time spans (of the order of 10^{2}–10^{3} years) (see Fig. 6). Most uplift rates in the literature show a general decrease with the increase of the measured time span. It is therefore likely that the 20 mm a^{-1} Ischia uplift rate, averaged over 30 ka, might have been higher during shorter time spans. Therefore, major prehistoric collapses occurred in response to fast uplift phases that oversteepened the Mt. Epomeo flanks to critical slope angles. For these conditions, the analysis of the historical seismicity of the island provides a possible triggering mechanism for landsliding (Violante *et al.* 2004).

Among volcano-related phenomena those involving caldera or sector collapse at coastal or island volcanoes have a major tsunamigenic potential. In particular, sector or flank collapse and their attendant debris avalanches entering the sea produce impulse tsunami waves. To date there have been few witnessed historical or contemporaneous accounts of such events: the Oshima-Oshima (Japan) volcano collapse in 1741 (Satake & Kato 2001); the avalanche at Unzen volcano (Japan) in 1792 (Hoshizumi *et al.* 1999); the debris avalanche at Mt. St. Augustine (Alaska) in 1883 (Kienle *et al.* 1987); the 1888 Ritter Island (Western Pacific) collapse (Ward & Day 2003), and recently a small avalanche at Stromboli volcano (Italy) on December 2002 (Tinti *et al.* 2003). In addition to these marine events the case of the Mt. St. Helens volcano (Washington, USA) is noteworthy; there a debris avalanche 'splashed' into the Spirit Lake in 1980 (Lipman & Mullineaux 1981).

In all these cases, apart from Stromboli, the collapses were entirely subaerial, of the order of 10^{-1}–10^{1} km^3, and all the failed masses displaced a certain amount of water over a very short time span, which varied from seconds to minutes. The wave heights reported for these events varied from hundreds of metres (Mt. St. Helens) to tens of metres (Oshima-Oshima, Unzen, Ritter island), to a few metres (Mt. St. Augustine). At Stromboli a much smaller (10^6 m^3) avalanche produced a 3–4 m high tsunami that partially inundated some sectors of the island.

In the case of Ischia three basic typologies of collapse-landslide may be inferred: (1) entirely subaerial; (2) entirely submarine; and (3) partially subaerial. There is little doubt that the northern and western DAs belong to the first category whereas the southern DA may be of the second or third type depending on the scenario adopted (Chiocci & de Alteriis 2006).

Conclusions

(1) Marine geophysical surveys and a wide range of sea-bottom observations and sampling carried out in

Table 3. *Major physical characteristics of Ischia underwater DAs based on marine geophysical and geological data*

	SDA	WDD	NDD_E and NDD_W
Hummocky area (km^2)	150	11	17
Runout (L) referred to farthest blocks (km)	45	7	10
Height of collapsed area (m)	1500	800	800
Apparent mobility H/L	0.03	0.11	0.12
Maximum thickness* (m)	30	45	60
Mean thickness (m)	5	5	20–30
Inferred volume range (km^3)	1.5–3	0.2–0.5	0.2–0.6
Depositional depth range (m)	500–1100	0–120	0–140
Overall pattern	Elongated and constrained within narrow chute	Spreading	Fairly elongated; NDD_E incised and remobilized by NDD_W
Inferred number of events	1 or 2	1	2
Mean block size	20 m × 20 m	10 m × 10 m	10 m × 10 m and 70 m × 70 m (bimodal)
Maximum block size and its runout distance	180 m × 200 m 30 km	30 m × 30 m 4 km	900 m × 200 m 3 km
Outrunner blocks	Yes	No	No
Excavation/levees	Yes/yes	No/no	Yes/yes
Transition to DF	Yes	?	?
Flow fabric (pressure ridges, flow lineations, shear patterns, etc.)	No	No	Yes
Post-avalanche sea-bottom features	Mostly erosional (minor landslides, gullies, creep, comet marks)	Mostly depositional (organogenic sands)	Mostly depositional (silts and muds)
Post-emplacement slow sliding movements	No	No	Likely

*Based on seismic data or cores, or inferred from topography.
DA, debris avalanche; DF, debris flow; H, height of failed material.

the past 10 years have revealed the importance of mass movements offshore Ischia (Table 2).

(2) At present four major debris deposits, namely the SDA (southern debris avalanche, Chiocci & de Alteriis 2006), the WDD (western debris deposit), the NDD_E (northeastern debris deposit, older event) and the NDD_W (northwestern debris deposit, younger event) have been identified, with volumes ranging from >1 km^3 to <0.5 km^3. These all spread out from the Mt. Epomeo uplifted block and are partly or entirely submarine. Whereas the SDA correlates with a clear land–sea amphitheatre scar occurring along the southern flank of Mt. Epomeo, the WDD and NDD lack clear failure scarps both on land and at sea.

(3) Most of these mass movements are debris avalanches and debris flows, which are found all around the island, except for the eastern sector, where no landslides were found. Their textural characteristics, and the geophysical–geological evidence, suggest a catastrophic rather than slow-moving emplacement dynamics, thus implying sudden failures of the volcano submarine and/or subaerial flanks. It is possible, however, that part of the failed mass has been remobilized as a slow-moving slide, as seems to be the case for the NDD_E.

(4) At present, a precise timing for the collapse events is still not possible because the dating of the DAs is not complete. Nevertheless, a relative chronology can be established on the basis of seismic stratigraphy, correlation with marine terraces and with onshore volcanic formations, and previously published data (Chiocci & de Alteriis 2006) (Table 3). These data suggest that the SDA, the WDD and the NDD_E events probably occurred during prehistory, with the NDD_W being more recent.

(5) With regard to the genetic link between the DAs, three possible scenarios can be hypothesized. In the first scenario, the WDD and NDD_E are genetically linked but are not related to the SDA. Their original scars have been mostly removed and/or masked by strong erosion and remodelling, following fast volcano-tectonic uplift and younger slope instabilities on the northwestern subaerial and submarine slopes of the island. In the second scenario, the WDD, NDD_E and SDA each corresponded to an episode of collapse, these collapses being independent of each other. In the third scenario, the WDD, NDD_E and SDA were emplaced after a major collapse of Mt. Epomeo, implying a large amphitheatre failure scar along its southern flank and summit 'decapitation' in response to fast volcano-tectonic uplift (see Fig. 24).

(6) For all these events, debris masses of the order of 0.1 to >1 km^3 have entered the sea both in shallow waters (to the west and north) and in deep waters (to the south). These are likely to have produced tsunami waves spreading across the Tyrrhenian Sea and possibly into the Bay of Naples, where wave energy amplification and constraining would occur. Such a possibility is the subject of continuing investigations that include further geological data and high-resolution computational grids for numeric simulation of the collapse–emplacement dynamics (Tinti *et al.* 2006).

F. L. Chiocci (University of Rome) first sensed, in 1998, the importance of catastrophic submarine landslides originating from Ischia and other Italian volcanic islands, and was the initial promoter of our research. We are grateful to R. Urgeles and J. Griffiths for their reviews and suggestions. Geolab s.r.l. is acknowledged for having made some seismic data available. The research was funded by the INGV (Istituto Nazionale di Geofisica e Vulcanologia) and DPC (Italian Department for Civil Protection) and by the Italian Ministry of Environment.

References

ACOCELLA, V. & FUNICIELLO, R. 1999. The interaction between regional and local tectonics during resurgent doming: the case of the island of Ischia, Italy. *Journal of Volcanology and Geothermal Research*, **88**, 109–123.

ALESSIO, G., ESPOSITO, E., FERRANTI, L., MASTROLORENZO, G. & PORFIDO, S. 1996. Correlazione tra sismicità ed elementi strutturali dell'Isola d'Ischia. *Quaternario*, **9**, 303–308.

BARRA, D., CINQUE, A., ITALIANO, A. & SCORZIELLO, R. 1992. Il Pleistocene superiore marino di Ischia: paleoecologia e rapporti con l'evoluzione tettonica recente. *Studi Geologici Camerti*, **1992**, 231–243.

BEGÉT, J. & KIENLE, J. 1992. Cyclic formation of debris avalanches at Mount St. Augustine volcano. *Nature*, **356**, 701–704.

BERRINO, G., CORRADO, G., LUONGO, G. & TORO, B. 1984. Ground deformation and gravity changes accompanying the Pozzuoli uplift. *Bulletin of Volcanology*, **47**, 187–200.

BRUNO, P. P., DE ALTERIIS, G. & FLORIO, G. 2002. The western undersea section of the Ischia volcanic complex (Italy, Tyrrhenian Sea) inferred from marine geophysical data. *Geophysical Research Letters*, **29**, 1029–1034.

BUCHNER, G. 1986. Eruzioni vulcaniche e fenomeni vulcano-tettonici di età preistorica e storica nell'isola di Ischia. *In*: CENTRE JEAN BÉRARD (ed.) *Tremblements de terre, éruptions volcaniques et vie des hommes dans la Campanie antique*. Napoli, **7**, 145–188.

BUCHNER, G., ITALIANO, A. & VITA-FINZI, C. 1996. Recent uplift of Ischia, Southern Italy. *In*: MCGUIRE, W. J., JONES, A. P. & NEUBERG, J. (eds) *Volcano Instability on the Earth and other Planets*. Geological Society, London, Special Publications, **110**, 249–252.

CANALS, M., LASTRAS, G. *ET AL*. 2004. Slope failure dynamics and impacts from seafloor and shallow sub-seafloor geophysical data: case studies from the COSTA project. *Marine Geology*, **213**, 9–72.

CAPRA, L., MACÌAS, J. L., SCOTT, K. M., ABRAMS, M. & GARDUÑO, M. 2002. Debris avalanches and debris flows transformed from collapses in the Trans-Mexican Volcanic Belt, Mexico—behaviour and implications for hazard assessment. *Journal of Volcanology and Geothermal Research*, **113**, 81–110.

CHEN, J. K., TAYLOR, F. W., EDWARDS, L., CHENG, H. & BURR, S. G. 1995. Recent emerged reef terraces of the Yenkhae resurgent block, Tanna Vanuatu: implications for volcanic, landslide and tsunami hazards. *Journal of Geology*, **103**, 577–590.

CHIOCCI, F. L. & DE ALTERIIS, G. 2006. The Ischia debris avalanche. First, clear submarine evidence in the Mediterranean of a volcanic island pre-historic collapse. *Terra Nova*, **18**, 202–209.

CHIOCCI, F. L., MARTORELLI, E., SPOSATO, A. & TIVOLI, RESEARCH GROUP. 1998. Prime immagini TOBI dei fondali del Tirreno centro-meridionale (settore orientale). *Geologica Romana*, **34**, 207–222.

CUBELLIS, E. & LUONGO, G. 1998. *Il terremoto del 28 luglio 1883 a Casamicciola nell'isola d'Ischia*. Servizio Sismico Nazionale, Instituto Poligrafico (ed.), Roma, 49–116.

D'ARGENIO, B., AIELLO, G. *ET AL.* 2004. Digital Elevation Model of the Bay of Naples and adjacent areas (Eastern Tyrrhenian Sea). *In*: PASQUARÈ, G. & VENTURINI, C. (eds) *Mapping Geology in Italy*. APAT-Dipartimento Difesa del Suolo Monography, Firenze, 21–28.

DE ALTERIIS, G., TONIELLI, R., PASSARO, S. & DE LAURO, M. 2006. *Isole Flegree (Ischia e Procida)*. Serie: Batimetria dei fondali marini della Campania. Liguori, Napoli.

DE VITA, S., SANSIVERO, F., ORSI, G. & MAROTTA, E. 2006. Cyclical slope instability and volcanism related to volcano-tectonism in resurgent calderas: the Ischia island (Italy) case study. *Engineering Geology*, **86**, 148–165.

DEPLUS, C., LE FRIANT, A. *ET AL.* 2001. Submarine evidence for large-scale debris avalanches in the Lesser Antilles Arc. *Earth and Planetary Science Letters*, **192**, 145–157.

DZURISIN, D., WICKS, C. & THATCHER, W. 1999. Renewed uplift at the Yellowstone caldera measured by leveling surveys and satellite radar interferometry. *Bulletin of Volcanology*, **61**, 349–355.

FUNCK, T. & SCHMINCKE, H.-U. 1998. Growth and destruction of Gran Canaria deduced from seismic reflection and bathymetric data. *Journal of Geophysical Research*, **103**, 15393–15407.

GILLOT, P.-Y., CHIESA, S., PASQUARÈ, G. & VEZZOLI, L. 1982. 33.000 yr K/Ar dating of the volcano-tectonic horst of the isle of Ischia, Gulf of Naples. *Nature*, **229**, 242–245.

GLICKEN, H. 1991. Sedimentary architecture of large volcanic-debris avalanches. *In*: FISHER, R. V. & SMITH, G. A. (eds) *Sedimentation in Volcanic Settings. SEPM Special Publication*, **45**, 99–106.

GLICKEN, H. 1998. Rockslide-debris avalanche of May 18, 1980, Mount St Helens volcano, Washington. *Geological Survey of Japan Bulletin*, **49**, 55–106.

GUADAGNO, F. M. & MELE, R. 1995. La fragile isola d'Ischia. *Geologia Applicata e Idrogeologia*, **30**, 177–187.

HILL, D. P. 1984. Monitoring unrest in a large silicic caldera, the Long Valley–Inyo craters volcanic complex in east–central California. *Bulletin of Volcanology*, **47**, 371–395.

HOSHIZUMI, H., UTO, K. & WATANABE, K. 1999. Geology and eruptive history of Unzen volcano, Shimabara Peninsula, Kyushu, SW Japan. *Journal of Volcanology and Geothermal Research*, **89**, 81–94.

KAIZUKA, S. 1992. Coastal evolution at a rapidly uplifting volcanic island: Iwo-Jima, western Pacific Ocean. *Quaternary International*, **15716**, 7–16.

KIENLE, J., KOWALIK, Z. & MURTY, T. S. 1987. Tsunamis generated by eruption from Mt. St. Augustine Volcano, Alaska. *Science*, **236**, 1442–1447.

LABAZUY, P. 1996. Recurrent landslide events on the submarine flank of Piton de La Fournaise volcano (Réunion Island). *In*: MCGUIRE, W. J., JONES, A. P. & NEUBERG, J. (eds) *Volcano Instability on the earth and other planets*. Geological Society, London, Special Publications, **110**, 295–306.

LIPMAN, P. W. & MULLINEAUX, D. R. (eds) 1981. *The 1980 eruptions of Mount St. Helens, Washington*. US Geological Survey Professional Papers, **1250**.

LIPMAN, P. W., NORMARK, W. R., MOORE, J. G., WILSON, J. B. & GUTMACHER, C. E. 1988. The giant submarine Alika debris slide, Mauna Loa, Hawaii. *Journal of Geophysical Research*, **93**, 4279–4299.

MAHOOD, G. A. & HILDRETH, W. 1983. Nested calderas and trapdoor uplift at Pantelleria, Strait of Sicily. *Geology*, **11**, 722–726.

MANZO, M., RICCIARDI, G. P. *ET AL.* 2006. Surface deformation analysis in the Ischia Island (Italy) based on spaceborne radar interferometry. *Journal of Volcanology and Geothermal Research*, **151**, 399–416.

MASSON, D. G. 1996. Catastrophic collapse of the volcanic island of Hierro 15 ka ago and the history of landslides in the Canary islands. *Geology*, **24**, 231–234.

MASSON, D. G., CANALS, M., ALONSO, B., URGELES, R. & HUHNERBACH, V. 1998. The Canary Debris Flow: source area morphology and failure mechanisms. *Sedimentology*, **45**, 411–432.

MASSON, D. G., WATTS, A. B., GEE, M. J. R., URGELES, R., MITCHELL, N. C., LE BAS, T. P. & CANALS, M. 2002. Slope failures on the flanks of the western Canary Islands. *Earth-Science Reviews*, **57**, 1–35.

MCKEE, C., MORI, J. & TALAI, B. 1989. Microgravity changes and ground deformation at Rabaul caldera, 1973–1985. *In*: LATTER, J. H. (ed.) *IAVCEI Proceeding, Volcanic Hazards*. Springer, Berlin, 399–428.

MELE, R. & DEL PRETE, S. 1998. Fenomeni di instabilità dei versanti in Tufo Verde del Monte Epomeo (Isola d'Ischia, Campania). *Bollettino della Società Geologica Italiana*, **117**, 93–112.

MILIA, A. & TORRENTE, M. M. 1999. Tectonics and stratigraphic architecture of a peri-Tyrrhenian half-graben (Bay of Naples, Italy). *Tectonophysics*, **315**, 301–318.

MOORE, J. G., NORMARK, W. R. & HOLCOMB, R. T. 1994. Giant Hawaiian Landslides. *Annual Review of Earth and Planetary Sciences*, **22**, 119–144.

MULDER, T. & CHOCONAT, P. 1996. Classification of offshore mass movements. *Journal of Sedimentary Research*, **66**, 43–57.

NORMARK, W. R., MCGANN, M. & SLITER, R. 2004. Age of Palos Verdes submarine debris avalanche, southern California. *Marine Geology*, **203**, 247–259.

ORSI, G., GALLO, G. & ZANCHI, A. 1991. Simple shearing block-resurgence in caldera depression. A model from Pantelleria and Ischia. *Journal of Volcanology and Geothermal Research*, **47**, 1–11.

RITTMANN, A. 1930. Geologie der Insel Ischia. *Zeitung Vulkanologie*, **6**, 1–265.

RITTMANN, A. & GOTTINI, V. 1980. L'isola d'Ischia: Geologia. *Bollettino del Servizio Geologico Italiano*, **101**, 131–274.

SATAKE, K. & KATO, Y. 2001. The 1741 Oshima-Oshima eruption: extent and volume of submarine debris avalanche. *Geophysical Research Letters*, **28**, 427–430.

SHANMUGAN, G. 1997. The Bouma sequence and the turbidite mind set. *Earth-Science Reviews*, **42**, 201–229.

TIBALDI, A. & VEZZOLI, L. 1998. The space problem of caldera resurgence: an example from Ischia island. Italy. *Geologische Rundschau*, **87**, 53–66.

TIBALDI, A. & VEZZOLI, L. 2000. Late Quaternary monoclinal folding induced by caldera resurgence at Ischia, Italy. *In*: CONSGROVE, J. W. & AMEEN, M. S. (eds) *Forced Folds and Fractures*. Geological Society, London, Special Publications, **169**, 103–113.

TINTI, S., MANUCCI, A., PAGNONI, G., OKAL, E. A., YALCINER, A. & SYNOLAKIS, C. E. 2003. Field survey of the 30 December 2002 Stromboli Tsunami. EOS transactions, AGU, 84(46), Fall Meeting Supplement, Abstract 0S21A-07.

TINTI, S., DE ALTERIIS, G. & ARMIGLIATO, A. 2006. Preliminary studies of tsunami generation by the Ischia Debris Avalanche (IDA), Tyrrhenian Sea, Italy. *Geophysical Research Abstracts*, **8**, 03961.

UI, T., TAKERADA, S. & YOSHIMOTO, M. 2000. Debris avalanche. *In*: SIGURDSSON, H., HOUGHTON, B., MCNUTT, S., RYMER, H. & STIX, J. (eds) *Encyclopedia of Volcanoes*. Academic Press, San Diego, CA, 617–629.

URGELES, R., CANALS, M., BARAZA, J., ALONSO, B. & MASSON, D. G. 1997. The most recent megalandslides of the Canary Islands: the El Golfo debris avalanche and the Canary debris flow, west El Hierro Island. *Journal of Geophysical Research*, **102**, 20305–20323.

URGELES, R., CANALS, M., BARAZA, J. & ALONSO, B. 1998. Seismostratigraphy of the western flanks of El Hierro and La Palma (Canary Islands): a record of the Canary volcanism. *Marine Geology*, **146**, 225–241.

VALLANCE, J. W. 2000. Lahars. *In*: SIGURDSSON, H., HOUGHTON, B., MCNUTT, S., RYMER, H. & STIX, J. (eds) *Encyclopedia of Volcanoes*. Academic Press, San Diego, CA, 601–616.

VEZZOLI, L. 1988. *Island of Ischia*. Quaderni de 'La Ricerca Scientifica' Progetto finalizzato 'Geodinamica', CNR Monografie finali, **10**.

VIOLANTE, C., BUDILLON, F., ESPOSITO, E., PORFIDO, S. & VITTORI, E. 2004. Submerged hummocky topographies and relations with landslides on the northwestern flank of Ischia island, southern Italy. *In*: PATRON EDITORE, *Proceedings of 'Occurrence and mechanisms of flow-like landslides in natural slopes and earthfills', Sorrento, 14–16 May 2003*. Associazione Geotecnica Italiana, Bologna, 309–315.

WARD, S. N. & DAY, S. 2003. Ritter Island Volcano—lateral collapse and tsunami of 1888. *Geophysical Journal International*, **154**, 891–902.

Evidence of slope instabilities and tsunami associated with the 3.5 ka Avellino eruption of Somma–Vesuvius volcano, Italy

ALFONSA MILIA[1]*, ARTURO RASPINI[2] & MAURIZIO M. TORRENTE[3]

[1]*Istituto per l'Ambiente Marino Costiero (IAMC), CNR, Calata Porta di Massa, Porto di Napoli, 80133, Napoli, Italy*

[2]*Istituto di Geoscienze e Georisorse (IGG), CNR, Via Giorgio La Pira, 4, 50121, Firenze, Italy*

[3]*Dipartimento di Studi Geologici ed Ambientali (DSGA), Università del Sannio, Via Portarsa 11, 82100, Benevento, Italy*

**Corresponding author (e-mail: alfonsa.milia@iamc.cnr.it)*

Abstract: The interpretation of high-resolution seismic profiles acquired close the eastern coast of the Bay of Naples offshore Somma–Vesuvius volcano, and calibrated by borehole data, allowed us to recognize a reflection-free seismic facies consisting of a volcaniclastic debris avalanche overlain by pyroclastic density current deposits. Both are associated with the 3.5 ka 'Avellino eruption' and are in turn overlain by a marine succession. The top of the volcaniclastic unit corresponds to a deep erosional surface covered by a coarse-grained bioclastic layer with rounded pumice. We argue that these features are related to a tsunami that was triggered by the sudden entrance into the sea of the gigantic (>1 km^3) volcaniclastic unit. The onshore–offshore correlation and mapping of this volcaniclastic unit have allowed us to evaluate its distribution west of Somma–Vesuvius. The pyroclastic density current deposits terminate seaward at 40 m water depth; at the same depth the debris avalanche is affected by a gravitational instability. Over the last 3.5 ka the slump has been in constant movement, as documented by the angular unconformities at the top of the chaotic seismic unit and within the overlying marine deposits.

Submarine slope failure has been documented in many submarine environments all over the world (e.g. Hampton *et al.* 1996). It occurs at active and passive continental margins and volcanoes and is one of the most important mechanisms in shaping them, transporting vast quantities of sediment downslope. It is only in the last two decades that subaerial and submarine volcanoes were recognized as having evolved over long periods of construction punctuated by short and sometimes violent destructive slope failures. These events induce slumps, debris avalanches or slides, with volumes of several hundred to a few thousand cubic metres (McGuire *et al.* 1993; Vallance *et al.* 1995; Glicken 1996; Munro & Rowland 1996; Mangeney *et al.* 2000; Siebert 2002), which move at high speed at distances exceeding 10 km. Oversteepening of the flank caused by cryptodome intrusion is one of the more commonly invoked triggering factors for flank collapses (Lipman *et al.* 1981; McGuire *et al.* 1991; Donnadieu & Merle 1998). Sometimes, as in 1980 on Mount St. Helens, such flank collapses can trigger the explosive eruption of a rising magma column, which suddenly depressurizes after the decapitation of the volcano.

Many of the largest lateral collapses are located adjacent to the margins of island and coastal volcanoes, and have generated horseshoe scars on the flanks (Holcomb & Searle 1991; Moore *et al.* 1994*a*, *b*; McGuire 2003; Cronin *et al.* 2004). All the Hawaiian volcanoes, for example, are surrounded by submarine aprons of allochthonous volcanic material that was emplaced by sliding or slumping, and similar deposits have also been recognized around many marine volcanoes (Holcomb & Searle 1991; Begét & Kienle 1992; Filmer *et al.* 1994; Smoot 1995; Labazuy 1996; Keating & McGuire 2000; Deplus *et al.* 2001; Tibaldi 2001; Milia *et al.* 2003; Ward & Day 2003; Chiocci & de Alteriis 2006; Pérez-Torrado *et al.* 2006; de Alteriis & Violante 2009).

Episodes of flank instability and major lateral failure have to be considered as a ubiquitous occurrence in the normal life-cycle of marine volcanoes and are estimated to have occurred world-wide

From: VIOLANTE, C. (ed.) *Geohazard in Rocky Coastal Areas*. The Geological Society, London, Special Publications, **322**, 105–119. DOI: 10.1144/SP322.4

at least four times per century over the past 500 years (e.g. Siebert *et al.* 1987; Ui *et al.* 2000; McGuire 2003). They represent a major geohazard for marine environments, offshore infrastructure (Campbell 1999) and nearby coastal areas, because when associated with debris avalanches they can move rapidly with velocities in excess of 360 km h^{-1} (McGuire 2003; see also Mehl & Schmincke 1999) and locally destroy everything in their paths, triggering tsunamis on entering the sea (e.g. Heinrich *et al.* 1998). One such event was witnessed on Stromboli volcano on 30 December 2002, when the occurrence of a subaerial landslide involving the underwater flank of the volcanic edifice generated a tsunami up to 10 m high that caused severe damage along the Stromboli coast, and reached the northern coast of Sicily *c.* 50 km to the south (Tinti *et al.* 2000; Chiocci *et al.* 2003).

Somma–Vesuvius, located in southern Italy (Fig. 1), has been one of the most active volcanoes in the world. More than 80 eruptions are listed in a recent worldwide directory of volcanic activity (Simkin & Siebert 1994) and the products of two debris avalanches have been documented offshore (Milia *et al.* 2003, 2007). Recent stratigraphic and structural studies offshore Somma–Vesuvius have documented volcanic products on the continental shelf (Milia *et al.* 1998, 2003, 2007). In detail, two debris avalanches were detected in the Late Quaternary succession of the Bay of Naples off Somma–Vesuvius and a link between these debris avalanches and the eruptions of the 'Pomici di Base' (18 ka BP) and the 'Avellino Pumice' (3.5 ka BP) was postulated. The first historical eruption was the major Plinian explosion of AD 79, which buried the major Roman cities of Herculaneum and Pompeii.

Because of the surrounding high level of urbanization (about 700 000 people live within a 10 km radius of the crater) Somma–Vesuvius has been classified as a very high risk volcano. For many years numerous volcanological studies dealing with the reconstruction of the Somma–Vesuvius eruption and the onland distribution of eruption products have been carried out (Delibrias *et al.* 1979; Rosi & Sbrana 1987; Barberi *et al.* 1990; Scandone *et al.* 1993; Andronico *et al.* 1995). Based on the results of these volcanological studies the Italian Government has devised an Emergency Management Plan for the population living in the area of the volcano (Santacroce 1996).

The aims of this paper are: (1) the detection and onshore–offshore mapping of the Avellino eruption deposits of Somma–Vesuvius (3.5 ka) in the eastern Bay of Naples and in the adjacent coastal and alluvial plain areas; and (2) the evaluation of the consequences of the voluminous volcaniclastic mass wasting (linked to the above eruption) entering the sea.

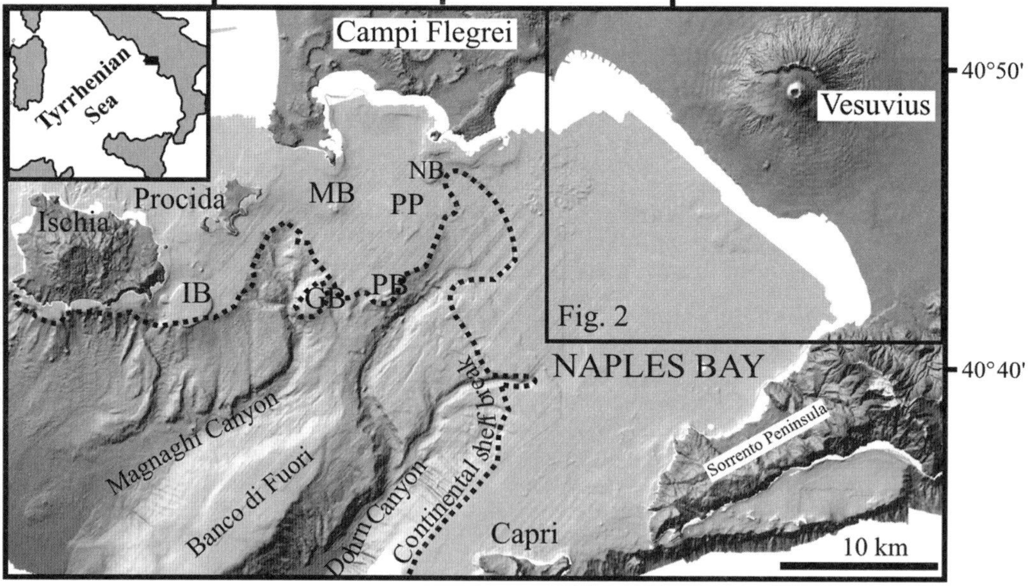

Fig. 1. Physiographic map of the Bay of Naples showing the continental shelf and two canyons across the slope. IB, Ischia Bank; GB, Gaia Bank; PB, Pia Bank; MB, Miseno Bank; PP, Penta Palummo Bank; NB, Nisida Bank.

Methods

Continuous seismic reflection profiles were acquired offshore Somma–Vesuvius in September 2000 (Fig. 2). The survey was conducted using a multi-tip sparker system equipped with multi-electrode arrays. A sparker source of 200 J with a broad frequency range (200–2000 Hz) was used. The ship position was determined by means of a differential global positioning system (GPS) that has a positioning accuracy of 1 m. All seismic-section data were recorded graphically with a vertical recording scale of 0.25 s. The best vertical resolution was *c.* 1 m. The total length of the seismic profiles was 110 km with a mean spacing of 1 km. This coverage of seismic data off Somma–Vesuvius facilitates confident stratigraphic correlations between profiles and is sufficient to reveal detailed structural and stratigraphic variations.

We investigated the region west of Vesuvius by means of a lithostratigraphic analysis of boreholes drilled by private companies and local authorities, and by the reinterpretation of data from the literature (D'Erasmo 1931; Bellucci 1998; Fig. 2).

The stratigraphic framework was reconstructed using a physical stratigraphy approach (e.g. Mitchum *et al.* 1977). Seismic units are groups of seismic reflections whose parameters (configuration, amplitude, continuity and frequency) differ from those of adjacent groups (Mitchum *et al.* 1977). Volcanic and sedimentary units were delineated on the basis of contact relations and internal and external configurations. The latter permit the interpretation of the environmental settings from seismic data. Using seismic characteristics, the stratigraphic relationships between the various units and the physical continuity of coastal outcrops we attempted to relate each seismic unit to a geological one.

Geological setting

Somma–Vesuvius volcano is located on the north-eastern coast of the Bay of Naples (Fig. 1). The Bay of Naples is a Middle Pleistocene half-graben filled by fourth-order (*c.* 100 ka) depositional sequences arranged in sequence sets that display long-term aggradational–progradational stacking patterns (Milia 1999; Milia & Torrente 1999). The Mesozoic–Cenozoic carbonate substrate, cropping out on the Sorrento Peninsula, dips towards the

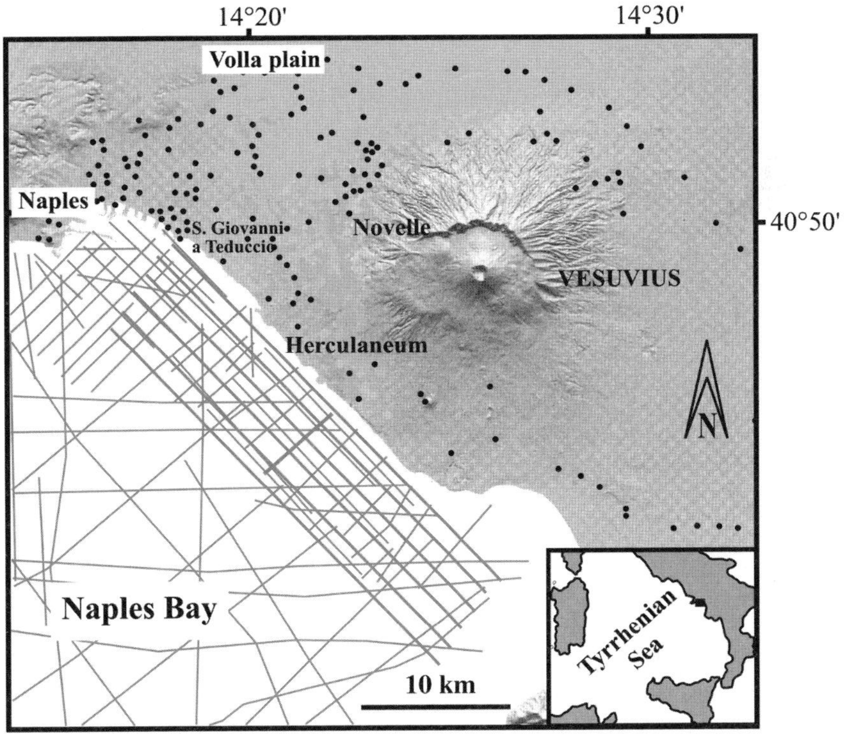

Fig. 2. Map showing location of sparker seismic lines in the Bay of Naples and onshore boreholes.

NW and is overlain by Quaternary sediments and volcanic products.

The physiography of the Bay of Naples is characterized by a wide continental shelf that extends to water depths of 100–180 m. The shelf width varies from a maximum of about 20 km in the central bay to about 2.5 km off the islands of Capri and Procida (Fig. 1). The northern area has an irregular continental shelf, which forms part of an extensive system of volcanic banks. In the central area, offshore Vesuvius, the continental shelf is wider and covers an area of *c.* 380 km^2.

The structure of Somma–Vesuvius corresponds to the breached crater of the Monte Somma volcano, which funnelled much of the material erupted from Vesuvius toward the sea.

The Somma–Vesuvius volcanic succession mainly consists of lava flows and minor scoria fall deposits (age $\leq$ 20–25 ka; Santacroce 1987), overlain by the deposits of four main Plinian eruptions (Andronico *et al.* 1995): the 'Pomici di Base' (18 ka BP); the 'Mercato Pumice' (8 ka BP); the 'Avellino Pumice' (3.5 ka BP); and the 'Pompeii Pumice' (AD 79). Only a few outcrops of the pyroclastic flows associated with the Pomici di Base eruption (18 ka) are documented on Monte Somma; the Mercato Pumice eruption (8 ka) produced pyroclastic flows, the major thicknesses of which occurred on the northern sector of Monte Somma; the two younger Plinian eruptions (3.5 ka–AD 79), in contrast, feature extensive surge and pyroclastic flows that moved both west and south of the crater (Sigurdsson *et al.* 1985; Mastrolorenzo *et al.* 2006).

Based on the interpretation of high-resolution seismic reflection profiles collected in the Bay of Naples, Milia *et al.* (2003) documented two debris avalanches originating from flank collapses of Somma–Vesuvius volcano and interlayered with marine sediments. In particular, the older debris avalanche was dated at 18 ka, being covered by the lowstand systems tract associated with the last glacial maximum, whereas the younger one is interbedded within the marine sediments of the highstand systems tract deposited over the last 5.5 ka. Milia *et al.* (2003) postulated the older debris avalanche to be associated with the 18 ka Pomici di Base eruption and the younger one with the 3.5 ka Avellino eruption. Their study was based on the stratigraphic relationships between the debris avalanches and the systems tracts of the Late Quaternary depositional sequence together with the ages of the dated Somma–Vesuvius Plinian eruptions. The 18 ka debris avalanche is present down to water depths of 100 m between Portici and Torre Annunziata and has a volume of 2.9 km^3, whereas the 3.5 ka debris avalanche is present to depths of 150 m and extends from Portici to Naples harbour, and has a volume of 1 km^3 (Milia *et al.* 2007, 2008).

Onshore the Avellino eruption was characterized by a fall phase followed by predominantly phreatomagmatic activity (Lirer *et al.* 1973; Rolandi *et al.* 1993, 1998) that gave rise to widespread surge deposits that covered the western sector of Monte Somma. Thick Avellino volcaniclastic debris-flow products underlying the Avellino fall and base surges have recently been discovered west of the volcano crater (Fig. 3a; Rolandi *et al.* 2004). The volcaniclastic deposits are grey, massive and lithic-rich, with lithic blocks up to 0.5 m in diameter within a sandy matrix. This eruption sent volcaniclastic debris flows westward and toward the sea. These products crop out directly above the ancient lava flows of Monte Somma. At the Herculaneum excavation site the AD 79 pyroclastic flows and surges cover pyroclastic surges attributed to the Avellino eruption (Lirer *et al.* 1973). The latter lie above a *c.* 10 m thick volcaniclastic unit featuring lava blocks (Fig. 3b; Bellucci 1998). In the Volla Plain, west of Vesuvius, boreholes encountered a thick tuff unit from Somma–Vesuvius interlayered with alluvial and marine sediments and locally underlain by lava breccias (Fig. 3c; Bellucci 1998).

Results

Slope instability was detected on the western flank of Somma–Vesuvius and on the continental shelf of the Bay of Naples by means of seismic profiles and borehole data. In detail, we focus on the offshore and onshore features of the 3.5 ka Avellino debris avalanche and tuffs, and document their recent mobilization in the submarine area as a result of slumping.

A strictly spaced grid of seismic reflection profiles permitted us to reconstruct the 3D features of the deposits associated with the Avellino eruption in the Bay of Naples. These deposits, which reach a water depth of 150 m, are characterized by chaotic seismic facies.

The interpretation of seismic profiles offshore from the western sector of Vesuvius reveals two superimposed reflection-free seismic units bounded at the top by a nonconformity that displays an irregular morphology (Fig. 4). Samples from boreholes close to the coast (Fig. 3d) reveal that the chaotic seismic unit lies above marine fine sands and is made up of a 15–20 m thickness of loose gravels and sands with subangular pumice, lava and scoria fragments overlain by some 5 m of lithoid yellow to green tuff including pumice, scoriae and mainly calcareous lithic fragments. This unit is abruptly covered by a 2 m layer consisting of loose sand and gravel with abundant fragments of marine shells and

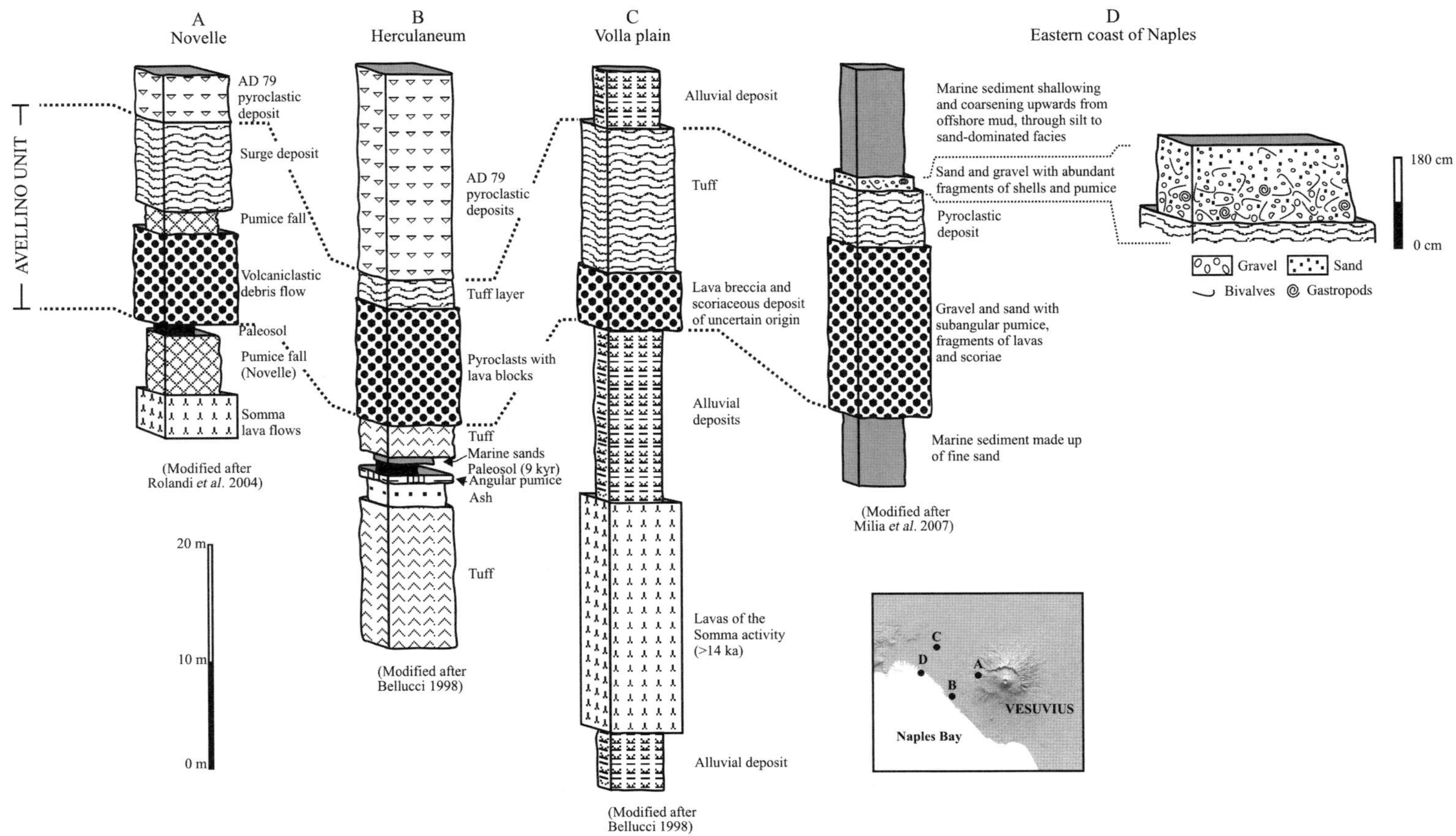

Fig. 3. Stratigraphic columns of the succession cropping out on the western flank of Vesuvius (column A); borehole logs in the southern Campana Plain (columns B and C) and on the eastern Naples coast (column D), with a close-up of the bioclast-rich layer.

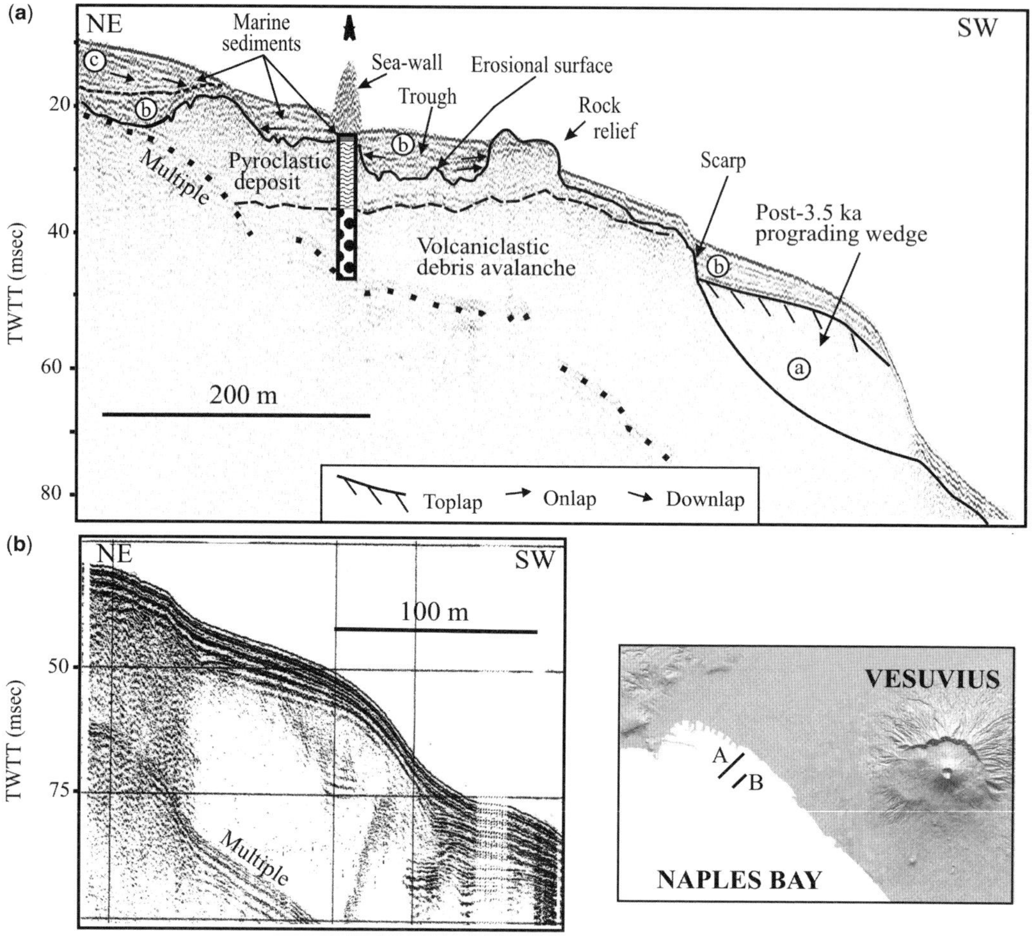

Fig. 4. (**a**) Sparker seismic line offshore S. Giovanni a Teduccio calibrated by a borehole displaying the Avellino eruption succession bounded at the top by an erosional surface and covered by the post 3.5 ka marine deposits. These deposits, from older to younger, consist of: (a) a prograding wedge; (b) a seismic unit with subparallel seismic reflectors; (c) a prograding unit close to the coast. Modified from Milia *et al.* (2007). (**b**) Sparker seismic line showing the prograding wedge.

rounded pumice. The succession terminates with shallowing and coarsening upward marine sediments consisting of muds, silts and sands.

Bearing the above considerations in mind, we demonstrate that the Avellino unit is made up of two main subunits that correspond to the thick basal coarse-grained volcaniclastic deposit (debris avalanche) and the overlying deposit related to pyroclastic gravity currents; and that the unit is sandwiched by marine sediments. The deposit of the Avellino unit related to pyroclastic gravity currents terminates at 40 m water depth and corresponds to a reflection-free seismic facies (Fig. 4), implying no change in the acoustic impedance. This is because such pyroclastic currents are generally poorly sorted, being generated by vertically directed explosions that occur when the density of the eruption column (or part of it) becomes greater than that of the surrounding atmosphere, such that the hot fragments collapse downward under gravity.

A comparable stratigraphic succession of the Avellino unit, as detected on the eastern coast of Naples and formed by the basal volcaniclastic debris avalanche overlain by surge deposits, is also present near the vent at Novelle (Figs 2 and 3a; Rolandi *et al.* 2004) and at the Herculaneum excavation site (Fig. 3b; Bellucci 1998).

The correlation between the offshore geological section, reconstructed by means of the interpretation of seismic reflection profiles, and the onshore geological section, reconstructed by the interpretation

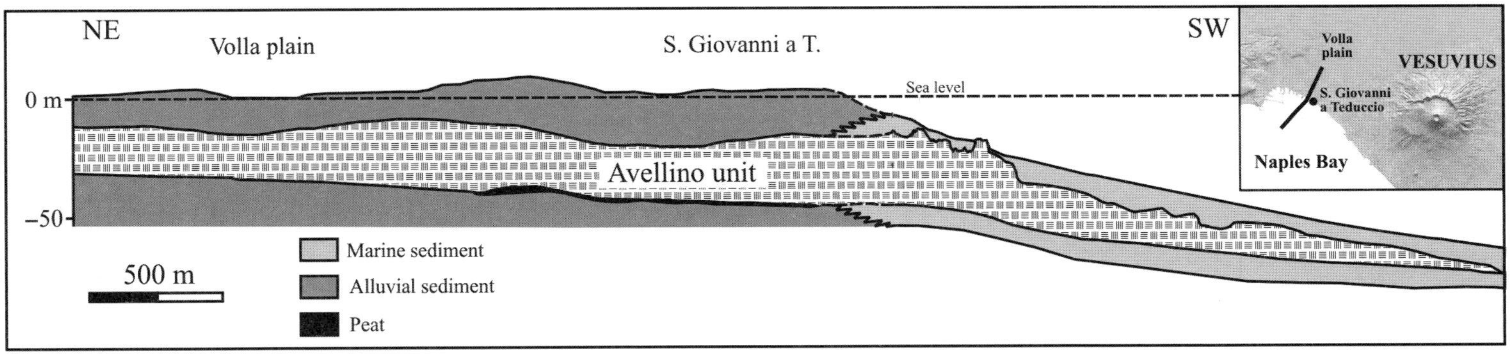

Fig. 5. Geological section from the Bay of Naples to the Volla plain displaying the physical continuity of the Avellino volcaniclastic unit.

of borehole data, reveals a physical continuity between the previously recognized thick tuff deposits, in places overlying breccia deposits in the Volla plain (Bellucci 1998), and the Avellino unit recognized offshore and consisting of a debris avalanche covered by deposits related to pyroclastic gravity currents (Fig. 5).

The isopach map of Figure 6 shows the distribution of the Avellino unit in marine and continental areas between Naples and Vesuvius with a maximum thickness of *c*. 70 m. The location of this unit suggests a westward flow direction of both pyroclastic gravity flows and debris avalanche.

In the area off San Giovanni a Teduccio the Avellino unit reaches its maximum thickness and is characterized by an irregular upper surface with up to 20 m high rock relief and up to 20 m deep troughs (Fig. 4). This surface is overlain by three seismic units: unit a, a prograding wedge displaying an oblique seismic configuration with the toplap surface located between 30 and 40 m water depth; unit b, a unit showing subparallel seismic reflectors that covers the prograding wedge and onlaps the troughs; unit c, a prograding unit close to the coast downlapping the seismic unit b. Based on the cumulative isopach map of these three seismic units, an irregular morphology of the nonconformable erosional surface was also inferred (Fig. 7). This buried morphology partly acted as a drainage system transferring eroded sediments from the coastal area to the sea, where they were deposited to form a prograding wedge (unit a in Fig. 4) that reaches a maximum thickness of 18 m (Fig. 7).

A seismic line across the continental shelf (Fig. 8) shows the morphology of the Avellino unit toward the sea. Seaward of a submarine scarp the top of the Avellino volcaniclastic debris avalanche is characterized by a staircase morphology, which features landward-dipping flat tops and sliding surfaces associated with the evacuation zone of a slump. In the distal area, in contrast, the top of the Avellino debris avalanche is characterized by mounds that can be interpreted as the accumulation zone of the slump. The occurrence of several unconformities within the marine deposits overlying the Avellino unit suggests that the slump has been moving over the last 3.5 ka and up to the present.

Discussion

To evaluate the slope instability of the coastal Somma–Vesuvius volcano during the Holocene,

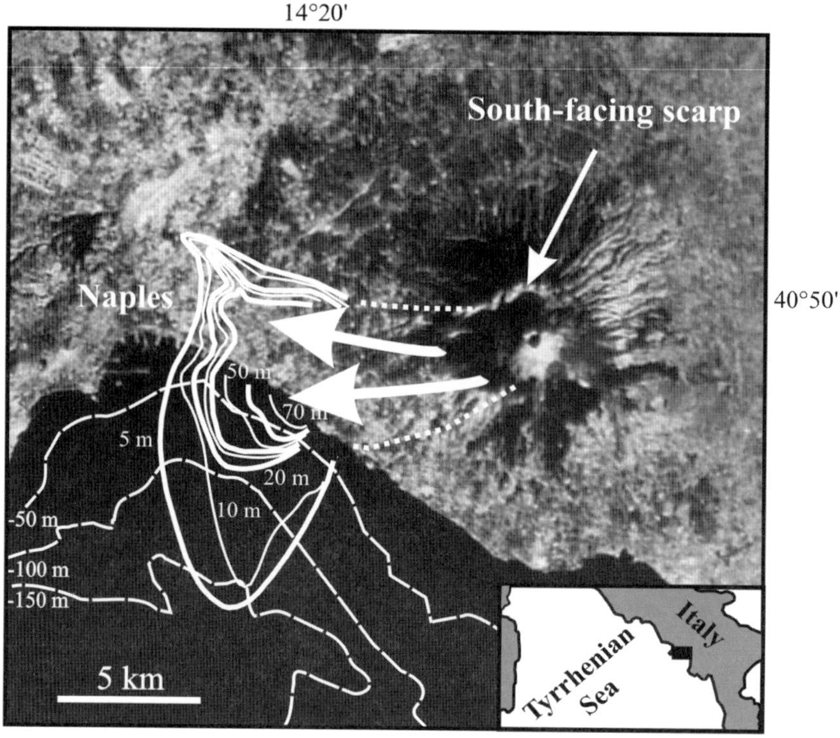

Fig. 6. Isopach map of the Avellino unit in the marine and continental area between Naples and Vesuvius. White arrows show the westward flow direction. Modified from Milia *et al.* (2007).

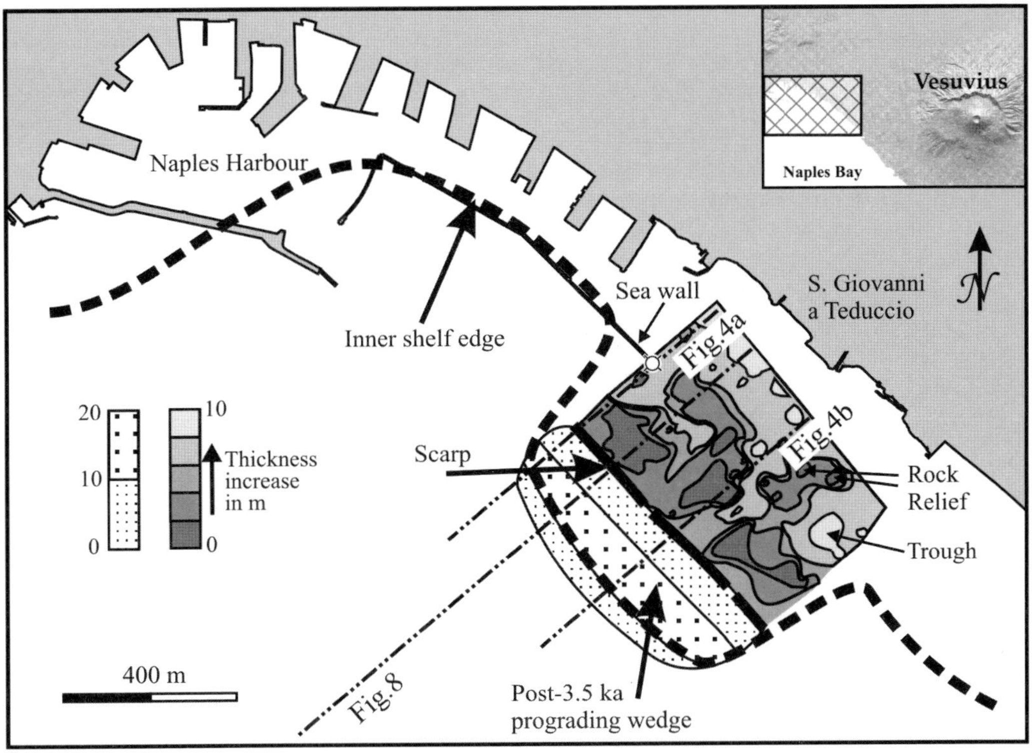

Fig. 7. Isopach map of the post-Avellino eruption marine deposits between 15 and 60 m water depth. The maximum and minimum thicknesses in the inner shelf zones correspond to, respectively, troughs and reliefs of the underlying erosional surface. Dotted areas correspond to a thick prograding wedge at the margin of the inner shelf. (Note the abrupt seaward shift of the inner shelf edge off San Giovanni a Teduccio.) Modified from Milia *et al.* (2007).

we used core data and high-resolution seismic profiles to reconstruct the architecture of the volcaniclastic unit relating to the Avellino eruption (3.5 ka BP) in the Bay of Naples and adjacent coastal and alluvial plain areas. In these areas this unit consists of a thick volcanic debris avalanche overlain by deposits related to pyroclastic gravity currents and has a total volume exceeding 1 km^3. The distribution of the Avellino unit documents a debris avalanche that moved over a distance of *c*. 10 km and the subsequent westward flow of pyroclastic currents. The emplacement of this voluminous unit induced an instantaneous inner shelf progradation (Fig. 7), which resulted in an abrupt shift of the inner shelf edge off San Giovanni a Teduccio, where the mound-shaped Avellino unit reaches thicknesses ranging from 50 to 70 m.

As mentioned above, a nonconformity bounds the Avellino unit at the top. This either corresponds to the depositional morphology of the deposits related to pyroclastic gravity currents or is an erosional surface. Pyroclastic currents are sediment gravity flows of volcanic fragments, often hot, that spread outwards around the volcano, partially filling valleys and mantling intervening ridges. Considering the highly irregular morphology, with troughs and highs of up to 20 m of relief, we therefore interpret the nonconformity as an erosional surface. This surface and the superimposed coarse-grained layer document a high-energy event in coastal areas. This event produced a deep erosion of the substratum to *c*. 30 m water depth (Fig. 4) and winnowing of fine sediments across the coastal setting before deposition of the 1.3 m thick layer consisting of sand and gravel with abundant marine shell fragments and rounded pumice (Fig. 3d). It is important to note that over the last 5500 years there is no documentary evidence of any comparable deep channelled surface with a coarse-grained bioclastic deposit resting on it in the entire highstand deposit of the Bay of Naples (Milia 1999; Milia & Torrente 1999; Sacchi *et al.* 2005), confirming that these deposits, although of limited extent (Fig. 7), represent an exceptional event.

In the study area such high-energy depositional conditions, which imply high flow velocities (e.g. Bryant 2001; Scheffers & Kelletat 2003), could be correlated to storms or tsunamis.

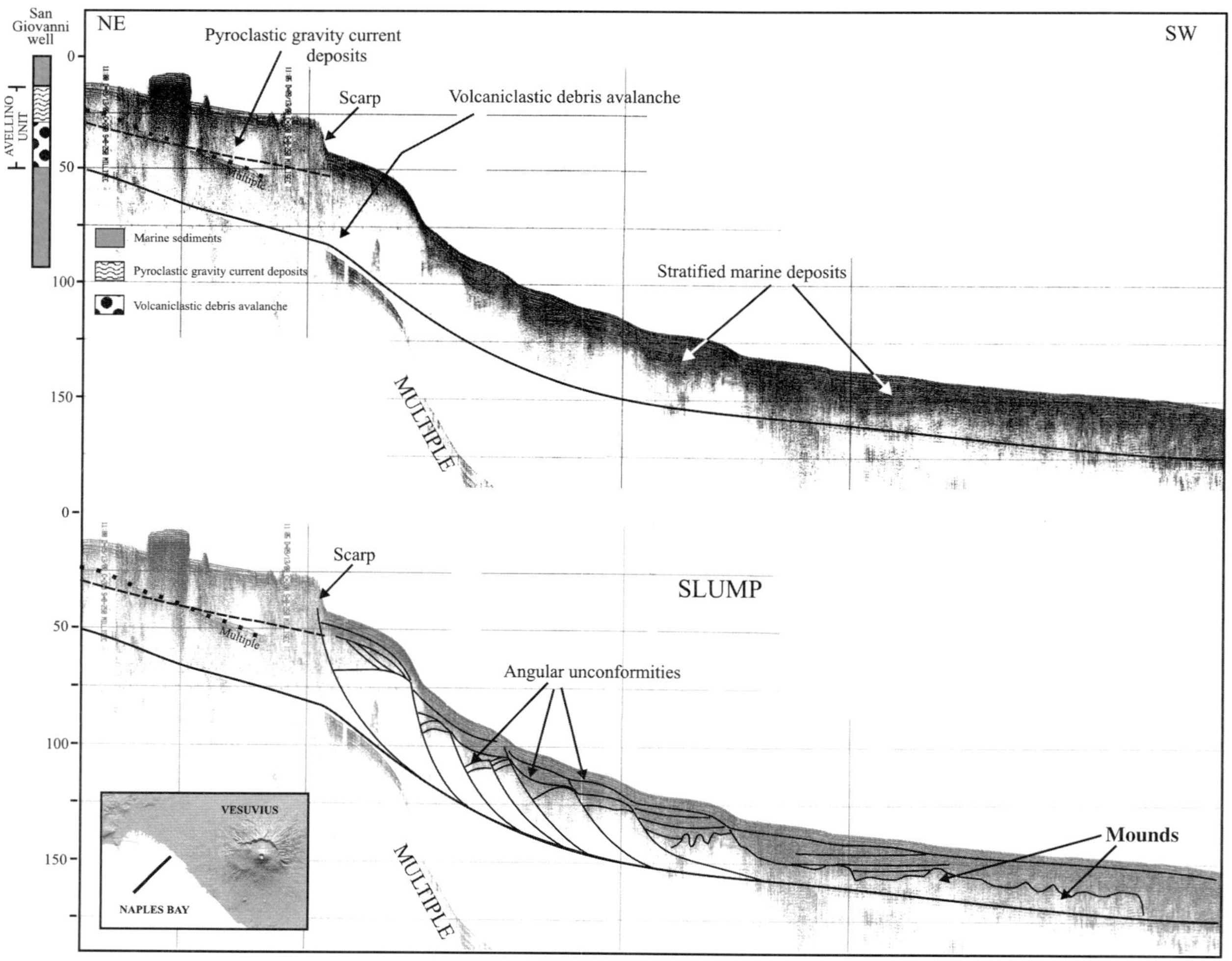

Fig. 8. Sparker seismic line showing a slump affecting the Avellino debris avalanche.

The relationship between storm processes and nearshore sediment transport and deposition is still poorly understood (Hill *et al.* 2004; see also Hartley & Jolley 1999). Nevertheless, considering the latitude of the study area, deposition by a major storm surge that deeply eroded the topmost part of the Avellino unit, and allowed the deposition of 1.3 m thick bioclastic sand and gravel appears unlikely for the following reasons: (1) Storm waves on the inner shelf and shoreface commonly result in the deposition of discrete sedimentary units that do not usually display features such as the large-scale, highly irregular erosional base and the great thickness (e.g. Morton 1981; Snedden & Nummedal 1991; Keen *et al.* 2004; Budillon *et al.* 2005) that is seen for the deposit overlying the Avellino unit. These sedimentological features require greater wave energy and primarily turbulent flow conditions (e.g. Davis *et al.* 1989; Walker & Plint 1992; Dawson & Shi 2000; Yamano *et al.* 2001; Tuttle *et al.* 2004); and (2) The 1.3 m thick bioclastic deposit resting on the Avellino unit is too uniform and locally extensive to be considered a storm layer (e.g. Scheffers & Kelletat 2003; Goff *et al.* 2004). The unique stratigraphic occurrence of the bioclastic layer resting on a deep irregular surface points to a truly exceptional erosional event, much rarer then would be expected from major storm activity.

Tsunamis may be produced by one of the following processes: shallow marine earthquakes, pyroclastic currents entering the sea and landslides. Tsunamigenic landslides can also be associated with volcanic eruption activity, as is the case with volcano flank collapses that produce debris avalanches.

Although described in areas particularly subjected to tsunamis in modern or recent times (Young & Bryant 1992; Nishimura *et al.* 1999; Dawson & Smith 2000; Fujiwara *et al.* 2000; Keating & McGuire 2000; Dominey-Howes 2004), tsunami deposits are rarely reported in the stratigraphic record because diagnostic criteria for their recognition are equivocal. In particular, tsunami deposits are difficult to detect in coastal environments and very few investigations have attempted to describe and estimate their impact on Pleistocene and Holocene coastal development (e.g. Minoura *et al.* 1996; Scheffers & Kelletat 2003; Tuttle *et al.* 2004). On the coastline a tsunami produces local accelerations and decelerations of the water column that can induce dramatic, widespread erosion and sediment deposition in the geological record, documenting changes in coastal topography (Yeh *et al.* 1993; Dawson 1994; Synolakis *et al.* 1995; Einsele 1998; Bryant 2001; Shuto 2001; Cantalamessa & Di Celma 2005). Extreme Holocene tsunami events were the dominant factor in the coastal development of southeastern Australia and Japan, and were responsible for the formation of barrier islands, cliffs, canyons and sculptured bedrock forms, as well as for cutting sand spits, tombolos and sandbars, and causing the deepening and widening of many channels (e.g. Bryant & Young 1996; Shuto 2001; Paine 2002). As a result of turbulence and a downward pressure exerted by the water, deep erosion of the substratum and the formation of wave-cut features are frequently associated with tsunamis as they approach the coast (Coleman 1978; Dawson & Smith 2000; Fujiwara *et al.* 2000; Waythomas 2000; Shuto 2001; Dominey-Howes 2004; see also Pratt 2002). Finally, it is important to consider that on entering the sea several debris avalanches produced by coastal volcano flank failures triggered tsunamis that resulted in severe damage and fatalities (Hayashi & Self 1992; Simkin & Siebert 1994; Ward & Day 2003).

Taking into account the above considerations, a plausible mechanism that can account for the occurrence of the channelled morphology and settlement of the 1.3 m thick bioclastic deposit on the Avellino unit is a tsunami related to the 3.5 ka Avellino eruption (Milia *et al.* 2007). A scenario can therefore be envisaged whereby the sudden entrance into the sea of the rapidly moving, voluminous Avellino unit, producing a fast energy transfer, significantly displaced water away from the shore, inducing a tsunami in the process. In this context, the erosional surface overlying the Avellino unit was due to high flow velocities, whereas the 1.3 m thick coarse-grained bioclastic layer settled during the waning phase following the passage of the tsunami. Tsunami-related backflows, following the local morphology, possibly increased their erosional capacity further as they moved across the coastal channelled area, allowing large amounts of debris eroded from shore settings to be transported seaward by some kind of sediment flow (e.g. Cantalamessa & Di Celma 2005), as was also observed in the 2004 Asian tsunami (e.g. http://www.dcita.gov.au/cca). This could be the case for the deposition of seismic unit a (Fig. 4), or part of it, which displays an oblique prograding wedge with the toplap surface located between 30 and 40 m water depth. Finally, 'normal' marine sedimentation returned in the eastern coastal settings of the Bay of Naples (Fig. 3d). As a whole, the proposed mechanism of tsunami formation is comparable with that of the tsunami that inundated the coasts of the Sunda Straits during the 1883 Krakatau eruption and was probably linked to the entrance of pyroclastic mass flows into the sea (Carey *et al.* 1996, 2000).

As stated above, we found evidence of a slump affecting the volcaniclastic debris avalanche and the overlying marine sediments below 40 m water depth. Parameters that generally favour or trigger slope instabilities include seismic activity, angle of slope of the margin, basement architecture, sea current and high pore-fluid pressure (Vendeville & Gaullier 2003). In this case the voluminous volcaniclastic debris avalanche was instantaneously emplaced under the sea, producing a rapid physiographic change and a potential area of disequilibrium. In such a scenario we cannot exclude that the slump may have been triggered by tsunami-induced pore-pressure changes.

This study is still in its preliminary stage and more data are required for a better understanding of both the impact of the 3.5 ka tsunami on the eastern coastal setting of the Bay of Naples and the sedimentological features of the related deposit. The results of this study illustrate, however, that, apart from the volcanic risk, tsunamis and active slumps are hazards that must be evaluated in the risk assessment of the Bay of Naples and the adjacent coastal and alluvial plain areas around the active Somma–Vesuvius volcano.

Conclusions

During the Holocene the geological evolution of the Vesuvius coastal area was characterized by many hazardous processes, such as the emplacement of deposits related to a debris avalanche and pyroclastic currents that travelled over a distance of *c.* 10 km, and the formation of a tsunami caused by the entrance of this volcaniclastic body (in excess of 1 km^3) into the sea and active submarine slumps.

In the Bay of Naples, and in the adjacent coastal areas, the collapse of the western seaward-facing flank of the Somma–Vesuvius volcano and the Avellino eruption (3.5 ka BP) that immediately followed allowed the emplacement of a thick debris avalanche overlain by a deposit related to pyroclastic gravity currents. As it entered the sea at a very high speed the voluminous (>1 km^3) volcanic material produced a tsunami that struck the coast of the Bay of Naples, deeply eroding the topmost part of the Avellino unit and forming a channelled morphology. During the waning phase that followed the passage of the tsunami surge a 1.3 m thick layer consisting of sand and gravel with abundant fragments of shells and rounded pumice settled.

Seaward of a scarp below 40 m water depth, the Avellino volcaniclastic debris avalanche displays typical morphologies suggesting a slumping process possibly triggered by tsunami-induced pore-pressure changes. Angular unconformities at the top of the debris avalanche and within the marine deposit covering the Avellino unit suggest that over the last 3.5 ka the slump was affected by repetitive movement.

The results of this study illustrate that, in addition to the volcanic risk, tsunami triggered by the voluminous mass wasting linked to volcanic eruptions (and possible gravitational instabilities of the related volcaniclastic products) entering the sea are potentially devastating hazards. They must therefore be seriously evaluated in the risk assessment and disaster management planning of the densely populated region around the coastal active Somma–Vesuvius volcano.

This paper benefited significantly from the constructive reviews provided by W. McGuire and M. Jaboyedoff. This research was partly funded by the Italian Ministero dell'Università e della Ricerca Scientifica e Tecnologica (FRA 2006, 2007, to M.M.T.).

References

ANDRONICO, D., CALDERONI, G., CIONI, R., SBRANA, A., SUPPLIZIO, R. & SANTACROCE, R. 1995. Geological map of Somma–Vesuvius Volcano. *Periodico di Mineralogia*, **64**, 77–78.

BARBERI, F., MACEDONIO, G., PARESCHI, M. T. & SANTACROCE, R. 1990. Mapping the tephra fallout risk: an example from Vesuvius, Italy. *Nature*, **344**, 142–144.

BEGÉT, J. E. & KIENLE, J. 1992. Cyclic formation of debris avalanches at Mount St. Augustine volcano. *Nature*, **356**, 701–704.

BELLUCCI, F. 1998. Nuove conoscenze stratigrafiche sui depositi effusivi ed esplosivi nel sottosuolo dell'area del Somma–Vesuvio. *Bollettino della Società Geologica Italiana*, **117**, 385–405.

BRYANT, E. A. 2001. *Tsunami. The Underrated Hazard.* Cambridge University Press, Cambridge.

BRYANT, E. A. & YOUNG, R. 1996. Bedrock-sculpturing by tsunami, south coast New South Wales. *Journal of Geology*, **104**, 565–582.

BUDILLON, F., ESPOSITO, E., IORIO, M., PELOSI, N., PORFIDO, S. & VIOLANTE, C. 2005. The geological record of storm events over the last 1000 years in the Salerno Bay (Southern Tyrrhenian Sea): new proxy evidences. *Advances in Geosciences*, **2**, 123–130.

CAMPBELL, K. J. 1999. Deep water geohazards: how significant are they? *Leading Edge*, **18**, 514–519.

CANTALAMESSA, G. & DI CELMA, C. 2005. Sedimentary features of tsunami backwash deposits in a shallow marine Miocene setting, Mejillones Peninsula, northern Chile. *Sedimentary Geology*, **178**, 259–273.

CAREY, S., SIGURDSSON, H., MANDEVILLE, C. & BRONTO, S. 1996. Pyroclastic flows and surges over water: an example from the 1883 Krakatau eruption. *Bulletin of Volcanology*, **57**, 493–511.

CAREY, S., SIGURDSSON, H., MANDEVILLE, C. & BRONTO, S. 2000. Volcanic hazards from pyroclastic flow discharge into the sea: examples from the 1883 eruption of Krakatau, Indonesia. *In*: MCCOY, F. W. & HEIKEN, G. (eds) *Volcanic Hazards and Disasters in Human Antiquity*. Geological Society of America, Special Papers, **345**, 1–14.

CHIOCCI, F. I. & DE ALTERIIS, G. 2006. The Ischia debris avalanche: first clear submarine evidence in the Mediterranean of a volcanic island prehistorical collapse. *Terra Nora*, **18**, 202–209.

CHIOCCI, F. L., BOSMAN, A., ROMAGNOLI, C., TOMMASI, P. & DE ALTERIIS, G. 2003. The December 2002 Sciara del Fuoco (Stromboli Island) submarine landslide: a first characterization. *Geophysical Research Abstracts*, **5**, 12069.

COLEMAN, P. J. 1978. Tsunami sedimentation. *In*: FAIRBRIDGE, R. W. & BOURGEOIS, J. (eds) *The Encyclopedia of Sedimentology*. Dowden, Hutchinson & Ross, Stroudsburg, PA, 828–831.

CRONIN, S. J., FERLAND, M. A. & TERRY, J. P. 2004. Nabukelevu volcano (Mt. Washington), Kadavu—A source of hitherto unknown volcanic hazard in Fiji. *Journal of Volcanology and Geothermal Research*, **131**, 371–396.

DAVIS, R. A., KNOWLES, S. C. & BLAND, M. J. 1989. Role of hurricanes in the Holocene stratigraphy of estuaries. Examples from the Gulf Coast of Florida. *Journal of Sedimentary Petrology*, **59**, 1052–1061.

DAWSON, A. G. 1994. Geomorphological effects of tsunami run-up and backwash. *Geomorphology*, **10**, 83–94.

DAWSON, A. G. & SHI, S. 2000. Tsunami deposits. *Pure and Applied Geophysics*, **157**, 875–897.

DAWSON, S. & SMITH, D. E. 2000. The sedimentology of Middle Holocene tsunami facies in northern Sutherland, Scotland, UK. *Marine Geology*, **170**, 69–79.

DE ALTERIIS, G. & VIOLANTE, C. 2009. Catastrophic landslides off Ischia volcanic island (Italy) during pre history. *In*: VIOLANTE, C. (ed.) *Geohazard in Rocky Coastal Areas*. Geological Society, London, Special Publications, **322**, 73–104.

DELIBRIAS, G., DI PAOLA, G. M., ROSI, M. & SANTACROCE, R. 1979. La storia eruttiva del complesso vulcanico Somma–Vesuvio ricostruita dalle successioni piroclastiche del M. te Somma. *Rendiconti della Società Italiana Mineralogia e Petrologia*, **35**, 411–438.

DEPLUS, C., LE FRIANT, A. *ET AL*. 2001. Submarine evidence for large-scale debris avalanches in the Lesser Antilles Arc. *Earth and Planetary Science Letters*, **192**, 145–157.

D'ERASMO, G. 1931. Studio geologico dei pozzi profondi della Campania. *Bollettino della Società dei Naturalisti in Napoli*, **43**, 15–130.

DOMINEY-HOWES, D. 2004. A re-analysis of the Late Bronze Age eruption and tsunami of Santorini, Greece, and the implications for the volcano-tsunami hazard. *Journal of Volcanology and Geothermal Research*, **130**, 107–132.

DONNADIEU, F. & MERLE, O. 1998. Experiments on the indentation process during cryptodome intrusions: new insights into Mount St. Helens deformation. *Geology*, **26**, 79–82.

EINSELE, G. 1998. Event stratigraphy: recognition and interpretation of sedimentary event horizon. *In*: DOYLE, P. & BENNET, M. R. (eds) *Unlocking the Stratigraphical Record—Advances in Modern Stratigraphy*. Wiley, Chichester, 145–193.

FILMER, P. E., MCNUTT, M. K., WEBB, H. F. & DIXON, D. J. 1994. Volcanism and archipelagic aprons in the Marquesas and Hawaiian Islands. *Marine Geophysical Review*, **16**, 385–406.

FUJIWARA, O., MASUDA, F., SAKAI, T., IRIZUKI, T. & FUSE, K. 2000. Tsunami deposits in Holocene bay mud in southern Kanto region, Pacific coast of central Japan. *Sedimentary Geology*, **135**, 219–230.

GLICKEN, H. 1996. *Rockslide-debris avalanche of May 18, 1980, Mount St. Helens volcano, Washington*. US Geological Survey, Open-File Report, **96-677**.

GOFF, J., MCFADGEN, B. G. & CHAUGUE-GOFF, C. 2004. Sedimentary differences between the 2002 Easter storm and the 15th-century Okoropunga tsunami, southeastern North Island, New Zealand. *Marine Geology*, **204**, 235–250.

HAMPTON, M. A., LEE, J. L. & LOCAT, J. 1996. Submarine landslides. *Reviews of Geophysics*, **34**, 33–59.

HARTLEY, A. J. & JOLLEY, E. J. 1999. Unusual coarse, clastic, wave-dominated shoreface deposits, Pliocene to middle Pleistocene, Northern Chile: implications for coastal facies analysis. *Journal of Sedimentary Research*, **69**, 105–114.

HAYASHI, J. N. & SELF, S. 1992. A comparison of pyroclastic flow and debris avalanche mobility. *Journal of Geophysical Research*, **97**, 9063–9071.

HEINRICH, P., MANGENEY, A., GUIBOURG, S. & ROCHE, R. 1998. Simulation of water waves generated by a potential debris avalanche in Monserrat, Lesser Antilles. *Geophysical Research Letters*, **25**, 3697–3700.

HILL, H. W., KELLEY, J. Y., BELKNAP, D. F. & DICKSON, S. M. 2004. The effects of storms and storm-generated currents on sand beaches in Southern Maine, USA. *Marine Geology*, **210**, 149–168.

HOLCOMB, R. T. & SEARLE, R. C. 1991. Large landslides from oceanic volcanoes. *Marine Geotechnology and Georesources*, **10**, 19–32.

KEATING, B. H. & MCGUIRE, W. J. 2000. Island edifice failures and associated tsunami hazards. *Pure and Applied Geophysics*, **157**, 899–955.

KEEN, T. R., BENTLEY, S. J., CHAD VAUGHAN, W. & BLAIN, C. A. 2004. The generation and preservation of multiple hurricane beds in the northern Gulf of Mexico. *Marine Geology*, **210**, 79–105.

LABAZUY, P. 1996. Recurrent landsliding events on the submarine flank of Piton de la Fournaise volcano (Réunion Island). *In*: MCGUIRE, W. J., JONES, A. P. & NEUBERG, J. (eds) *Volcano Instability on the Earth and other Planets*. Geological Society, London, Special Publications, **110**, 293–306.

LIPMAN, P. W., MOORE, J. G. & SWANSON, D. A. 1981. Bulging of the north flank before the May 18 eruption—Geodetic data. *In*: LIPMAN, P. W. & MULLINEAUX, D. R. (eds) *The 1980 eruptions of Mount St. Helens, Washington*. US Geological Survey, Professional Papers, **1250**, 143–155.

LIRER, L., PESCATORE, T. S., BOOTH, B. & WALKER, G. P. L. 1973. Two Plinian pumice-fall deposits from Somma–Vesuvius, Italy. *Geological Society of America Bulletin*, **84**, 759–772.

MANGENEY, A., HEINRICH, P., ROCHE, R., BOUDON, G. & CHEMINEE, J. L. 2000. Modeling of debris avalanche and generated water waves: application to real and potential events in Montserrat. *Physics and Chemistry of the Earth, Part A*, **25**, 741–745.

MASTROLORENZO, G., PETRONE, P., PAPPALARDO, L. & SHERIDAN, M. F. 2006. The 3780-yr-B.P.

catastrophes as a worst case scenario for a future eruption at Vesuvius. *Proceedings of the National Academy of Science*, **103**, 4366–4370.

McGuire, W. J. 2003. Volcano instability and lateral collapse. *Revista*, **1**, 33–45.

McGuire, W. J., Murray, J. B., Pullen, A. D. & Saunders, S. J. 1991. Ground deformation monitoring for dyke emplacement and slope instability. *Journal of Volcanology*, **38**, 325–344.

McGuire, W. J., Saunders, S. J. & Pullen, A. D. 1993. Rifting at Mount Etna. *WOVO News*, **4**, 16–18.

Mehl, K. W. & Schmincke, H.-U. 1999. Structure and emplacement of the Pliocene Roque Nublo debris avalanche deposit, Gran Canaria, Spain. *Journal of Volcanology and Geothermal Research*, **94**, 105–134.

Milia, A. 1999. Aggrading and prograding infill of a peri-Tyrrhenian Basin (Naples Bay, Italy). *Geo-Marine Letters*, **19**, 237–244.

Milia, A. & Torrente, M. M. 1999. Tectonics and stratigraphic architecture of a peri-Tyrrhenian half-graben (Bay of Naples, Italy). *Tectonophysics*, **315**, 301–318.

Milia, A., Mirabile, L., Torrente, M. M. & Dvorak, J. J. 1998. Volcanism offshore of Vesuvius Volcano in Naples Bay. *Bulletin of Volcanology*, **59**, 404–413.

Milia, A., Torrente, M. M. & Zuppetta, A. 2003. Offshore debris avalanches at Somma–Vesuvius volcano (Italy): implications for hazard evaluation. *Journal of the Geological Society, London*, **160**, 309–317.

Milia, A., Raspini, A. & Torrente, M. M. 2007. The dark nature of Somma–Vesuvius volcano: evidence from the ~3.5 ka B.P. Avellino eruption. *Quaternary International*, **173–174**, 57–66.

Milia, A., Raspini, A. & Torrente, M. M. 2008. The dark nature of Somma–Vesuvius volcano: evidence from the ~3.5 ka B.P. Avellino eruption. Reply. *Quaternary International*, **192**, 110–115.

Minoura, K., Gusiakov, V. G., Kurbatov, A., Takeuti, S., Svendsen, J. I., Bondevik, S. & Oda, T. 1996. Tsunami sedimentation associated with the 1923 Kamchatka earthquake. *Sedimentary Geology*, **106**, 145–154.

Mitchum, R. M., Vail, P. R. & Sangree, J. B. 1977. Seismic stratigraphy and global changes of sea level, part 6: stratigraphic interpretation of seismic reflection patterns in depositional sequences. *In*: Payton, C. E. (ed.) *Seismic Stratigraphy—Applications to Hydrocarbon Exploration*. American Association of Petroleum Geologists, Memoirs, **26**, 117–133.

Moore, J. G., Bryan, W. B. & Ludwig, K. R. 1994*a*. Chaotic deposition by a giant wave, Molokai, Hawaii. *Geological Society of America Bulletin*, **106**, 962–967.

Moore, J. G., Normark, W. R. & Holcomb, R. T. 1994*b*. Giant Hawaiian landslides. *Annual Review of Earth and Planetary Sciences*, **22**, 119–144.

Morton, R. A. 1981. Formation of storm deposits by wind-forced currents in the Gulf of Mexico and the North Sea. *In*: Nio, S.-D., Shuttenhelm, R. T. E. & van Weering, Tj. C. E. (eds) *Holocene Marine Sedimentation in the North Sea Basin*. International Association of Sedimentologists, Special Publications, **5**, 385–396.

Munro, D. C. & Rowland, S. K. 1996. Caldera morphology in the western Galapagos and implications for volcano eruptive behavior and mechanism of caldera formation. *Journal of Volcanology and Geothermal Research*, **72**, 85–100.

Nishimura, Y., Miyaji, N. & Suzuki, M. 1999. Behavior of historic tsunamis of volcanic origin as revealed by onshore tsunami deposits. *Physics and Chemistry of the Earth, Part A*, **24**, 985–988.

Paine, M. 2002. Tsunami book gives a better understanding of ancient floods on Mars. *Science of Tsunami Hazards*, **20**, 50–56.

Pérez-Torrado, F., Paris, R. *et al.* 2006. Tsunami deposits related to flank collapse in oceanic volcanoes: The Agaete Valley evidence, Gran Canaria, Canary Islands. *Marine Geology*, **227**, 135–149.

Pratt, B. R. 2002. Storm versus tsunamis: dynamic interplay of sedimentary, diagenetic, and tectonic processes in the Cambrian of Montana. *Geology*, **30**, 423–426.

Rolandi, G., Mastrolorenzo, G., Barrella, A. M. & Borrelli, A. 1993. The Avellino plinian eruption of Somma–Vesuvius (3760 y. B.P.): the progressive evolution from magmatic to hydromagmatic style. *Journal of Volcanology and Geothermal Research*, **58**, 67–88.

Rolandi, G., Petrosino, P. & McGeehin, J. 1998. The interplinian activity at Somma–Vesuvius in the last 3500 years. *Journal of Volcanology and Geothermal Research*, **82**, 19–52.

Rolandi, G., Bellucci, F. & Cortini, M. 2004. A new model for the formation of the Somma Caldera. *Mineralogy and Petrology*, **80**, 27–44.

Rosi, M. & Sbrana, A. (eds). 1987. *Phlegrean Fields*. Quaderni de 'La Ricerca Scientifica', CNR, Progetto Finalizzato Geodinamica, **114**, 1–175.

Sacchi, M., Insinga, D., Milia, A., Molisso, F., Raspini, A., Torrente, M. M. & Conforti, A. 2005. Stratigraphic signature of the Vesuvius 79 AD event off the Sarno prodelta system, Naples Bay. *Marine Geology, Special Issue C*, **222-223**, 443–469.

Santacroce, R. (ed.). 1987. *Somma–Vesuvius*. CNR Quaderni de 'La Ricerca Pubblicazone, **114**.

Santacroce, R. 1996. Preparing Naples for Vesuvius. *IAVCEI News*, **1/2**, 5–7.

Scandone, P., Arganese, G. & Galdi, F. 1993. The evaluation of volcanic risk in the Vesuvian area. *Journal of Volcanology and Geothermal Research*, **22**, 163–197.

Scheffers, A. & Kelletat, D. 2003. Sedimentologic and geomorphologic tsunami imprints worldwide—A review. *Earth-Science Reviews*, **63**, 83–92.

Shuto, N. 2001. Tsunami-induced topographical change recorded in documents in Japan. *In*: Whitney, R. L., Dunlap, M. & Ventenbergs, K. (eds) *Proceedings of the International Tsunami Symposium 2001, Seattle, Session 3*, 513–522.

Siebert, L. 2002. Landslides resulting from structural failure of volcanoes. *In*: Evans, S. G. & De Graff, J. W. (eds) *Catastrophic Landslides: Effects, Occurrence and Mechanisms*. Geological Society of America, Reviews in Engineering Geology, **15**, 209–235.

SIEBERT, L., GLICKEN, H. & UI, T. 1987. Volcanic hazards from Bezymianny- and Bandai-type eruptions. *Bulletin of Volcanology*, **49**, 435–459.

SIGURDSSON, H., CAREY, S., CORNELL, W. & PESCATORE, T. S. 1985. The Eruption of Vesuvius in A.D. 79. *National Geographic Research*, **1**, 332–387.

SIMKIN, T. & SIEBERT, L. 1994. *Volcanoes of the World*. Geosciences Press, Tucson, AZ.

SMOOT, N. C. 1995. Guyot and seamount morphology and tectonics of the Hawaiian–Emperor elbow by multi-beam sonar. *Marine Geology*, **64**, 203–215.

SNEDDEN, J. W. & NUMMEDAL, D. 1991. Origin and geometry of storm-deposited sand beds in modern sediments of the Texas continental shelf. *In*: SWIFT, D. J. P., OERTEL, G. F., TILMANN, R. W. & THORNE, J. A. (eds) *Shelf Sand and Sandstone Bodies*. International Association of Sedimentologists, Special Publications, **14**, 283–308.

SYNOLAKIS, C. E., IMAMURA, F., TSUJI, Y., MATSUTOMI, H., TINTI, S., COOK, B. & USHMAN, M. 1995. Damage conditions of east Java tsunami of 1994 analyzed. *EOS Transactions, American Geophysical Union*, **76**, 257–262.

TIBALDI, A. 2001. Multiple sector collapses at Stromboli volcano, Italy: how they work. *Bulletin of Volcanology*, **63**, 112–125.

TINTI, S., BORTOLUCCI, E. & ROMAGNOLI, C. 2000. Computer simulations of tsunamis due to flank collapse at Stromboli, Italy. *Journal of Volcanology and Geothermal Research*, **96**, 103–128.

TUTTLE, M. P., RUFFMAN, A., ANDERSON, T. & JETER, H. 2004. Distinguishing tsunami from storm deposits in Eastern North America: the 1929 Grand Banks Tsunami versus the 1991 Halloween Storm. *Seismological Research Letters*, **75**, 117–131.

UI, T., TAKARADA, S. & YOSHIMOTO, M. 2000. Debris avalanches. *In*: SIGURDSSON, H., HOUGHTON, B., MCNUTT, S. R., HAZEL, R. & STIX, J. (eds) *Encyclopedia of Volcanoes*. Academic Press, New York, 617–626.

VALLANCE, J. W., SIEBERT, L., ROSE, W. I., GIRON, J. R. & BANKS, N. K. 1995. Edifice collapse and related hazards in Guatemala. *Journal of Volcanology and Geothermal Research*, **66**, 337–355.

VENDEVILLE, B. C. & GAULLIER, V. 2003. Role of pore-fluid pressures and slope angle in triggering submarine mass movements: natural examples and pilot experimental models. *In*: LOCAT, J. & MIENERT, J. (eds) *Submarine Mass Movements and their Consequences*. Kluwer, Dordrecht, 137–144.

WALKER, R. G. & PLINT, A. G. 1992. Wave- and storm-dominated shallow marine systems. *In*: WALKER, R. G. & JAMES, N. P. (eds) *Facies models—Response to Sea-level Changes*. Geological Association of Canada, Toronto, Ontario, 219–238.

WARD, S. N. & DAY, S. 2003. Ritter Island Volcano—Lateral collapse and the tsunami of 1888. *Geophysical Journal International*, **154**, 891–902.

WAYTHOMAS, C. F. 2000. Re-evaluation of tsunami formation by debris avalanche at Augustine Volcano, Alaska. *Pure and Applied Geophysics*, **157**, 1145–1188.

YAMANO, H., KAYANNE, H. & YONEKURA, N. 2001. Anatomy of a modern coral reef flat: a recorder of storms and uplift in the Late Holocene. *Journal of Sedimentary Research*, **71**, 295–304.

YEH, H., IMAMURA, F., SYNOLAKIS, C. E., TSUJI, Y., LIU, P. & SHI, S. 1993. The Flores Island tsunamis. *EOS Transactions, American Geophysical Union*, **74**, 369–373.

YOUNG, R. & BRYANT, E. A. 1992. Catastrophic wave erosion on the south-eastern coast of Australia. Impact of the Lanai tsunami *ca*. 105 ka? *Geology*, **20**, 199–202.

Landslides in coastal areas of Italy

C. IADANZA, A. TRIGILA*, E. VITTORI & L. SERVA

ISPRA (Institute for Environmental Protection and Research)/Geological Survey of Italy, Via Vitaliano Brancati, 48, 00144 Roma, Italy

**Corresponding author (e-mail: alessandro.trigila@isprambiente.it)*

Abstract: Of the more than 8000 km of coasts in Italy, about 35% consist of cliffs and rocky shores. On them many villages are located, some of them dating back to Roman or Greek times. Typical natural hazards along these coastal areas are rock falls, slides and debris flows. They induce locally extremely high levels of risk, as shown by the many disasters that have occurred in recent time. This paper provides a first portrait over the whole of Italy of (1) the number and distribution of coastal landslides; (2) their typologies; (3) their state of activity; (4) the most common damage and exposed elements; and (5) the most typical mitigation measures. This information has been largely derived from the IFFI archive (Italian Landslides Inventory). Geographic information system processing has allowed us to identify 4135 landslides affecting a large portion of the Italian cliffed coasts. The damage evaluation, based on the AVI archive (inventory of information on sites historically affected by landslides and floods in Italy for the period 1918–2000), has shown the extent of the problem: there have been 408 casualties in the 20th century and 713 significant disruptions of the rail and road networks. A more detailed analysis has been carried out in three sample areas to illustrate the basic factors controlling the genesis and evolution of slope instabilities in the various coastal settings.

The Italian peninsula and islands are bounded by *c.* 8100 km of coasts; nearly 2850 km are steep-sloping, mostly rocky, cliffs or bluffs directly plunging into the sea, or separated from it by a thin strip of sandy or gravel beach or a reef of fallen blocks. The remainder are gently sloping beaches, often backed by low-lying flood plains and locally by coastal lakes or marshes, which are what remains now of the many more that existed before the extensive land reclamation works of the 19th and 20th centuries.

In Italy the major concern is generally devoted to the erosion of beaches, most of which are important tourist resorts, and to the flooding of coastal plains, which are increasingly occupied by industrial developments, urban settlements and infrastructure. Nevertheless, the hazard on rocky coasts posed by rock falls, rotational or translational slides and debris flows channelled along steep and deeply incised streams is relevant in several regions of Italy (Almagià 1910; Guerricchio 1988), where disasters have occurred in the past, with the loss of many lives.

Indeed, many villages are located on the cliffed coastal areas of Italy, some of them dating back to pre-Roman times; up to the 16th century such a panoramic location offered better protection from pirates, and also from floods and slides. Many of these villages are now a preferred choice of summer and mid-season tourism, when the number of residents may become 10 times and more that of the winter. Congested roads, as well as houses and boats, are threatened by rock falls and slides that are sometimes triggered by heavy storms, which are not infrequent where mountains neighbour the sea, or by earthquakes, especially along the Campania, Calabria and Sicily coasts. Therefore the overall risk along such coastal sectors is very high, and requires appropriate studies and countermeasures.

This paper aims at providing an outline of the present knowledge on coastal landslides in Italy, based principally on the IFFI database (Inventario dei Fenomeni Franosi in Italia; Italian Landslide Inventory) developed by ISPRA (Institute for Environmental Protection and Research), which will be used to highlight the size of the problem (i.e. the number of landslides and the area affected) and the most typical hazardous settings, emphasizing the leading parameters in the triggering and subsequent dynamics of the phenomena. Examples drawn from representative sites illustrate the various kinds of hazard and the engineering countermeasures taken by regional authorities.

In addition, useful information on damage caused by coastal landslides has been obtained from the AVI database (inventory of information on sites historically affected by landslides and floods in Italy for the period 1918–2000), produced by GNDCI (Italian Group for Hydrogeological Disaster Prevention), which summarizes information collected from newspapers and historical documents.

From: VIOLANTE, C. (ed.) *Geohazard in Rocky Coastal Areas*. The Geological Society, London, Special Publications, **322**, 121–141. DOI: 10.1144/SP322.5
0305-8719/09/$15.00

Brief outline of geology and geography of Italian coast

The distribution of cliffed and low-lying stretches of coast is a direct consequence of the complex tectonic history of the Italian peninsula, related to the convergence of the African and Euro-Asian plates. The subduction and rotation of minor plates and the opening of the oceanic crust-floored Tyrrhenian and Balearic seas led during the Neogene and Quaternary to mountain building (mainly the NE-verging fold-and-thrust belt of the Apennines) and filling by sediment of NE-migrating fore-deeps. Since the Pliocene, extension has occurred, with the overprinting of a horst-and-graben architecture on the western side of the fold-and-thrust system (Malinverno & Ryan 1986; Royden *et al.* 1987). Only Sardinia (part of the European plate), the Hyblean plateau (southeastern Sicily) and the Salento peninsula in Apulia have been left substantially untouched (forelands) by the compressive wave. Vertical movements are still active in several coastal sectors (Bordoni & Valensise 1998; Antonioli *et al.* 1999; Lambeck *et al.* 2004; Vannoli *et al.* 2004). Tectonic uplift is particularly prominent in Calabria and northeastern Sicily, whereas subsidence, partly caused by active tectonics and partly by sediment diagenesis, affects most of the major coastal flood plains. A further contributing factor is the still-debated effect of isostatic rebound related to the Last Glacial Maximum (Lambeck *et al.* 2004).

Some segments of the Tyrrhenian Sea coast from Tuscany to Campania have been influenced by Plio-Quaternary volcanism. Volcanic islands have emerged from the sea from Latium to Sicily: the Pontine Islands, Ischia and Procida in the Bay of Naples, the Aeolian Islands (including Alicudi, Filicudi and Ustica), Pantelleria and Linosa, the last two being in the Mediterranean Sea south of Sicily. All these islands are characterized by a dominantly high rocky coastline with minor pocket beaches. Volcanic activity still affects part of the Campania coast (Phlegrean Fields, Vesuvius and Ischia), and the eastern (Etna) and northern Sicilian coasts (Aeolian Islands). Positive and negative volcano-tectonic movements are particularly evident in the Phlegrean Fields and Ischia (Dvorak & Mastrolorenzo 1991).

Along the coast, counter-clockwise from the Gulf of Trieste in the northern Adriatic Sea, are the following high-standing sectors, generally made of Mesozoic–Cenozoic carbonates except where otherwise stated: (1) the cliffs near Trieste; (2) the Conero promontory near Ancona, with cliffs of marly limestone and bluffs of terrigenous deposits north of it; (3) the coast from south of Francavilla to Vasto (Abruzzo), made partly of active cliffs cut in Quaternary clastic deposits, and partly of gentle slopes mostly modelled by landslides, and the modest spur of Termoli in Molise; (4) the Gargano promontory; (5) the nearly undeformed but isostatically uplifted Murge and Salento peninsula from Barletta to north of Gallipoli in the Ionian Sea; (6) modest outcrops of clastic flyschoid sediments (sandstones and shales) north of Capo Spulico; (7) large stretches of the Calabrian coast, principally made of flyschoid and metamorphic rocks, locally of calcarenites (e.g. from Cirò Marina to Crotone) and limestone only north of Diamante in the Tyrrhenian Sea, interrupted by coastal plains, of which the main ones are the Sibari plain, Catanzaro–Lamezia graben and Gioia Tauro; (8) the Cilento region, made of limestone, calcarenite and sandstone units; (9) the Sorrento peninsula where the sea cliffs are partly made of pyroclastic rocks (east of Sorrento); (10) the coastline of Vesuvius and the Phlegrean Fields (mainly pyroclastic rocks); (11) most of the coast from Formia to Terracina; (12) the Circeo promontory; (13) the modest spurs of Anzio (Pleistocene calcarenites); (14) Santa Marinella–Civitavecchia (sandstones, shales and calcarenites), respectively south and north of the Tiber river delta; (15) Ansedonia and the Argentario promontory; (16) Talamone, Castiglione, the Piombino promontory and Castiglioncello (ophiolite) just south of Livorno, which break up the low-lying sandy coast of Tuscany; (17) the Ligurian rugged coast from the Magra river to the French border (limestone in the La Spezia–Cinque Terre area, then shales, ophiolite and metamorphic rocks), which is in general very steep and affected by frequent flash floods and debris flows, but locally has narrow or pocket beaches and minor alluvial plains.

Most of the smaller islands are characterized by steep slopes locally interrupted by narrow pocket beaches. Sardinia has extensive sectors of high coast, mostly carved in granite, metamorphic rocks of various grade and volcanic rocks, except in the east (Gulf of Orosei), where limestone crops out, with very clear evidence of the stage 5e sea-level highstand (Antonioli *et al.* 1999).

A variety of settings characterize Sicily, with granitoids and metamorphosed rocks in the NE, calcarenites near Messina, limestone at Taormina, lavas and conglomerates in the Etna region, and limestone from Augusta to Syracuse and in the SE (Hyblean plateau, with some basaltic lavas near Capo Passero). From Gela to Trapani mainly Plio-Pleistocene calcarenites with some marls and chalk crop out, and north of Trapani to Palermo limestone cliffs characterize the landscape, interrupted by several coastal plains (Castellammare,

Mondello). East of Palermo mainly clastic flysch deposits are present, except for the low-lying coast between Termini Imerese and Cefalù, which rests scenically on a limestone spur. East of Capo d'Orlando mainly granitoids crop out, interrupted by the low-lying Milazzo plain (Fig. 1).

The IFFI Project

The IFFI Project aims at identifying and mapping slope instabilities over the whole Italian territory, based on standardized criteria (Amanti *et al.* 2000; Trigila 2007; Trigila & Iadanza 2008). So far,

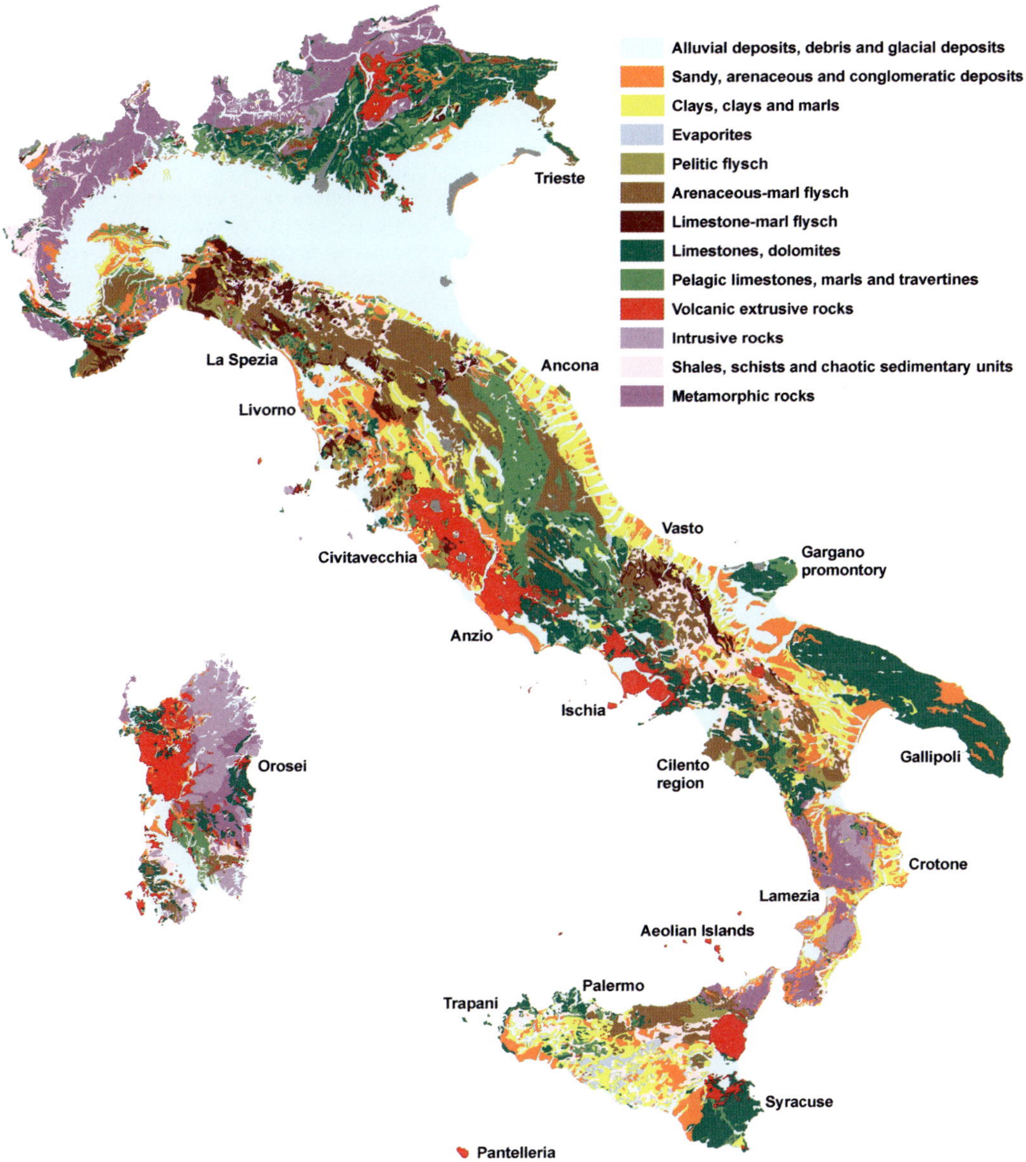

Fig. 1. Lithological map of Italy (derived from Italian Geological Map at 1:500 000 scale).

about 480 000 landslides are included in the database. The IFFI website (http://www.sinanet.apat.it/progettoiffi), allows researchers to explore the data and obtain detailed information on landslides.

The project has been financed with €4.1 million by the Committee of Ministries for Soil Protection, established by Law no. 183/89. The institutions involved in the accomplishment of the IFFI Project are: (1) ISPRA–Geological Survey of Italy, which has the task of coordinating the activities, verifying the data conformity, and building up the national landslide database and a WebGIS application (Iadanza *et al.* 2006); (2) Regions and Self-Governing Provinces, which are charged to collect and organize the historical and archive data and to map the areas affected by landslides.

The chosen method is based on aerial photo-interpretation, search of published sources and historical documents, and field surveys. The working scales vary between 1:10 000 and 1:25 000. The geodatabase includes vector layers of landslides and an alphanumeric archive of landslide attributes.

The logic structure of the archive follows the landslide data sheet devised by APAT, which is organized in three information levels: the first level contains the basic data on landslide location, type and state of activity; the second level provides data on morphometry, geological setting, lithology, land use and main causes of activation; the third level gives detailed information on damage, investigations and remedial measures for risk reduction. Based on classifications by Varnes (1978), Cruden & Varnes (1996) and Amanti *et al.* (1996, 2001), landslide movements have been divided into 11 types: rock fall–topple, rotational–translational slide, lateral spread, slow earth flow, rapid debris flow, sinkhole, complex landslide, deep-seated gravitational slope deformation, area affected by numerous rockfalls–topples, area affected by numerous sinkholes, and area affected by numerous shallow landslides. The last three types have been introduced to group landslide phenomena too small to be mapped that affect the same slope.

Landslides are represented by a georeferenced point, located at the highest point of the crown, with a polygon when the surface is wider than 10 000 m^2 or with a line when the width is very narrow, as in the case of debris flows.

Methods and data processing

The input data used for this study, in addition to the IFFI inventory, are: the AVI archive (Guzzetti 2000*a*, *b*); the CORINE Land Cover 2000 database (APAT 2005); the road and railway networks; a digital elevation model (DEM; 20 m × 20 m grid cells) and its derivative products, such as the slope angle and the upslope contributing area (total upslope area that drains into a grid cell); the Geological Map of Italy scale 1:500 000 and the lithological map derived from it; the topographic map scale 1:25 000 by IGM (Geographic Military Institute); the digital colour orthophoto of Italy with 1 m resolution (Terraitaly™ it2000).

The sectors of rocky shores have been discriminated using the Spatial Analyst (ESRI ArcGIS™) slope function applied to the 20 m DEM, choosing those sectors displaying a slope angle >4°. Such a modest angle has proved adequate to this purpose, based on tests in well-known coastal areas, compared with mapping carried out by other means. Of the Italian coasts, 2874 km (35.4%) can be classified as rocky shores (Fig. 2a). These results have been compared with a high-definition mapping of Italian coastlines developed by another Department of ISPRA (Barbano *et al.* 2005). Based on the aerial photo-interpretation of digital colour orthophotos (Terraitaly™ it2000), 2824 km of coast were classified as rocky coasts. The small difference obtained by the two approaches can be attributed to the 20 m resolution of the DEM, which does not permit pocket and very narrow beaches to be detached.

According to the aim of this work, all the landslides located within 1 km of the coastline have been classified as 'coastal landslides'. A statistical survey on some of the most significant coastal sectors has shown that this distance allows us to include almost all the landslides directly affecting the coast and coastal settlements and infrastructure. A total of 4135 coastal landslides have been extracted from the IFFI database according to this criterion (Fig. 2b).

Comparing the two maps in Figure 2, it emerges that most of the Italian rocky coasts, as expected, are affected by landslide phenomena. In Figure 3 an example of a retreating cliff is shown (Mattinata–Gargano promontory, Apulia region). A peculiar case exists in some sectors of the Adriatic coast, where landslides involve coastal hills or abandoned cliffs that are now located a few hundred metres from the present-day low-lying sandy coast.

Statistical analyses have been performed on the following parameters: type of movement, state of activity, damage and exposed elements.

Type of movement and state of activity

The most frequent types of movement are rock falls or topples (31.2%), followed by rotational and translational slides (24.7%), and complex landslides (13.5%) (Figs 4 and 5). The kinematics of coastal landslides depends on: (1) lithology, stratigraphy, structural setting, ground-water level, weathering of the rock mass, morphology of the slope; (2) sea

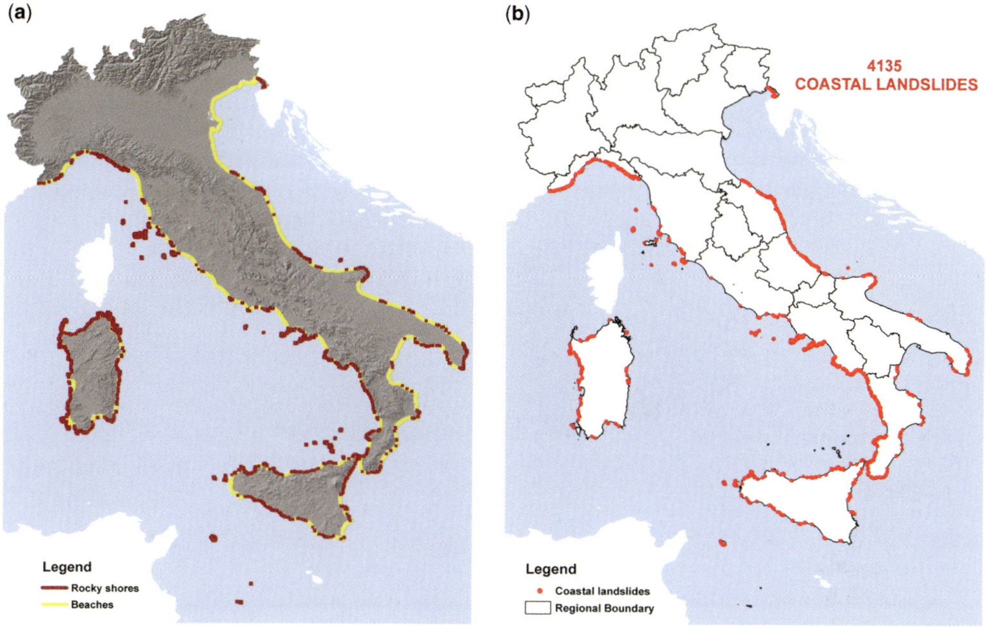

Fig. 2. (**a**) Distribution of rocky coasts and low-lying sandy shore in Italy. Red line, rocky shores; yellow line, beaches. (**b**) Distribution of coastal landslides in Italy.

Fig. 3. Rocky cliff, Mattinata (FG), Puglia (southern Italy).

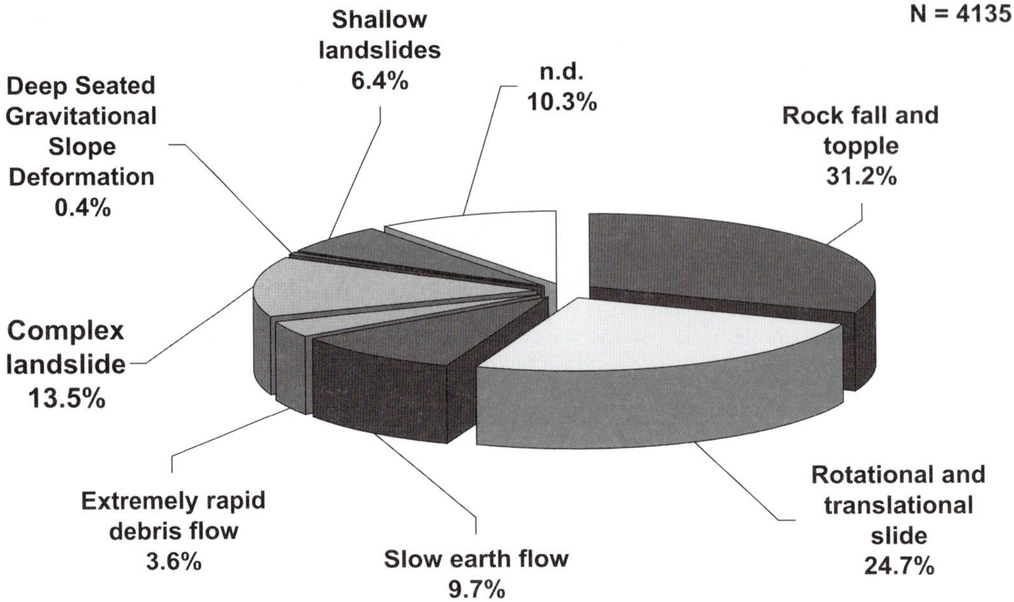

Fig. 4. Types of movement expressed as percentage of the total coastal landslides reported by the IFFI Project.

Fig. 5. Rock topple occurring in volcanic deposits along the coast of Dorgali (NU), Sardinia (central Italy).

wave action, which is often of primary relevance, especially for high vertical cliffs. The energy with which waves strike the coast depends on the height of waves at the base of a cliff. The peak erosive energy is attained when the waves break just at the foot of the slope (Sunamura 1992; Budetta *et al.* 2000).

Landslides occurring along rocky coasts may significantly contribute to their evolution. The boulders and rock debris accumulating at the foot of a cliff can effectively reduce, at least temporarily, further wave action; sometimes they may also allow the development of narrow shingle beachs. Catastrophic debris flows occurring along the paths of coastal streams can develop fan-delta systems (Sacchi *et al.* 2009; Violante 2009) with temporary coastline progradation. A typical example is provided by Vietri sul Mare on the Costiera Amalfitana (Esposito *et al.* 2004*a*; Violante *et al.* 2009).

About 45% of the coastal landslides are either at present 'active', 'suspended' (have moved during the last seasonal cycle), or 'reactivated'. Landslides now 'dormant', but expected to move again in the future make up another 45% of the total. Landslides artificially or naturally stabilized make up 4%, whereas only 1% are 'relict' landslides; that is, landslides that were active under climatic and morphological conditions different from the present ones (WP/WLI 1993; Cruden & Varnes 1996). The state of activity is not defined for 5% of landslides.

Figure 6 shows the complex landslide of Guvano (municipality of Vernazza, province of LaSpezia, northern Italy), characterized by rock falls in the crown and translation movement in the main body, which finally grades into earth flow reaching the coastline. This landslide was stabilized many years ago during the construction of the Genova–Roma railway. When the railway track was transferred inside the mountain via a tunnel, the landslide started moving again, as a result of the abandonment of the stabilization works and their slow demolition by the sea waves (De Stefanis *et al.* 1978; Federici *et al.* 2001; Bottero *et al.* 2004).

Damage and exposed elements

Historically, coastal landslides have produced significant damage to urban and tourist settlements, port facilities, and road and railway networks

Fig. 6. The landslide of Guvano (SP), Liguria (northern Italy).

Fig. 7. Damage produced by coastal landslides. (**a**) Railway track displacement produced by slow landslide movement at Petacciato (CB) Molise (after Guerricchio *et al.* 1996); (**b**) damage to state road 16 and to railway track, Ancona, Marche (after Cotecchia & Simeone 1998).

(Fig. 7), with a high number of fatalities and people made homeless.

An estimate of damage can be obtained through the AVI archive. In this database, updated to the year 2000 and including 16 630 georeferenced damaging events, 1453 landslides are located within 1 km from the coastline. Table 1 gives a summary of damage reported for these events. Figure 8 shows the high cliff overhanging the beach of Chiaia di Luna on the island of Ponza (Pontine Islands, central Italy) where a rock fall from the rim of the cliff killed a tourist in 2001 (D'Angelo *et al.* 1999).

In recent decades urban settlements have experienced a major expansion along most of the Italian coastal areas. According to the Environmental Data Yearbook (APAT 2006), between 1975 and 1992 this growth was particularly strong in the provinces of Viterbo (Latium), Matera (Basilicata), Catania and Siracusa (Sicily), Macerata and Ascoli Piceno (Marche), Udine (Friuli) and Brindisi (Puglia), with percentages exceeding 60%. In the period 1990–2000, according to the data of the project CORINE Land Cover (APAT 2005), the greatest urban sprawl has been recorded in Sardinia and Calabria, and in the provinces of Chieti (Abruzzo) and Livorno (Tuscany).

In this paper, the elements exposed to landslide risk have been identified by combining the GIS layer of the landslides with the layers of the infrastructure and urban settlements (Fig. 9). The most critical points along the rail network are represented in Figure 10. Sadly, it must be noted that every new severe meteorological event in Italy adds new infrastructure to this list, revealing a systemic weakness, clearly related to insufficient design awareness.

Table 1. *Damage caused by coastal landslides (AVI Archive 1918–2000)*

Damage	Number
Fatalities	408
Injured people	344
Evacuated people	3656
People homeless	1047
Civil buildings	310
Highways	23
Roads	518
Railways	172
Monuments	18
Industrial buildings	14
Network infrastructure	78
Public infrastructure	40
Public buildings	27
Agricultural areas	65

Case studies

To better investigate the relationship between landslides and some factors, such as geological setting, lithology, slope angle and wave action, a more detailed analysis has been carried out on three study areas, chosen as representative of the most typical Italian geological–geomorphological rocky coastal settings (Fig. 11).

Adriatic coast of Marche, Abruzzo and Molise regions (central Italy)

Between Pesaro and Termoli over a length of nearly 370 km, the coastal strip is mostly characterized by relief composed of Plio-Pleistocene overconsolidated Grey–Blue Clay, overlain by coarser sediments (Esu & Martinetti 1965; Colosimo & Crescenti 1972; Colleselli & Colosimo 1977; Esu

Fig. 8. Rocky cliff of Chiaia di Luna, Ponza (Pontine Islands).

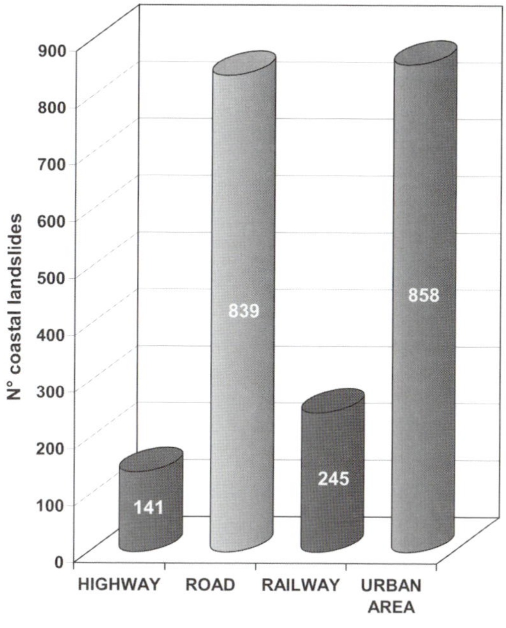

Fig. 9. Number of coastal landslides that affect highways, roads, railways and urban settlements.

2000), which form a succession of coastal hills, and active and abandoned cliffs (Hutchinson 1973; Emery & Khun 1982; D'Alessandro *et al.* 2001). Only in the Conero promontory, just south of Ancona, Mesozoic–Palaeogene marls and marly limestones crop out, making a nearly vertical cliff.

There have been 766 landslides affecting the Adriatic coastal strip between Pesaro and Termoli. The most frequent types of movement are rotational and translational slides (*c.* 41%), slow earth flows (18%), complex landslides (19%), falls and topples (7%), and shallow landslides (7%), as shown in Figure 12.

The main cause of failure in the active cliffs is sea-wave erosion at their base, as observed in the promontories of Gabicce and Conero. In contrast, the stability of abandoned cliffs is influenced by the same factors that control the stability of inland slopes, including heavy rainfall and changes of pore pressure, and the main predisposing factor is the reduction of the shear resistance of clay by processes of creep, softening and weathering.

Active cliffs affected by enhanced sea-wave erosion are preferred sites of rock falls or topples and translational or rotational slides, whereas coastal hills and abandoned cliffs are mainly affected by slides, earth flows, and shallow and complex landslides.

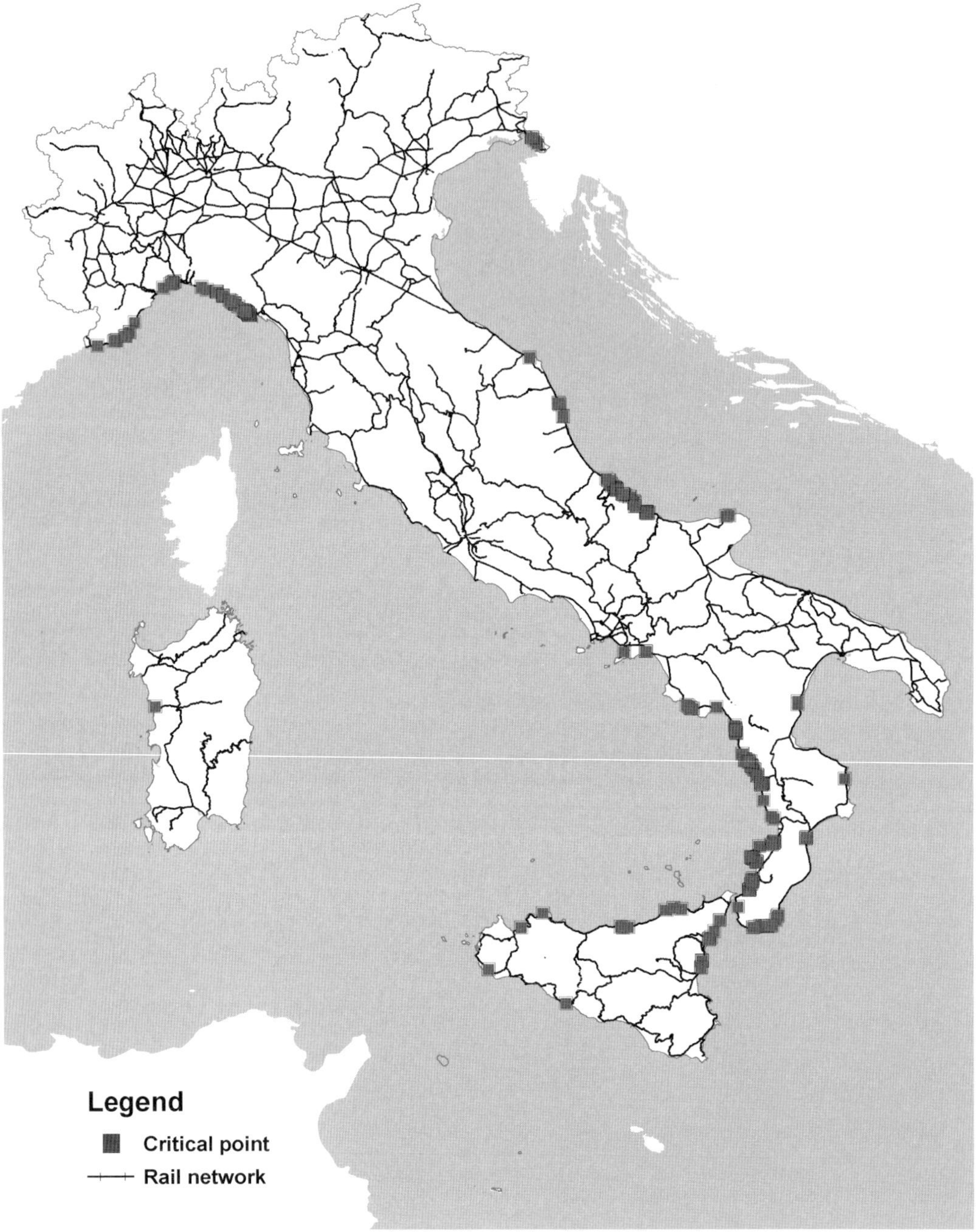

Fig. 10. Most critical areas along the Italian coastal railways.

The histogram of the distance of the landslide crown from the shoreline shows a main peak at *c*. 150 m from the sea related to landslides on active cliffs (Fig. 13). The slope angles in the detachment zone resulting from DEM analysis are generally modest, starting from 8° for rotational and translational slides, slow earth flows and shallow landslides. In contrast, as expected, the slope angle for rock falls or topples is much higher (25°; Fig. 14). Obviously, these angles are not true angles, but mean values resulting from the spatial resolution of the DEM.

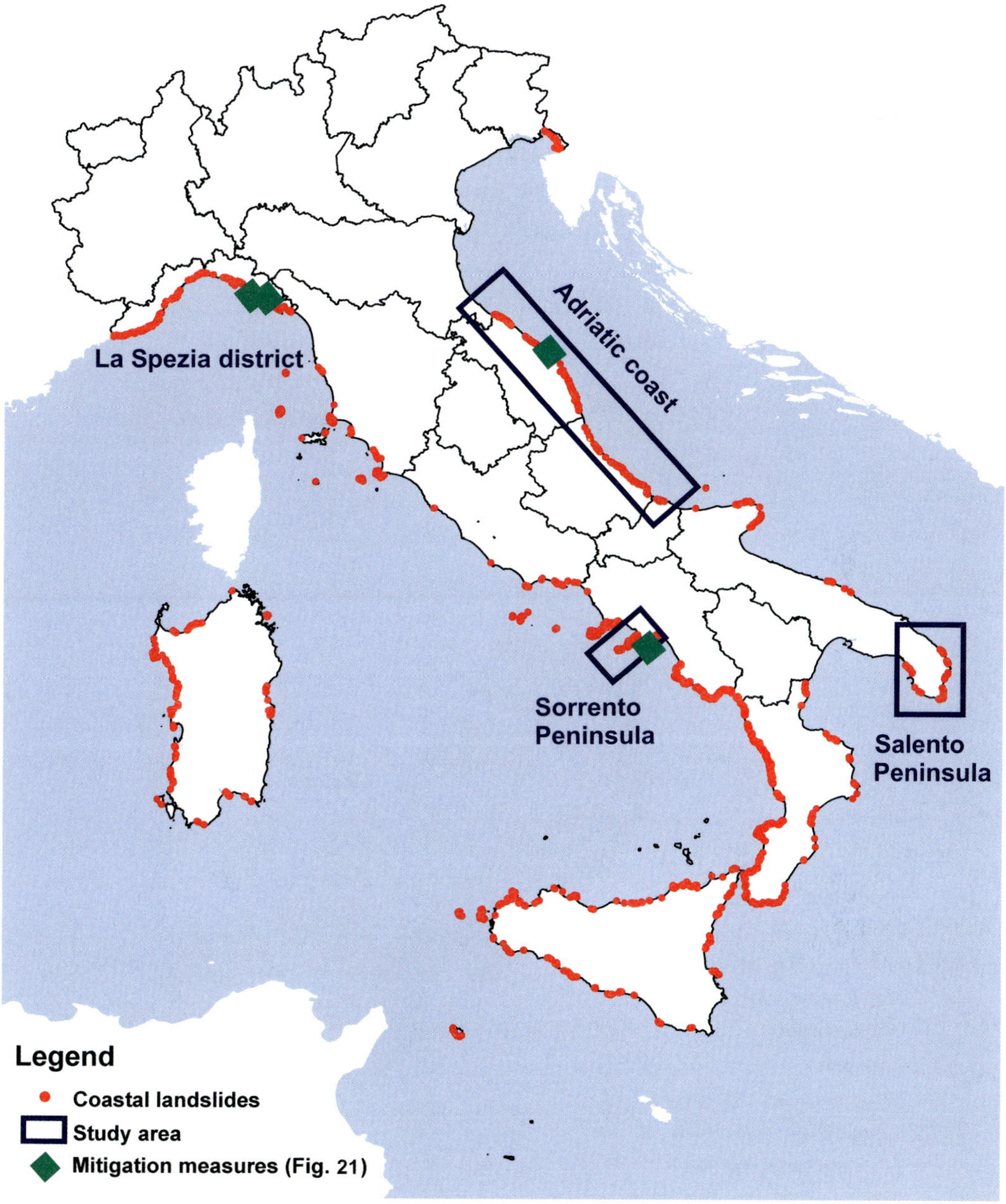

Fig. 11. Location of study areas and remedial works described in this paper.

The rotational–translational slides in the Plio-Pleistocene Grey–Blue Clay Fm. along the Adriatic coast, which have been extensively described in the scientific literature, are a very widespread phenomena, affecting often whole slopes from hilltop to the coast, with shear surfaces sometimes being very deep and even continuing below sea level, to produce bulges in the sea floor. The reactivation is seasonal, being most common from October to March, following heavy and persistent rainfall (Fiorillo & Guadagno 2000). These landslides have caused damage to urban settlements, roads and railways (Cancelli *et al.* 1984; Gori & Mezzabotta 1995; Guerricchio & Melidoro 1998*a*, *b*; Guerricchio *et al.* 1996; Melidoro & Mezzabotta 1998).

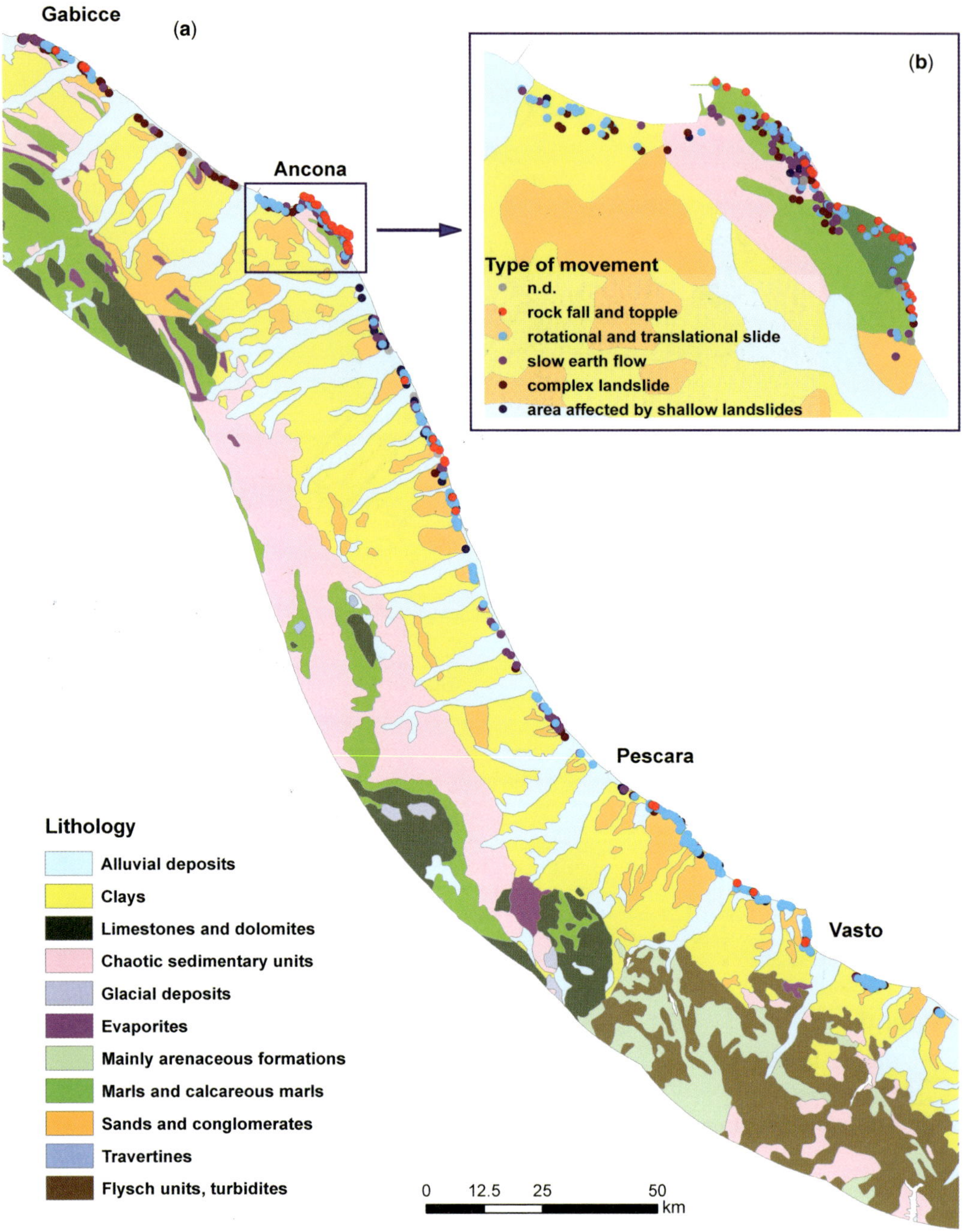

Fig. 12. Lithological map and distribution of landslide types (**a**) along the Adriatic coast and (**b**) in the Conero promontory.

Sorrento peninsula (Campania region, southern Italy)

The bedrock of the mountainous Sorrento peninsula consists of Triassic–Jurassic dolomitic limestones on the southeastern side (Costiera Amalfitana–Bay of Salerno) and Cretaceous limestone on the northwestern side (Bay of Naples). Large portions of the calcareous slopes are mantled by thick (up to several metres) blankets of loose, altered,

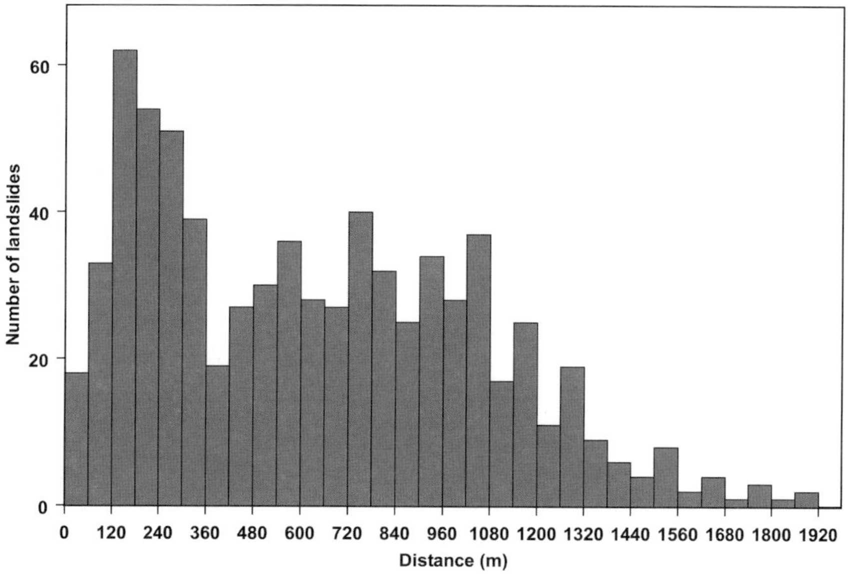

Fig. 13. Histogram of the distance of the landslide crown from the shoreline along the studied Adriatic coast.

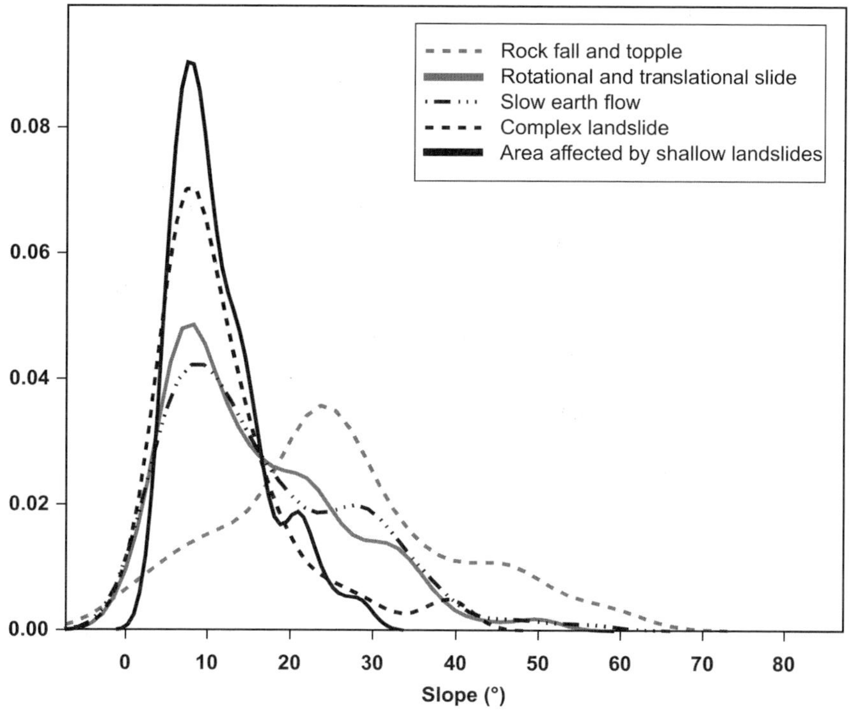

Fig. 14. Probability density function of slope angle in the source area of Adriatic coastal landslides.

pyroclastic deposits, in a great part coming from the Vesuvian AD 79 eruption (Fig. 15). In this geological–geomorphological setting, the most frequent and dangerous types of landslides, involving the volcanic material, are extremely rapid debris flows, triggered by heavy rainfall, after a particularly wet period (Fig. 16; Civita *et al.* 1975; Budetta *et al.* 1997; De Falco *et al.* 1997; Di Crescenzo & Santo 1999; Vallario 1998; Calcaterra *et al.* 2004; Calcaterra & Santo 2004; Esposito *et al.* 2004*b*). Sometimes even modest rock falls or slides on steep slopes (slope angle often >35°) have a significant impact, triggering much wider translational slides that often turn into extremely rapid debris flows. The shear plane is generally located within pumiceous layers, often inter-bedded with low-permeability ash layers, or at the contact with the limestone bedrock. When these pumiceous layers are saturated, a decrease of shear resistance occurs, often enhanced by pore overpressure.

Debris flows are classified as (1) 'channelled flows', with narrow, elongated shape, which commonly run along a network of gullies converging into a main stream; and (2) 'unchannelled flows', with an approximately triangular shape, which occur over an open slope with no gullies. The channelled flows are more fluid and travel longer distances. In the past, they had a tremendous destructive power and caused many fatalities (Esposito *et al.* 2004*b*; Porfido *et al.* 2009).

The main predisposing factors (i.e. slope angle and shape of drainage network) have been analysed along the Sorrento peninsula, where 281 rapid debris flows are listed in the IFFI database. The relation between debris-flow source area and slope angles reveals the highest frequency for angles from 35 to 40° (Fig. 17). The decreasing number of events for higher gradients is clearly due to the absence of volcanic deposits on steeper slopes.

The contributing area upslope of debris-flow source areas can be estimated from the 20 m resolution DEM by the flow accumulation number, which represents the number of cells draining into a given cell of the grid, and consequently provides an estimate of the drainage area by multiplying this number by the surface area of a cell (400 m^2). Using this approach, the upslope contributing area is found to be smaller than 10 000 m^2 in 89% of cases, meaning that the majority of flows are triggered not far from the watershed, on steep slopes draining relatively small areas upstream.

The travel-distance potential of extremely rapid debris flows has been evaluated plotting the vertical height (difference in elevation from source area to

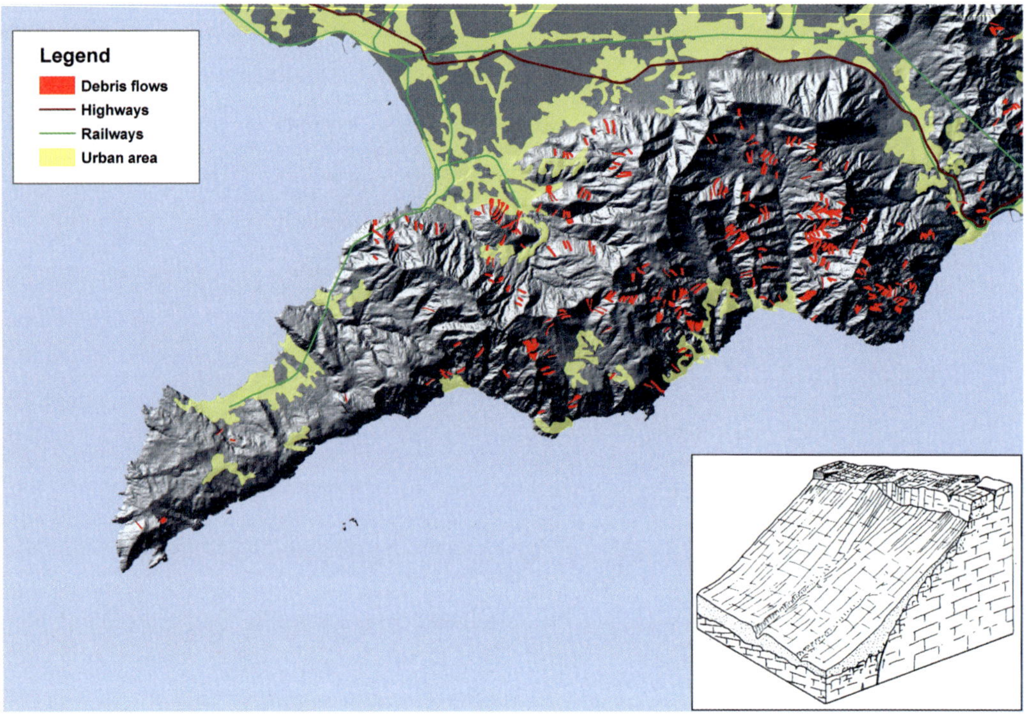

Fig. 15. Landslide distribution and urban areas in the Sorrento peninsula. Inset: geological sketch showing pyroclastic deposits over calcareous bedrock.

Fig. 16. The Pozzano landslide, Castellammare di Stabia (NA), Campania. Digital colour orthophoto TerraItaly™ – © BLOM Compagnia Generale Ripreseaeree S.p.A – Parma (http://www.terraitaly.it).

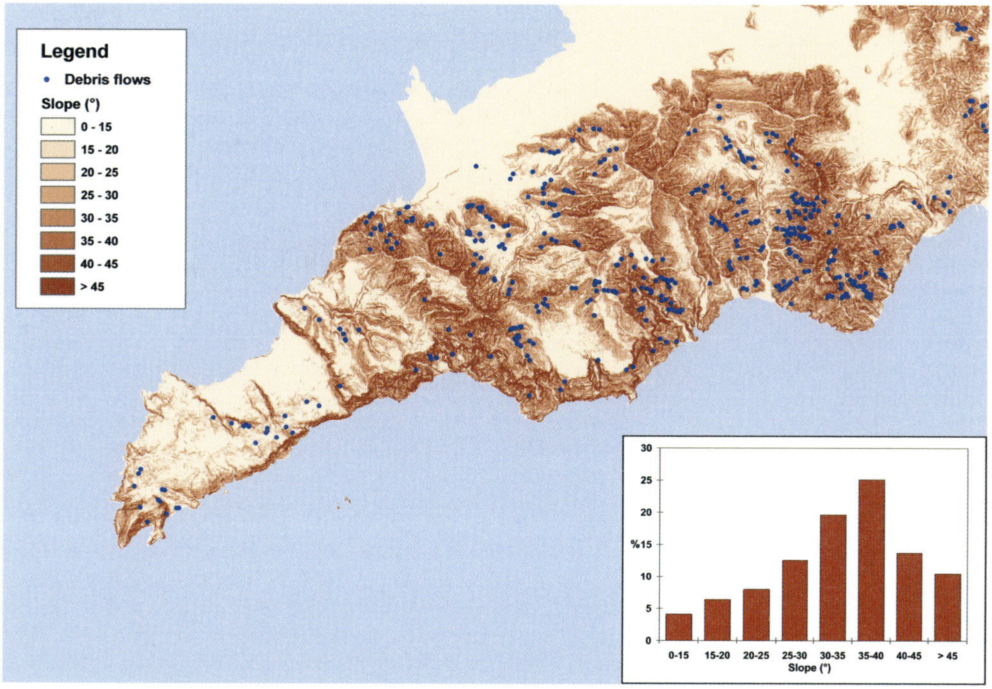

Fig. 17. Slope angle and debris-flow distribution along the Sorrento peninsula (southern Italy). Inset: percentage distribution of slope angle in debris-flow source areas.

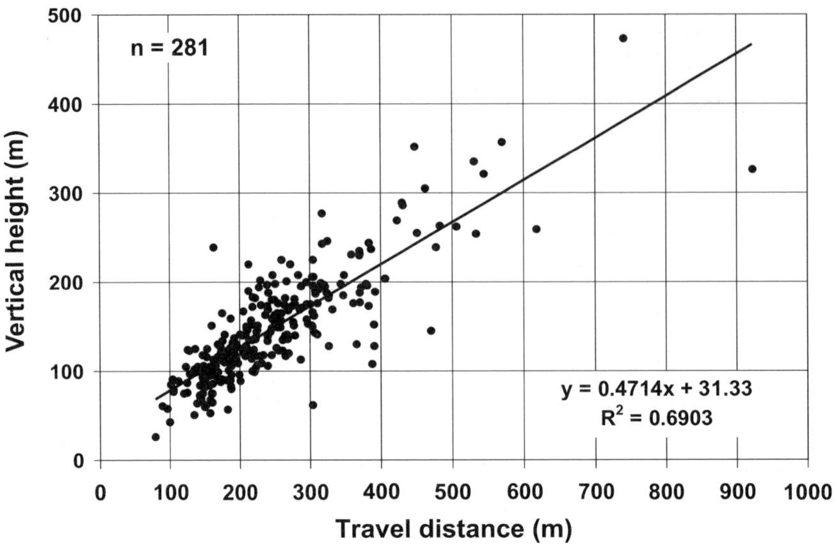

Fig. 18. Travel-distance potential of extremely rapid debris flows, based on 281 events occurring in the Sorrento peninsula (southern Italy).

depositional area) versus the maximum distance travelled by the flow. The plot in Figure 18 shows a linear direct relationship for the 281 flows, with a coefficient of 0.47, in agreement with the value reported in the literature (Calcaterra & Santo 2004). The lower values of the ratio H/L (angle of reach) are typical of channelled flows compared with those on open slopes.

Salento peninsula (Puglia region, southern Italy)

The Salento peninsula is the 'heel' of the boot-shaped Italian peninsula. It is one of the few outcropping pieces of foreland, although it has been affected by minor tectonic pulses since the Late Messinian; the bedrock is a thick succession (over 6000 m) of Mesozoic–Cenozoic limestone to dolostone rocks, covered by Plio-Pleistocene calcarenites (Bruno & Zezza 1992). A slight, but still significant, Pleistocene uplift (Bordoni & Valensise 1998, and references therein) has produced a coast largely made of rocky cliffs. They are affected by a number of landslides, 31 according to IFFI (Fig. 19), most of which (90%) are represented by rock falls and topples of sometimes large blocks (Fig. 20).

The main predisposing factors are the relatively dense network of subhorizontal bedding and subvertical discontinuities, weathering and karstic cavities. The main triggering cause is wave erosion at the cliff base. As a result of these mechanisms, the distance from the sea over which the detachments take place is less than 50 m.

The evolution of high rocky coasts in this area presents a typical cyclic behaviour (Sunamura 1992). The repeated rockfalls smooth the acclivity of the cliff, making it more stable; moreover, the accumulation of debris and fallen blocks at the foot of the cliff exerts a breakwater action, which can slow down or even halt the basal erosion. Without taking into account the long-term sea-level fluctuation (which can be of the order of many metres as a result of climatic changes), the sea waves are able to erode away more or less rapidly the debris and undercut the cliff, so that there is a return to a state of instability.

Mitigation measures

To mitigate the landslide risk along the coasts of Italy, remedial measures commonly taken are: rock scaling and slope redesign; retention structures (walls, piles, reinforced soil embankments, gabions); passive structures (rockfall barriers, wire mesh nets, rockfall tunnels); dowels and anchors; soil reinforcement and erosion control measures; subsurface and deep drainage; protection measures at the foot of cliffs against sea wave action (e.g. breakwaters).

After the disastrous rapid debris flows that occurred in 1998 in the Sarno area (Campania region), when more than 150 people lost their life, the Italian Government issued the Law 267/98 (commonly referred to as the 'Sarno Law') on prevention measures to reduce the risk of landslide and flooding. Two of the main focal points are: (1)

Fig. 19. Coastal landslides in the Salento peninsula, Puglia, according to the IFFI Project.

the identification, description and mapping of hazard and risk areas; (2) risk mitigation through land planning, land regulation and stabilization works. In this perspective, ISPRA–Geological Survey of Italy has been charged with monitoring the progress of such remedial measures (3215 projects throughout Italy, financed with €2.4 billion), from project submission to final realization. Below, some examples of these works are described (locations are shown in Fig. 11).

In the Ligurian Cinque Terre (La Spezia district), the safety of the Genova–Roma railway has required a rockfall tunnel, retaining walls and breakwater reefs (Fig. 21a). In Levanto, after a careful geological–geomorphological–geomechanical study and rock-fall simulation, the remedial measures for the very steep slope (>40°) cut in gabbros have been based on a selective rock scaling and vegetation removal, followed by the installation of double twisted wire mesh nets and rockfall barriers with high deformation capacity (Fig. 21b). Figure 21c shows the stabilization of a calcareous rockslope overhanging the wharf, used as a parking area, in the port of Cetara (Sorrento peninsula, Salerno district), by controlled removal of unstable blocks and installation of wire nets and anchors. Figure 21d shows a semi-circular breakwater in the Conero promontory aimed at preventing

Fig. 20. Rock fall along the cliff of Porto Miggiano, S. Cesarea Terme (LE), Puglia.

Fig. 21. Remedial works on coastal landslides: (**a**) railway rockfall tunnel (Cinque Terre, SP); (**b**) rockfall barriers (Levanto, SP); (**c**) wire mesh nets with anchors (Cetara, SA); (**d**) breakwater in the Conero promontory (AN).

sea-wave erosion of the cliff composed of relatively soft marls and marly limestone. An example of non-structural measures is the limited access to hazardous areas, as in the case of the Chiaia di Luna beach on the island of Ponza, where access is forbidden to the still unprotected section of the beach.

It is to be stressed that the most relevant non-structural measures have been introduced with the Law 267/98 cited above, which has required regional authorities to map all the hazardous areas and to avoid new constructions where the level of hazard is too high.

Final remarks

The summary review presented in this paper offers a reliable portrait of the coastal landslides and related risk in Italy. Although not exhaustive at the local scale (which was not its aim), it illustrates the most typical situations, their distribution and the commonest mitigation measures devised by the national authorities.

This review has been based on the most detailed datasets so far available, specifically the IFFI archive (Italian Landslides Inventory) developed by ISPRA, supported by information on damage from the AVI database (inventory of information on sites historically affected by landslides and floods in Italy for the period 1918–2000). The IFFI database has proven to offer the great advantage of covering the whole of Italy with homogeneous data and at a spatial resolution useful for reliable hazard evaluation at the regional scale and even at the local scale.

Commonly, along the Italian coasts the greatest attention has been devoted to the risk of coastal erosion, flooding, subsidence, tsunamis, and earthquakes in the flat areas mostly corresponding to alluvial plains. Such areas host numerous large urban settlements, industrial facilities, tourist resorts, airports, and important roads and railways. The urban sprawl of recent decades, as portrayed, for example, by the CORINE Land Cover data, well illustrates the parallel risk growth. However, it has been shown here that the landslide risk along high coasts is also relevant. This is even more true during summer, when the population may increase 10 times and more in the most renowned tourist resorts, with many people concentrated on reefs, pocket beaches and boats floating near rocky shores. Rock falls or topples affect many sea-cliffs, connected to the structural setting and fracture network of the rock mass, to basal wave erosion and sometimes to seismic events, especially in southern Italy. Moreover, where a layer of soil, debris or loose volcanic rocks covers steep slopes of bedrock plunging into the sea, rapid debris flows, triggered during short and intense rainfall, represent the main hazard (e.g. along wide coastal sectors of Campania and Calabria).

A future development of this research could be a more refined analysis at the local scale, based on the IFFI database and more detailed DEM and land use maps.

Special thanks go to D. Berti (ISPRA) for the useful contribution to the realization of the Lithological Map of Italy (scale 1:500 000).

References

Almagià, R. 1910. *Studi geografici sulle frane in Italia.* Memorie della Società Geografica Italiana, **14**.

Amanti, M., Casagli, N., Catani, F., D'Orefice, M. & Motteran, G. 1996. *Guida al censimento dei fenomeni franosi ed alla loro archiviazione.* Miscellanea del Servizio Geologico Nazionale, **VII**.

Amanti, M., Bertolini, G., Cara, P., Chiessi, V., Martini, M. G., Ramasco, M. & Ventura, R. 2000. *Allegato tecnico per l'attuazione del Progetto IFFI.* Servizio Geologico Nazionale, Roma.

Amanti, M., Bertolini, G., Chiessi, V., De Nardo, M. T. & Ramasco, M. 2001. Guida alla compilazione della scheda frane IFFI—Allegato 1. Servizio Geologico Nazionale, Roma. World Wide Web Address: http://www.sinanet.apat.it/progettoiffi

Antonioli, F., Silenzi, S., Vittori, E. & Villani, C. 1999. Sea level changes and tectonic mobility: precise measurements in three coastlines of Italy considered stable during the last 125 ky. *Physics and Chemistry of the Earth, Part A*, **24**, 337–342.

APAT 2005. La realizzazione in Italia del progetto europeo Corine Land Cover 2000 (Rapporti 61/2005). APAT, Roma.

APAT 2006. Annuario dei dati ambientali (Environmental Data Yearbook). APAT. World Wide Web Address: http://www.apat.gov.it

Barbano, A., Corsini, S., Paone, M. & La Mantia, C. 2005. Definizione della linea di costa italiana e delle opere di difesa. *In*: *Atti della 9ª Conferenza Nazionale ASITA, Catania, 15–18 novembre 2005*. Federazione delle Associazioni Scientifiche per le Informazioni Territoriali e Ambientali, 231–236.

Bordoni, P. & Valensise, G. 1998. Deformation of the 125 ka marine terrace in Italy: tectonic implications. *In*: Stewart, I. & Vita Finzi, C. (eds) *Late Quaternary Coastal Tectonics*. Geological Society, London, Special Publications, **146**, 71–110.

Bottero, D., Cavallo, C., De Stefanis, E., Gorziglia, G. & Poggi, F. 2004. *Progetto Inventario dei Fenomeni Franosi in Italia: Primi risultati in Liguria.* Regione Liguria.

Bruno, G. & Zezza, F. 1992. Typologies of instability and methods of stability verification for the coastal cliffs of Salento (Apulia, Italy). *Bollettino di Oceanologia Teorica e Applicata*, **X**, 123–131.

Budetta, P., Calcaterra, D. *et al.* 1997. Fenomeni Franosi connessi all'evento pluviometrico del gennaio 1997 in Penisola Sorrentina—M.ti Lattari: primo contributo. *In*: *Atti del IX Congresso Nazionale*

Geologi, Roma, 17–20 aprile 1997. Consiglio Nazionale dei Geologi, Roma, 223–231.

BUDETTA, P., GALIETTA, G. & SANTO, A. 2000. A methodology for the study of the relation between coastal cliff erosion and the mechanical strength of soils and rock masses. *Engineering Geology*, **56**, 243–256.

CALCATERRA, D. & SANTO, A. 2004. The January 10, 1997 *Pozzano landslide, Sorrento Peninsula, Italy. Engineering Geology*, **75**, 181–200.

CALCATERRA, D., DE RISO, R., EVANGELISTA, A., NICOTERA, M. V., SANTO, A. & SCOTTO DI SANTOLO, A. 2004. Slope instabilities in the pyroclastic deposits of the Phlegraean district and the carbonate Apennine (Campania, Italy). *In*: PICARELLI, L. (ed.) *International Workshop on Occurrence and Mechanisms of Flows in Natural Slopes and Earthfills, Sorrento, 14–16 May 2003*. PATRON Editore, Bologna, 61–76.

CANCELLI, A., MARABINI, F., PELLEGRINI, M. & TONNETTI, G. 1984. Incidenza delle frane sull'evoluzione della costa adriatica da Pesaro a Vasto. *Memorie della Società Geologica Italiana*, **27**, 555–568.

CIVITA, M., DE RISO, R., LUCINI, P. & NOTA D'ELOGIO, E. 1975. Studio delle condizioni di stabilità dei terreni della Penisola Sorrentina (Campania). *Geologia Applicata e Idrogeologia*, **10**, 129–188.

COLLESELLI, F. & COLOSIMO, P. 1977. Comportamento di argille plio-pleistoceniche in una falesia del litorale adriatico. *Rivista Italiana di Geotecnica*, **5**, 5–21.

COLOSIMO, P. & CRESCENTI, U. 1972. Carta geolitologica a orientamento geotettonico e della franosità della zona del Monte Conero (Comuni di Ancona, Numana e Sirolo). *Memorie della Società Geologica Italiana*, **12**, 317–334.

COTECCHIA, V. & SIMEONE, V. 1998. Studio dell'incidenza degli eventi di pioggia sulla grande frana di Ancona del 13 dicembre 1982. *In*: LUINO, F. (ed.) *Atti del Convegno Internazionale 'La prevenzione delle catastrofi idrogeologiche: il contributo della ricerca scientifica', Alba (CN), 5–7 novembre 1996*. CNR–GNDCI, *I*, 19–29.

CRUDEN, D. M. & VARNES, D. J. 1996. Landslide types and processes. *In*: TURNER, A. K. & SCHUSTER, R. L. (eds) *Landslides Investigation and Mitigation*. US Transportation Research Board, Special Report, **247**, 36–75.

D'ALESSANDRO, L., GENEVOIS, R. & MARINO, A. 2001. Dinamica recente della costa alta tra Ortona e Vasto (Abruzzo centro-meridionale). *Memorie della Società Geologica Italiana*, **56**, 53–60.

D'ANGELO, S., GISOTTI, G. & LEMBO, P. 1999. La spiaggia e la falesia di Chiaia di Luna (isola di Ponza): un geotopo con problemi di pericolosità geologica e di fruizione turistica. *Memorie Descrittive della Carta Geologica d'Italia*, **LIV**, 433–438.

DE FALCO, M., DE RISO, R. & DUCCI, D. 1997. La piovosità della Penisola Sorrentina e dei M.ti Lattari in relazione all'evento del gennaio 1997. *In*: *Atti del IX Congresso Nazionale Geologi, Roma, 17–20 aprile 1997*. Consiglio Nazionale dei Geologi, Roma, 321–329.

DE STEFANIS, A., MARINI, M., TERRANOVA, R. & DE LUIGI, G. 1978. I movimenti franosi di Guvano e di Rodalbia nelle Cinque Terre e i loro riflessi sulla morfologia della costa ligure e sugli insediamenti. *Memorie della Società Geologica Italiana*, **19**, 161–167.

DI CRESCENZO, G. & SANTO, A. 1999. Le frane da scorrimento-colata rapida in depositi piroclastici nella Penisola Sorrentina: nuovi dati geomorfologici. *In*: *Atti del Convegno 'Il Rischio Idrogeologico e la Difesa del Suolo', Roma, 1–2 ottobre 1998*. Accademia Nazionale di Lincei, Roma, 127–136.

DVORAK, J. J. & MASTROLORENZO, G. 1991. *The mechanism of recent vertical crustal movements in Campi Flegrei caldera, Southern Italy*. Geological Society of America, Special Papers, **263**.

EMERY, K. O. & KHUN, G. G. 1982. Sea cliffs: their processes, profiles, and classification. *Geological Society of America Bulletin*, **93**, 644–654.

ESPOSITO, E., PORFIDO, S. & VIOLANTE, C. (eds) 2004*a*. *Il nubifragio dell'ottobre 1954 a Vietri Sul Mare—costa di Amalfi, Salerno*. Pubblicazione Gruppo Nazionale per la, Difesa dalle Catastrofi Idrogeologiche, **2870**.

ESPOSITO, E., PORFIDO, S., VIOLANTE, C., BISCARINI, C., ALAIA, F. & ESPOSITO, G. 2004*b*. Water events and historical flood recurrences in the Vietri sul Mare coastal area (Costiera Amalfitana, southern Italy). *In*: RODDA, G. & UBERTINI, L. (eds) *The Basis of Civilization—Water Science?* International Association of Hydrological Sciences Publications, **286**, 1–12.

ESU, F. 2000. La stabilità dei pendii costieri. *In*: CANUTI, P. & FANTI, R. (eds) *CNR-Gruppo Nazionale per la Difesa dalle Catastrofi Idrogeologiche. Rapporto 1986–1995: Linea 2, Previsione e prevenzione eventi franosi a grande rischio*, CNR–GNDCI, 149–158.

ESU, F. & MARTINETTI, S. 1965. Considerazioni sulle caratteristiche tecniche delle argille Plio-Pleistoceniche della fascia costiera adriatica tra Rimini e Vasto. *Rivista Italiana di Geotecnica*, **12**, 164–185.

FEDERICI, P. R., BALDACCI, F., PETRESI, A. & SERANI, A. 2001. *Atlante dei Centri Abitati Instabili della Liguria. I. Provincia di La Spezia*. Regione Liguria CNR–GNDCI.

FIORILLO, F. & GUADAGNO, F. M. 2000. Analysis of rainfall patterns triggering reactivations of a large landslide in pleistocene clay in molise (Italy). *In*: BROMHEAD, E., DIXON, N. & IBSEN, M. L. (eds) *Landslides in Research, Theory and Practice*. Thomas Telford, London, 553–558.

GORI, U. & MEZZABOTTA, M. 1995. Sul cinematismo della frana costiera dell'abitato di Petacciato (CB). *Geologia applicata e Idrogeologia*, **30**, 659–670.

GUERRICCHIO, A. 1988. Aspetti geologici sull'erosione dei litorali e loro influenza nel campo applicativo. *Geologia Applicata e Idrogeologia*, **XXIII**, 29–78.

GUERRICCHIO, A. & MELIDORO, G. 1998*a*. Rischi da grandi frane nella fascia costiera adriatica. *In*: LUINO, F. (ed.) *Atti del Convegno Internazionale 'La prevenzione delle catastrofi idrogeologiche: il contributo della ricerca scientifica', Alba (CN), 5–7 novembre 1996*. CNR–GNDCI, *I*, 317–330.

GUERRICCHIO, A. & MELIDORO, D. 1998*b*. Deformazioni gravitative dei versanti costieri di Vasto

(Abruzzo). *In*: LUIND, F. (ed.) *Atti del Convegno Internazionale 'La prevenzione delle catastrofi idrogeologiche: il contributo della ricerca scientifica', Alba (CN), 5–7 novembre 1996.* CNR–GNDCI, *I*, 369–381.

GUERRICCHIO, A., MELIDORO, G. & SIMEONE, V. 1996. Le grandi frane di Petacciato sul versante costiero adriatico (Molise). *Memorie della Società Geologica Italiana*, **51**, 607–632.

GUZZETTI, F. 2000*a*. Progetto AVI. Rapporto sull'attività svolta nel corso del 1999. CNR–GNDCI. World Wide Web Address: http://avi.gndci.cnr.it/docs/attivita/Attivita_1999it.pdf

GUZZETTI, F. 2000*b*. Landslide fatalities and the evaluation of landslide risk in Italy. *Engineering Geology*, **58**, 89–107.

HUTCHINSON, J. N. 1973. The response of London Clay cliffs to differing rates of toe erosion. *Geologia Applicata e Idrogeologia*, **8**, 221–239.

IADANZA, C., TRIGILA, A. & CASABURI, M. 2006. L'applicazione WebGIS del Progetto IFFI (Inventario dei Fenomeni Franosi in Italia). *MondoGIS*, **52**, 31–34.

LAMBECK, K., ANTONIOLI, F., PURCELL, A. & SILENZI, S. 2004. Sea-level change along the Italian coast for the past 10,000 yr. *Quaternary Science Reviews*, **23**, 1567–1598.

MALINVERNO, A. & RYAN, W. B. F. 1986. Extension in the Tyrrhenian Sea and shortening in the Apennines as result of arc migration driven by sinking of the lithosphere. *Tectonics*, **5**, 227–245.

MELIDORO, G. & MEZZABOTTA, M. 1998. Monitoraggio ultrasecolare delle deformazioni gravitative costiere adriatiche. *In*: LUINO, F. (ed.) *Atti del Convegno Internazionale 'La prevenzione delle catastrofi idrogeologiche: il contributo della ricerca scientifica', Alba (CN), 5–7 novembre 1996.* CNR–GNDCI, *I*, 343–356.

PORFIDO, S., ESPOSITO, E., ALAIA, F., MOLISSO, F. & SACCHI, M. 2009. The use of documentary sources for reconstructing flood chronologies on the Amalfi rocky coast (southern Italy). *In*: VIOLANTE, C. (ed.) *Geohazard in Rocky Coastal Areas*. Geological Society, London, Special Publications, **322**, 171–185.

ROYDEN, L., PATACCA, E. & SCANDONE, P. 1987. Segmentation and configuration of subducted lithosphere in Italy: an important control on thrust-belt and foredeep-basin evolution. *Geology*, **15**, 714–717.

SACCHI, M., MOLISSO, F. *ET AL.* 2009. Insights into flood-dominated fan-deltas: very high-resolution seismic examples off the Amalfi cliffed coast, eastern Tyrrhenian Sea. *In*: VIOLANTE, C. (ed.) *Geohazard in Rocky Coastal Areas*. Geological Society, London, Special Publications, **322**, 33–72.

SUNAMURA, T. 1992. *Geomorphology of Rocky Coasts*. Wiley, Chichester.

TRIGILA, A. (ed.) 2007. Rapporto sulle frane in Italia, Il Progetto IFFI—Metodologia, risultati e rapporti regionali (Rapporti 78/2007). APAT, Roma. World Wide Web Address: http://www.apat.gov.it/site/it-IT/APAT/Pubblicazioni/Rapporti/

TRIGILA, A. & IADANZA, C. 2008. *Landslides in Italy*. Special Report 2008. (Rapporti 83/2008). ISPRA, Roma.

VALLARIO, A. 1998. *Il dissesto idrogeologico in Campania*. SIGEA, Roma, Geologia dell'Ambiente, **3**.

VANNOLI, P., BASILI, R. & VALENSISE, G. 2004. New geomorphic evidence for anticlinal growth driven by blind-thrust faulting along the northern Marche coastal belt (central Italy). *Journal of Seismology*, **8**, 297–312.

VARNES, D. J. 1978. Slope movements types and processes. *In*: SCHUSTER, R. L. & KRIZEK, R. J. (eds) *Landslides: Analysis and Control*. US Transportation Research Board, Special Report **176**, 11–33.

VIOLANTE, C. 2009. Rocky coasts: geological constraints for hazard assessment. *In*: VIOLANTE, C. (ed.) *Geohazard in Rocky Coastal Areas*. Geological Society, London, Special Publications, **322**, 1–31.

VIOLANTE, C., BISCARINI, C., ESPOSITO, E., MOLISSO, F., PORFIDO, S. & SACCHI, M. 2009. The consequences of hydrologic events on steep coastal watersheds: the Costa d'Amalfi, eastern Tyrrhenian Sea. *In*: LIEBSHER, H. J. *ET AL.* (eds) *The Role of Hydrology in Water Resource Management. Capri Italy, 2008.* IAHS Publication, **327**, 102–113.

WP/WLI (International Geotechnical Societies UNESCO Working Party on World Landslide Inventory) 1993. A suggested method for describing the activity of a landslide. *IAEG Bulletin*, **47**, 53–57.

Slope instability on rocky coast: a case study of Le Grazie landslides (eastern Liguria, northern Italy)

P. BRANDOLINI*, F. FACCINI, A. ROBBIANO & R. TERRANOVA

University of Genoa, DISAM, via Balbi 2, 16122 Genova, Italy

**Corresponding author (e-mail: brando@unige.it)*

Abstract: The stretch of rocky coast between Rapallo and Chiavari (Tigullio Gulf) is characterized by significant geomorphological instability phenomena, mainly contained between the shoreline and the Via Aurelia. Research was focused particularly on the geomorphological features and the states of activity of the landslides that are situated below the Santuario di N. S. delle Grazie. The dynamics can be partially attributed to natural causes, including the geological–tectonic layout, the geomechanical properties of the rock mass and wave undercutting, and partially to human activity related to the opening of limestone quarries and the construction of roadway, railway and communications networks. The recent evolution of the landslides, involving buildings, structures and infrastructure, requiring a variety of consolidation measures and work, has been observed. The latter were implemented following in-depth analysis of the subsoil geology and monitoring activity.

The study considers the well-known landslides below the Santuario di N. S. delle Grazie, situated towards the western margin of the town of Chiavari (Liguria, Italy) (Fig. 1). The gravitational movements occur on the southern slope of Mt. Cucco, affecting the areas between the shore and the main state road no. 1 ('Aurelia'). They also affect, to a lesser extent, the slope between the road and the watershed line.

Several landslide-prone zones were identified and defined; they appear to be interconnected by the presence of a deep-seated gravitational slope deformation. Proceeding from east to west, five active landslides were identified: the Belvedere landslide, the Colonia Piaggio landslide, the Via Aurelia landslide, the Santuario landslide and the Liggia landslide (Figs 2 and 3).

These mass movements have been known throughout historical times and it is certain that periods of activity have alternated with periods of dormancy, and that they have always represented a problem for public administrations and the local population. The activity of these movements constitutes a geomorphological risk for the road and railway networks, the stability of buildings and the safety of people, including swimmers and those in boats in the seaward sector.

The landslides were examined previously in scientific studies conducted in the second half of the 19th century (Issel 1892), but more detailed observation was carried out by Rovereto (1922, 1939). He associated the mass movements with the opening of the quarries from which the stone was extracted for the construction of most of Chiavari's old town centre. Some of the movements that took place between 1922 and 1940 are reported in the explanatory notes accompanying the meteorological observations of the Observatory of the Diocesan Seminary of Chiavari. The Santuario delle Grazie landslides, which led to serious problems along the Via Aurelia requiring the construction of the present tunnel, were described by Sanguineti (1949).

Cortemiglia & Terranova (1970) conducted a complete study of the characteristics of the landslides on the slope below the Sanctuary, the causes of the movement and its mechanisms, in addition to providing guidelines regarding remedial and protective measures. However, the latest scientific work on this subject is more than 10 years old (Brandolini & Terranova 1994). It focused on a description of several examples of geomorphological instability on slopes in Liguria and the implications for soil conservation.

Geographical and climatological outline

The coastal sector of Le Grazie is situated in the central part of the physiographic unit of the Tigullio Gulf. The unit is delimited by the promotories of Sestri Levante to the east and Portofino to the west.

The study area is characterized by a typically Mediterranean climate with warm summers and mild winters, whereas the Mt. Cucco and Mt. Anchetta watershed ridge represents a preliminary orographic barrier that blocks the arrival of warmer winds from the sea. The temperature and precipitation profile was defined using data recorded at the Chiavari station in the period 1883–2002 (Fig. 4). As regards rainfall distribution, a peak is identifiable in October and a minimum in July.

From: VIOLANTE, C. (ed.) *Geohazard in Rocky Coastal Areas*. The Geological Society, London, Special Publications, **322**, 143–154. DOI: 10.1144/SP322.6

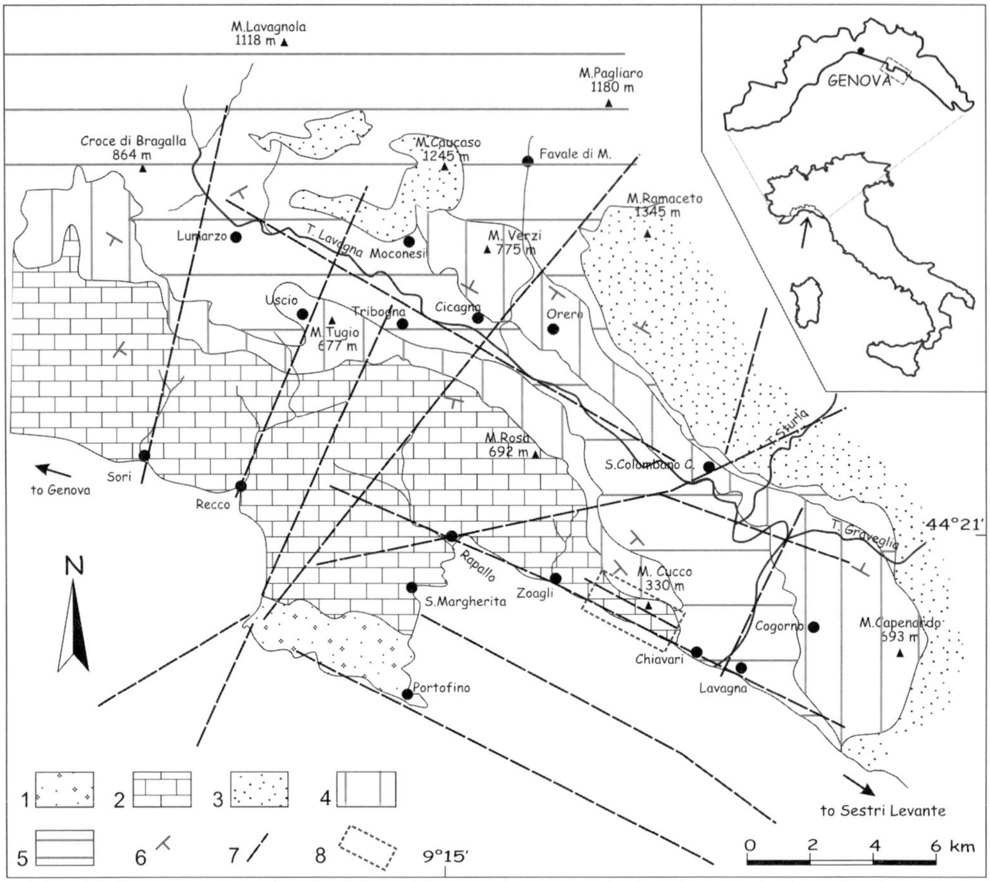

Fig. 1. Geological sketch map of the Tigullio Gulf. 1, conglomerate of Portofino (Oligocene); 2, flysch of Mt. Antola (Upper Cretaceous–Palaeocene); 3, sandstones of Mt. Gottero (Upper Cretaceous–Palaeocene); 4, slate of Mt. Verzi (Upper Cretaceous); 5, shales of Lavagna Valley (Upper Cretaceous); 6, attitude; 7, main tectonic lineations; 8, location of the study area.

Annual mean precipitation amounts to 1150 mm, and the annual mean temperature slightly exceeds 15 °C, showing a peak in the summer months (July–August) with temperatures slightly below 25 °C and a minimum in the winter months (December and January) at 10 °C.

The area affected by adverse winds, where swell acts upon the coast, has a window of about 122°, involving quadrants 2 and 3 in terms of winds and tides (Cortemiglia & Fierro 1965). The prevailing winds are from the SE whereas the strongest are from the SW (Fig. 5). Strong southwesterly sea storms are characterized by trains of waves with lengths reaching 120 m, heights of over 5 m and a velocity of up to 30 knots (15.4 m/s), in the open-sea environment of the Mediteranean, resulting in marked erosion of the sea floor and high-energy impact along the rocky coast.

Geological setting

The bedrock in the study area consists of flysch of Mt. Antola (Upper Cretaceous–Palaeocene), which characterizes the coastal sector between Genoa and Chiavari and the hinterland of Genoa as far as the boundary between the regions of Liguria and Emilia (see Fig. 1). The geological survey led to the identification of a sequence of rock types in the Mt. Antola Formation, typical of turbidite sedimentation (Marini 1981) grouped into the following lithofacies:

(1) compact grey marly limestone, sometimes showing conchoidal fracture, with thickness ranging from a few decimetres to several metres;

(2) ash-grey scaly marly limestone, sometimes showing coarse foliation, in layers generally not exceeding 2–3 m in thickness;

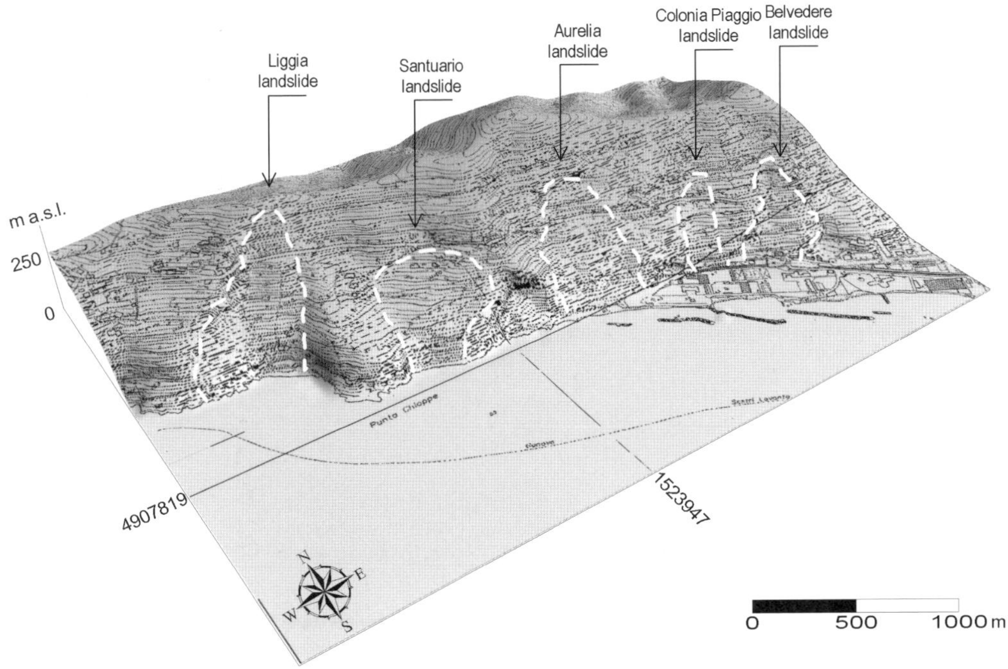

Fig. 2. A 3D view of the investigated coastal sector.

(3) slate, coarsely divisible into minute flakes and prisms, yellowish in colour owing to alteration, generally present in thin layers only several decimetres thick, occasionally in strata several metres thick;

(4) dark grey calcarenite, with veins of calcite;

(5) blackish grey shale, sometimes associated with clayey–arenaceous schist, generally occurring as thin layers 20–30 cm thick and occasionally as millimetre-scale intercalations.

The structural arrangement of the Mt. Antola Formation is generally monoclinal dipping towards south to SW, with the overall dip of the bedding more or less parallel to the slope (an unfavourable condition for slope stability), whereas locally, in the western sector, there is a different dip with respect to the slope (resulting in a fair or good condition). The bedding have a dip of 35–40° along the watershed ridge, and a lower dip (10–15°) in the western coastal sector; dip direction varies between 240° and 260°. The tectonic setting is characterized by at least two displacement systems that have approximately orthogonal trends, marked by neotectonic activity. The first system, to which the present orientation of the coast can be attributed, is oriented WNW–ESE, whereas the second one, along which the landslide masses are set into motion and the secondary drainage network is arranged, is oriented NNE–SSW.

The tectonic style is responsible for the generalized macro-fracturing of the bedrock: the latter is affected by discontinuities that are capable of isolating rock prisms of various sizes, creating conditions of slope instability.

Geomorphological features and landslide evolution

The geomorphological features of the study area have been represented on a map on which structural data, landforms and processes related to gravity, running waters, marine and karst processes, and human activity are indicated (Fig. 6).

The slope investigated is characterized by a complex morphology, considering the structural and tectonic arrangement of the Mt. Antola Formation, the thickness of the debris cover, the geomorphological evolution of the landslides, and the quarrying activity that led to the opening of vertically cut quarry fronts (Figs 7–9). The bedrock generally crops out in the watershed ridge sector, whereas in the middle and lower portions of the slope, thick debris cover overlies the bedrock and undergoes continuous erosion by wave motion (Fig. 10). At the foot of the slope, erosion of the rock cliff has led to the formation of a talus cone

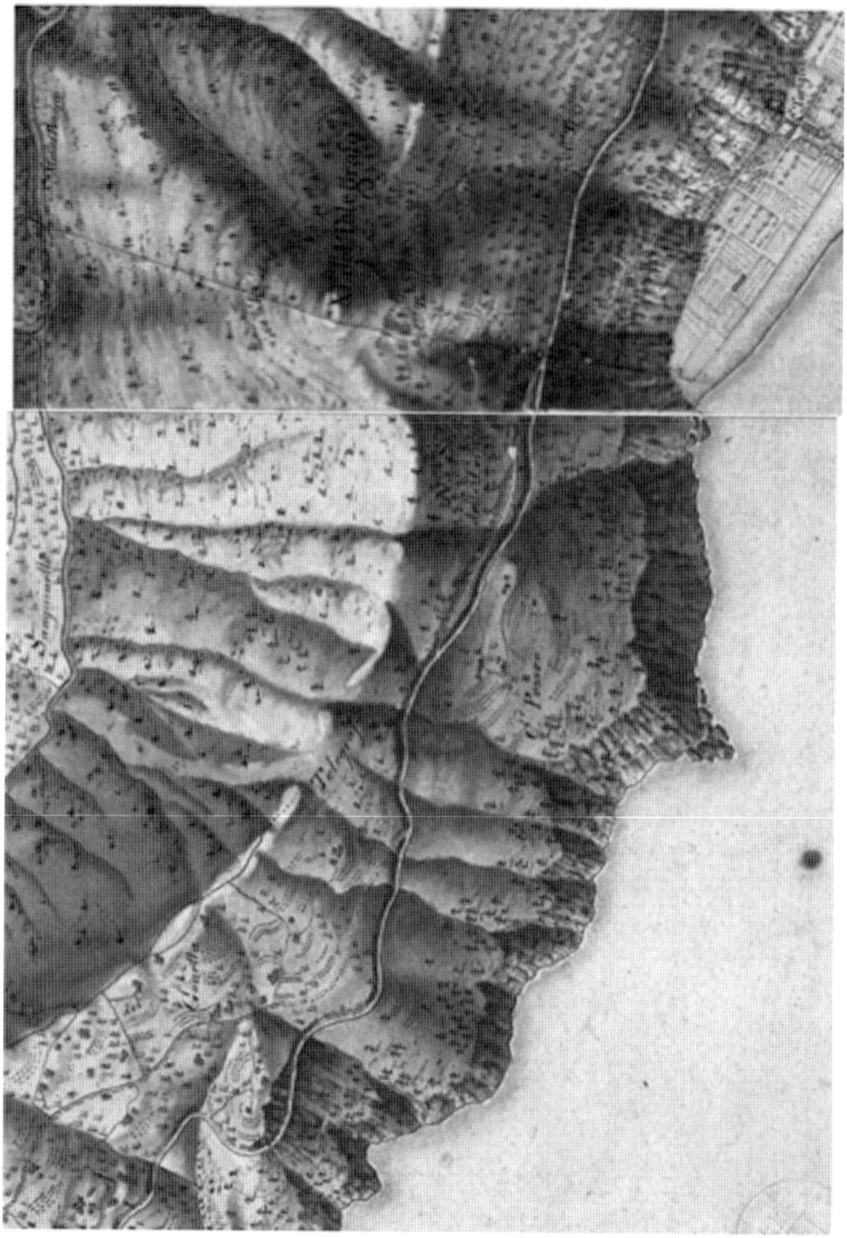

Fig. 3. Historical map of the investigated coastal area (beginning of the 19th century; from IGMI archives).

and scree slope, which, in turn, ensure a supply of material to the gravelly–pebbly pocket beaches occurring at the base of the Liggia and Via Aurelia landslides.

The landslides, which have been described in detail by Cortemiglia & Terranova (1970), are classifiable as complex movements caused by translational rock slides with components of rockfalls. The interpretation of aerial photographs, comparisons with historical maps and detailed field investigations have made it possible to define the state of activity and the morphological variations of the landslide masses, which have undergone very recent changes. The principal elements indicating the state of activity of the landslides characterized by an intermittent type of evolution, are as follows.

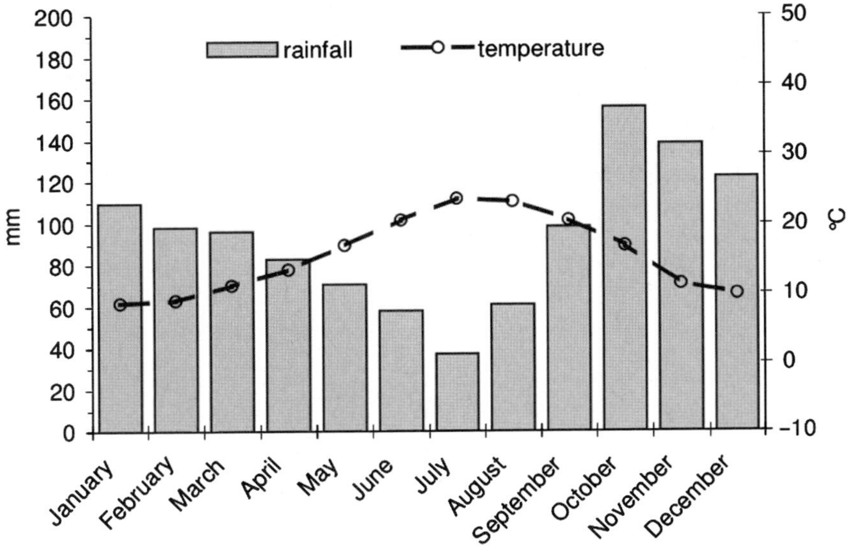

Fig. 4. Temperature and rainfall for the Chiavari weather station, 5 m above sea level (1883–2002 period).

(1) There are marked increases in the deformations visible along the Via Aurelia and the railway line, together with structural damage to the drystone walls of terraces and to buildings, including the famous Santuario di N. S. delle Grazie. In this regard, it may be noted that at various times the main road and the railway line have both undergone severe damage requiring changes to the original plans. These changes involved work to relocate portions of the original route, which are

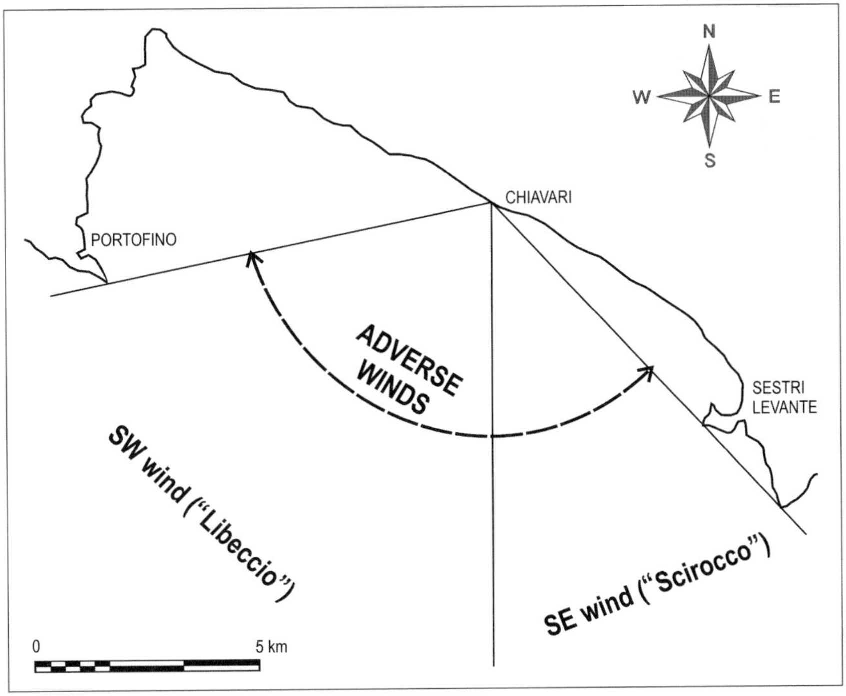

Fig. 5. Prevailing wind sector of the studied coastal area.

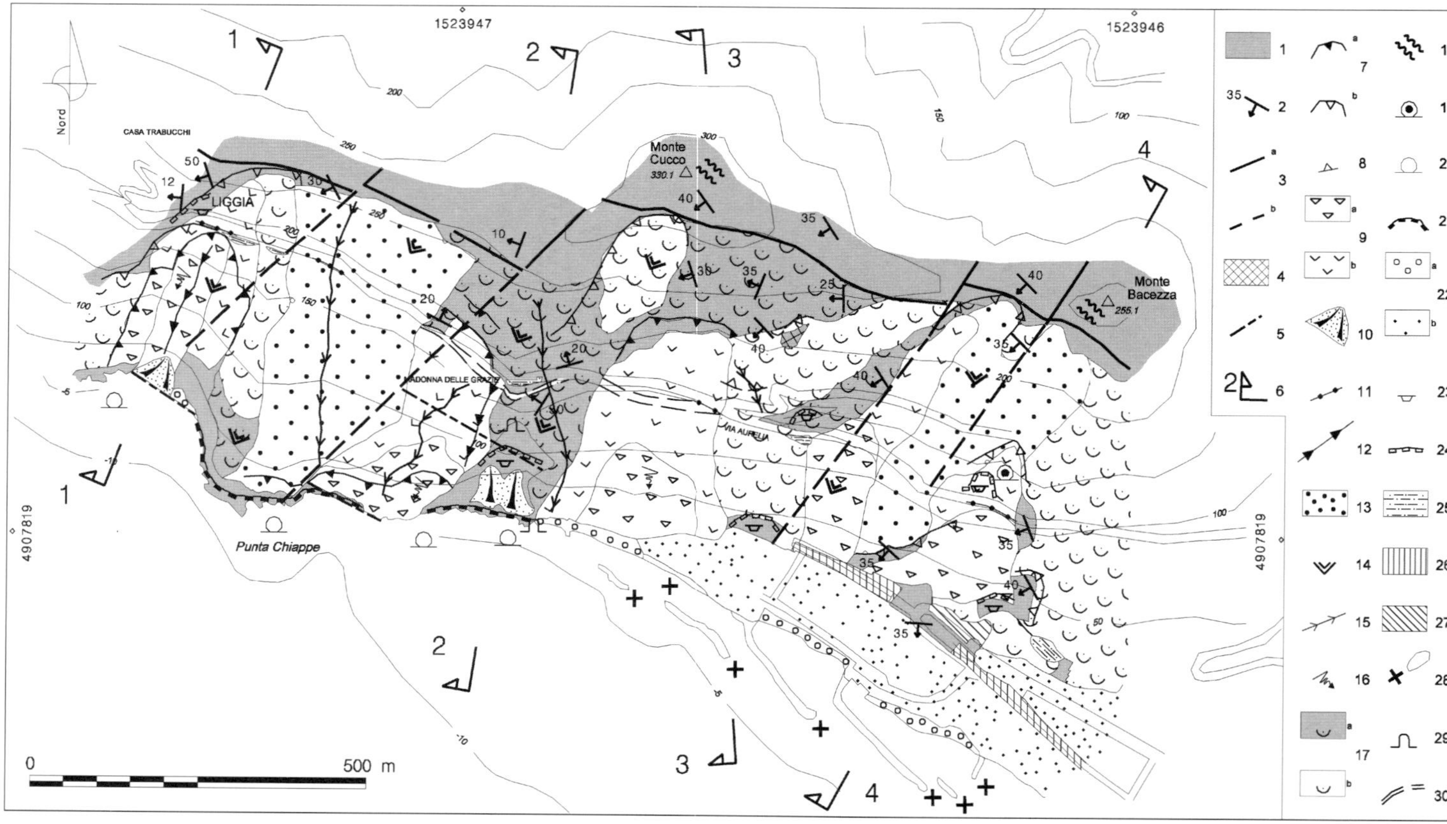

Fig. 6. Geomorphological map of the study area. **Structural data**: 1, flysch of Mt. Antola (Upper Cretaceous–Palaeocene); 2, attitude; 3, fault (a, certain; b, assumed); 4, strongly fractured bedrock; 5, fracture or tectonic lineation; 6, trace of geological cross-section. **Slope landform, processes and deposits due to gravity**: 7, edge of landslide scarp (a, active; b, dormant); 8, reverse slope; 9, landslide as a result of slide (a, active; b, dormant); 10, talus cone; 11, tension crack; 12, rock defile with occasional debris fall; 13, debris slope cover; 14, area affected by deep-seated gravitational slope deformation. **Landform, processes and deposits due to running water**: 15, gully; 16, area affected by rill wash; 17, colluvial deposit (a, thin; b, significant). **Karst landform**: 18, limestone outcrops with corrosion microforms; 19, cave. **Marine landform, processes and deposits**: 20, sea cave; 21, edge of scarp due to wave erosion; 22, beach deposits: (a, pebbles; b, sand). **Anthropogenic landform; processes and deposits**: 23, abandoned quarry; 24, edge of artificial scarp due to mining activities; 25, fill; 26, railway embankment; 27, artificial denudation area; 28, protective structure along the shoreline; 29, abandoned tunnel entry; 30, abandoned road track.

Fig. 7. Panoramic view of the central sector of the study area, showing on the left the Sanctuary landslide and on the right the ancient quarrying area.

still partially visible, and to create by-passes, some of which also involve tunnels.

(2) There are increases in the extension of the active parts of the landslide masses, as compared with data reported in the literature over the last 40 years (Cortemiglia & Terranova 1970; Brandolini & Terranova 1994).

(3) There are morphological variations observable in the masses, especially in the Liggia and Santuario landslides. The variations are related to

Fig. 8. Liggia landslide area in the western sector of the investigated area.

Fig. 9. Panoramic view, taken from the east, of the Aurelia landslide.

the removal of significant amounts of material at the foot by wave cutting. The other landslides present a similar evolutionary trend in terms of the geomorphological aspects, although the determining factor is related to precipitation, with rainfall of marked intensity occurring especially over the last few years (Faccini *et al.* 2005).

Landforms and processes related to running waters are observable throughout the sector surveyed. In terms of geomorphological dynamics, the

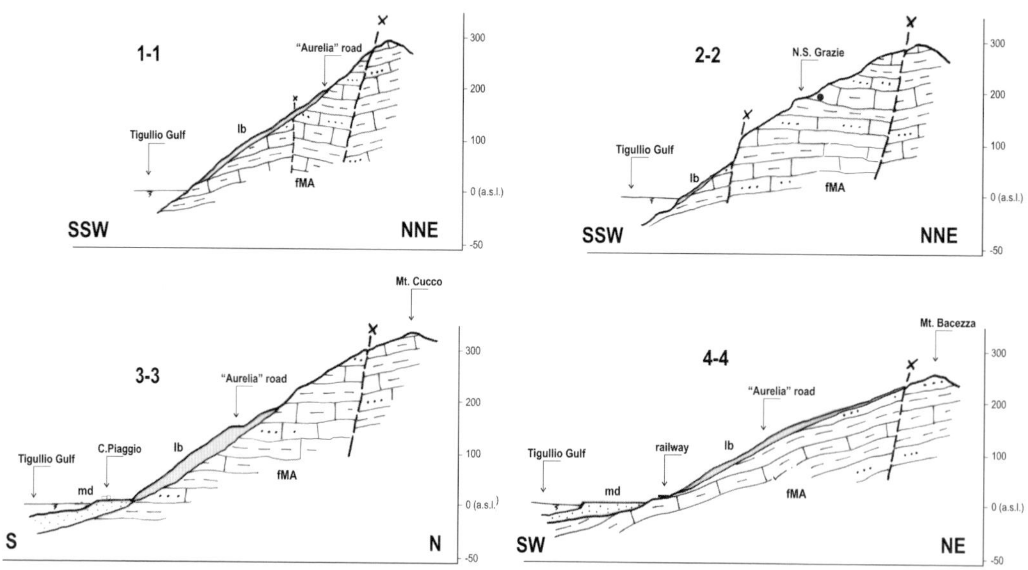

Fig. 10. Geological cross-sections: fMA, flysch of Mt. Antola; lb, landslide body; md, marine deposit. (See Fig. 6 for location.)

gullies in the central and western sectors of the study area are of particular significance. Moreover, colluvial deposits were observed especially in the uppermost slope portions, with thicknesses sometimes reaching several metres, and in all cases, they have been reworked by terracing for agricultural use.

Another element examined, the cave named Dugu II situated in the Belvedere landslide, also takes on importance in terms of a correlation between karstic morphogenesis and gravitational deformations. The cave is located on a fracture that runs parallel to the coastline. The cavity is a discontinuity with an aperture of some metres, which indicates the slope's instability.

Wave action takes on marked importance in this sector, because of the exposure of the coast with respect to the dominant winds, and because of the heterogeneous nature of the flysch and its geomechanical characteristics. Wave erosion has led to the formation of a cliff exceeding 25 m in height and of interbedded sea caves excavated by differential erosion.

Finally, human activity appears to have played a major role in this case. The opening and exploitation of numerous quarries scattered throughout the study area, and the construction of the surface and underground road and railway systems, are responsible for triggering instability phenomena and variation in the geohydrology. Both have jointly contributed to aggravating a situation characterized by natural instability.

Geomechanical characterization of the flysch

Rock mass classification

Structural geological surveys conducted at several measurement stations (Fig. 11) led to the identification of the main families of discontinuities that contribute to the determination of the quality of the rock mass and can influence slope stability. Three principal systems of discontinuity were surveyed: the bedding joint, which has a dip of 20–30° and a dip direction of 240–260°, fracture j1, with a dip between 60 and 80° and a dip direction between 130 and 180°, and fracture j2, which has a dip of 50–60° and a wider range of dip direction values, from 45 to 130°.

The method selected for characterization of the rock mass was the rock mass rating (RMR) system proposed by Bieniawski (1989). The parameters described below were determined by means of field studies (Fig. 12).

(1) *Joint compressive strength (JCS).* Schmidt hammer test results suitably processed provided

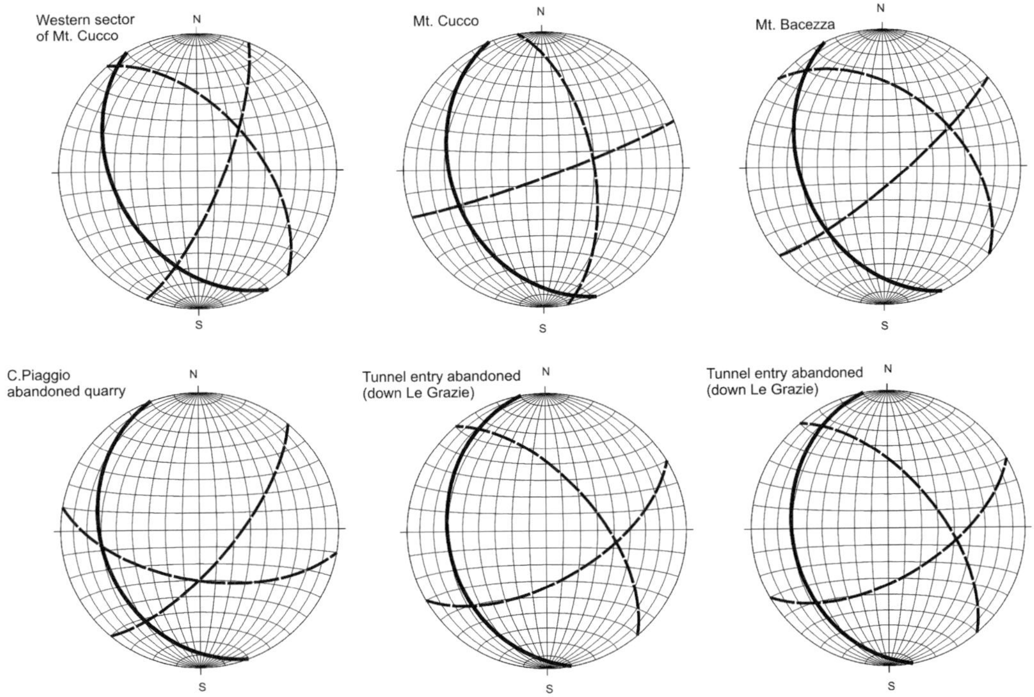

Fig. 11. Equal area projection, lower hemisphere great circle: bold line, bedding; dashed line, main fracture.

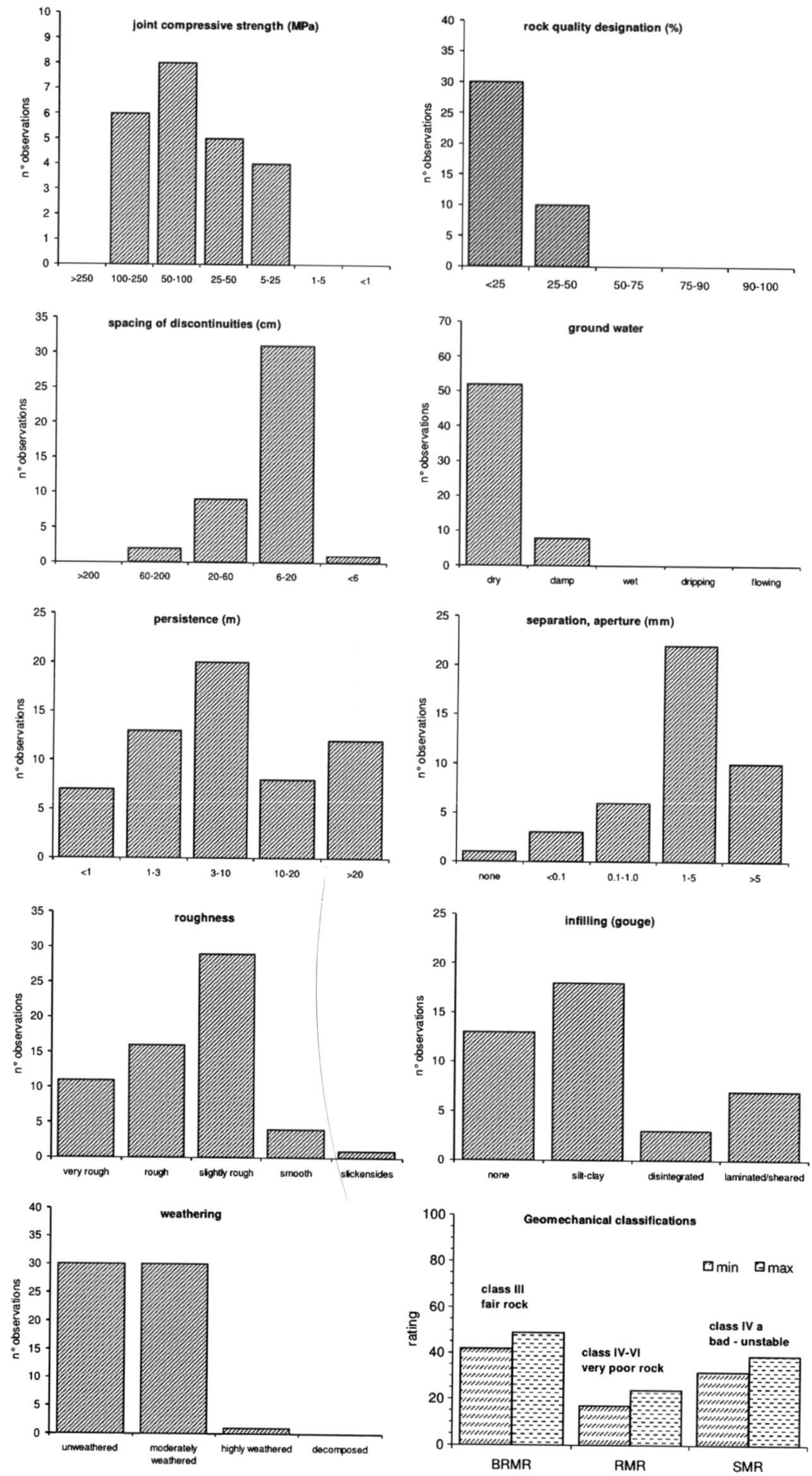

Fig. 12. Geomechanical data and frequency histograms.

rather variable JCS values. The mode range was between 50 and 100 MPa, with the dispersion being attributable to lithological heterogeneity and to problems involved in testing on planar surfaces. Considering the moderate alteration of the joints and the aims of this study, the JCS values were adopted as indicative values of uniaxial compressive strength.

(2) *Rock quality designation (RQD).* A modal value of less than 25% was obtained on the basis of the drilling results obtained around the Santuario di N. S. delle Grazie, a value that is in keeping with *in situ* observations.

(3) *Spacing of discontinuities.* The histogram frequency diagram shows a peak in the 6–20 cm range, which is consistent with the RQD value obtained.

(4) *Conditions of discontinuities.* Bearing in mind that fracturing is marked and widespread in the marly limestone and that dense schistosity is observable in the intercalations of clayey shale, the maximum persistence value (20 m) was held to be representative, as was the maximum separation value (1–5 mm). The investigations conducted also led to a definition of the joints as undulating and slightly rough. When present, the infilling (gouge) consists of fine-grained sediment, and less frequently of calcite or brecciated intervals. The sides of the discontinuities do not show significant alteration.

(5) *Ground water.* The discontinuities in the rock are mostly dry or more infrequently damp.

Taking into account the values obtained for the geomechanical survey parameters, the BMR (basic mass rating) value ranged between 42 and 48, thus ranking the rock mass as class III (fair rock).

Rating adjustment for the rocky slope

A preliminary estimate of rocky slope stability conditions can be made using a rating for the orientation of discontinuities applied to the BMR. In this case, we referred to an interpretation proposed by Pozzi & Clerici (1985), based on the application of the Markland Test (Markland, 1972), and to the interpretation proposed by Romana (1991), adopting the SMR (slope mass rating) method. The latter utilizes numerical coefficients based on the geometric relations between the discontinuities and the slope and excavation methods. By means of these methods, kinematic and static compatibility is analysed in relation to the planar and wedge slide mechanisms. The input parameters required for determination of the RMR and SMR indices according to the indicated procedures were obtained from the survey of the structural conditions.

Steepness was used for the inclination of the rock escarpments, whereas a value of 30° was adopted for the angle of shear strength of the generic discontinuity, taking into account the nature of the rock, joint roughness, infill and alteration. This latter value is a precautionary value, which can be ascribed to residual or basic conditions, but it resolves the problems involved in the adoption of peak values directly connected to a given value for normal stress, and also for cohesion intercept (Barton 1976).

The values resulting for the indices thus defined show a reduction of the RMR ranging from 18 to 22 (classes IV and V, poor and very poor rock), whereas the SMR varies between 30 and 35 (class IVa, bad–unstable), thus confirming the poor quality of the rock mass in terms of slope stability.

Final remarks

The rocky coast between Rapallo and Chiavari (Tigullio Gulf) is characterized by instability phenomena affecting the Via Aurelia, the railway line and buildings, including the Santuario di N. S. delle Grazie.

The geomorphological features of the coastal area located west of the city of Chiavari affected by historical landslides have been described. The factors causing the landslides are in part natural (geological–tectonic setting, geomorphological features, hydrology, geomechanical properties of the rock mass, wave action and rainfall) and in part anthropogenic (quarrying activity, construction of railway and road networks). The geomorphological dynamics observed in recent years, as in the past, indicate an intermittent type of activity characterizing these landslides.

The study is a contribution to encourage efforts to monitor the evolution of this area by public and private agencies. The overall objective would include the planning of general stabilization measures, or at least measures to reduce the evident geomorphological risk. The latter is defined as the combination between the hazard assessed and discussed above, and the level of vulnerability caused by tourist influx along the coast owing to the presence of the beaches and the navigation of pleasure craft. Moreover, along the subaerial coastal slope, in addition to the settlements and infrastructure mentioned above, there are numerous hiking trails, mainly leading to the sanctuary.

For the purpose of monitoring instability in this area, further direct and indirect types of investigation are considered to be indispensable. They should aim to identify the characteristics of the landslides, the geotechnical properties of the soils and bedrock, and the depth and dynamics of the water table. Moreover, drilling work would be of value, to install monitoring equipment such as inclinometers to survey the deformation and piezometers to measure pore pressure.

Such work is also justified by the significant environmental value of this coastal sector, which has been recently proposed for inclusion in the plan of the Natural Park of Portofino.

This work was carried out within the framework of the Ministry of Education, University and Research project 'Geomorphological heritage as a resource for a sustainable tourism' (COFIN 2004), National Coordinator: M. Panizza, and the local project of the Genoa University Unit on 'Liguria's geomorphological heritage as a resource for a sustainable tourism', Coordinator M. Piccazzo. The authors are grateful to D. Castaldini and V. Agnesi for useful suggestions and improvements to the text and figures.

References

BARTON, N. 1976. The shear strength of rock and rock joints. *International Journal of Rock Mechanics, Mining Science and Geomechanics, Abstracts*, **13**, 1–24.

BIENIAWSKI, Z. T. 1989. *Engineering Rock Mass Classifications*. Wiley, Chichester.

BRANDOLINI, P. & TERRANOVA, R. 1994. Esempi di dissesti geomorfologici dei versanti liguri e loro riflessi sulla conservazione del suolo. *Memorie Accademia Lunigianense di Scienze 'G. Capellini'*, **LXIV–LXV**, 55–77.

CORTEMIGLIA, G. C. & FIERRO, G. 1965. Variazioni del fondo marino e della spiaggia di Chiavari. *Annali Ricerche e Studi di Geografia*, **XXI**, 1–12.

CORTEMIGLIA, G. C. & TERRANOVA, R. 1970. Le frane della Collina delle Grazie nel comune di Chiavari e loro rapporti con la viabilità e gli insediamenti. *Bollettino della Società Geologica Italiana*, **89**, 277–298.

FACCINI, F., BRANDOLINI, P., ROBBIANO, A., PERASSO, L. & SOLA, A. 2005. Fenomeni di dissesto e precipitazioni in rapporto alla pianificazione territoriale: l'evento alluvionale del novembre 2002 nella bassa Val Lavagna (Liguria orientale). *Geografia Fisica e Dinamica Quaternaria*, **VII**, 145–153.

ISSEL, A. 1892. *Liguria geologica e preistorica*. Donath, Genova.

MARINI, M. 1981. Analisi geologica-strutturale ed interpretazione paleogeografia e tettogenetica dei Calcari del Monte Antola (Appennino Ligure). *Ofioliti*, **6**, 119–150.

MARKLAND, J. T. 1972. *A useful technique for estimating the stability of rock slopes when the rigid wedge slide type of failure is expected*. Imperial College, Rock Mechanics Research Reprints, **19**.

POZZI, R. & CLERICI, A. 1985. L'applicazione della classificazione geomeccanica di Z.T. Bieniawski alla stabilità di scarpate in roccia. *Costruzioni*, **XXXIV**, 32–35.

ROMANA, M. 1991. SMR classification. *In*: WITTKE, W. (ed.) *Proceedings of the 7th International Congress on Rock Mechanics, ISRM, Aachen*. Balkema, Rotterdam, 955–960.

ROVERETO, G. 1922. Note al rilevamento geologico dei fogli Rapallo e Chiavari. *Bollettino della Società Geologica Italiana*, **41**, 139–160.

ROVERETO, G. 1939. *Liguria Geologica*. Memorie della Società Geologica Italiana, **2**.

SANGUINETI, G. 1949. *Chiavari vista geologicamente*. Atti della Società Economica di Chiavari.

Alluvial and coastal hazards caused by long-range effects of Plinian eruptions: the case of the Lattari Mts. after the AD 79 eruption of Vesuvius

ALDO CINQUE[1]* & GAETANO ROBUSTELLI[2]

[1]*Università degli Studi Federico II, Dipartimento di Scienze della Terra, Largo S. Marcellino 10, 80138 Naples, Italy*

[2]*Università della Calabria, Dipartimento di Scienze della Terra, Via Pietro Bucci, 87036 Arcavacata di Rende, Cosenza, Italy*

**Corresponding author (e-mail: cinque@unina.it)*

Abstract: The Lattari Mountains (a limestone ridge about 20 km south of Vesuvius) received 1–2.5 m of fallout from the famous Plinian eruption of AD 79. As demonstrated by many residual outcrops of thick volcanoclastic debris-flow and alluvial deposits (referred to here as the Durece unit), the pyroclastic fall was soon followed by rapid erosion and landsliding that produced (1) decametre-scale aggradation of some narrow valley floors; (2) reactivation of alluvial fans; and (3) growth of new fan-deltas (extending as far as 500 m) at the coast. This response was primarily due to the steep topography of the area and the high erodability of the pyroclastic materials (light and cohesionless pumice fragments). Several geo-archaeological data indicate that the accelerated sedimentation had a duration of the order of decades and was followed by rapid dissection of the Durece unit deposits and fast dismantling by wave action of the newly created fan-deltas. This case highlights the need to consider the possibly catastrophic reaction of fluvial and coastal systems to large explosive eruptions, even in non-volcanic terrains at some distance from the volcano.

It is well known that coastal areas lying at the base of active volcanoes are exposed to a wide range of geological hazards that includes not only those related to coastal dynamics, but also those deriving from the activity and the geomorphological evolution of the adjacent volcano. These additional hazards are sometimes the direct and immediate consequence of volcanic activity (bradyseisms, earthquakes, lava flows, tephra fall, primary pyroclastic flows, etc.) or may be due to secondary events representing the reaction of the system to the geomorphological perturbation introduced by an eruption (Blong 2000). The latter hazards include outbreak floods from lakes as a result of volcanic damming and cold lahars that may accompany and/or follow an explosive eruption. Moreover, there are a many examples of large-scale eruptions resulting in large and intense sedimentary response over wide areas and in distant places (Walton 1986; Smith 1987, 1991*a*, *b*; Scott 1988; Buesch 1991; Smith & Lowe 1991; Mack *et al.* 1996; Major *et al.* 1996; Nakayama & Yoshikawa 1997; Riggs *et al.* 1997; Montgomery *et al.* 1999; Manville 2001; Manville *et al.* 2005).

In this paper we give an example of how major explosive eruptions may produce geomorphological hazard even in coastal areas at a significant distance from the volcano. With this aim we consider the case of the AD 79 eruption of Vesuvius, looking at its effects on the morphodynamics of the steep catchments that dissect the limestone ridges of the Lattari Mts. and descend to the coast of the Sorrento peninsula, about 20 km south of Vesuvius.

Geological and geomorphological setting

As shown in Figure 1, Vesuvius (1256 m high) rises abruptly from the almost flat Campania Plain bordering the Bay of Naples. Both the plain and the bay are the geomorphological expression of a deep Quaternary graben disturbing the western flank of the Southern Apennine chain, which is also broken by the graben hosting the Bay of Salerno and other Quaternary extensional features (Brancaccio *et al.* 1991).

The ridge of the Lattari Mts. is a WSW–ENE elongated horst that separates the two above-mentioned grabens. It is mainly composed of allochthonous Mesozoic carbonate rocks, which are well exposed wherever slope angles exceed 30–35°. On gentler slopes the carbonate rocks are covered by a carpet of Late Quaternary loose pyroclastic deposits of variable thickness (generally less than 2 m, but exceptionally up to 10 m or more) in which eruption units alternate with palaeosols and buried erosion surfaces. This cover was fed by

From: VIOLANTE, C. (ed.) *Geohazard in Rocky Coastal Areas*. The Geological Society, London, Special Publications, **322**, 155–171. DOI: 10.1144/SP322.7

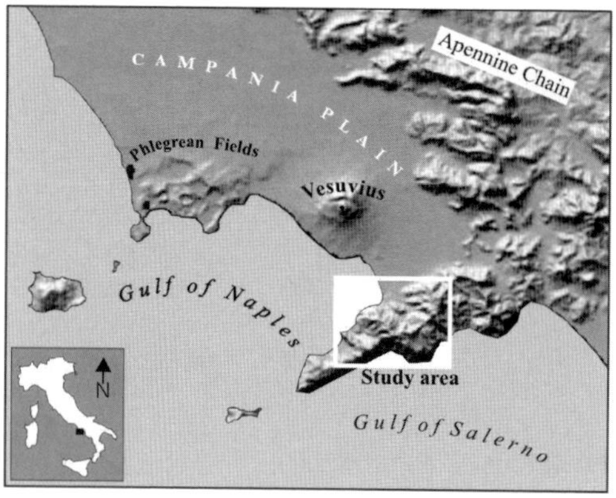

Fig. 1. Location of the study area in central Campania.

several explosive eruptions of the volcanic centres of the nearby Campania Plain (i.e. Vesuvius, Phlegrean Fields and other, older volcanoes that have been lowered by subsidence and buried under the graben sedimentary fill). The youngest phase of this carpeting complex belongs to the AD 79 eruption of Vesuvius, and often appears thinned by erosion but never much weathered. It rests either directly on the Mesozoic rocks or on older eruption units whose ashes and pumice are weathered to orange–brown loamy soils. The age of these weathered pyroclastic deposits ranges from Middle to Late Pleistocene. Between 18 ka ago and AD 79 the Lattari Mts. did not receive any significant fallout deposit because during that period Plinian eruptions of Vesuvius were dispersed in other directions (NE to east; Santacroce 1987).

Being an area of recent uplift and block-faulting, the Lattari Mts. are characterized by steep fault scarps (up to 1 km high) and steep V-shaped valleys. Suspended above these Quaternary features are remnants of much gentler erosional landscapes (Pliocene to Early Pleistocene in age), which have frequent subhorizontal elements alternating with hillslopes with an inclination of no more than 26–28°. Both the fault scarps and the valley side slopes that formed during the Quaternary are, in contrast, characterized by high mean gradients (Fig. 2) and by transverse profiles that typically alternate subvertical cliffs and elements inclined at 30–35°.

The AD 79 pyroclastic blanket on the Lattari Mts.

The AD 79 eruption of Vesuvius was responsible for the burial (and preservation) of the Roman towns of Herculaneum, Pompei and Stabiae, and it was carefully described by Pliny the Younger (see his letters in Tacitus, *Historiae*). The fallout of the AD 79 eruption was dispersed towards the SSE by the stratospheric winds acting at the time. Consequently, notwithstanding the 20 km distance from the crater, the Lattari ridge was mantled with a remarkable thickness of loose pyroclastic deposits. As the isopachs in Figure 3 show, the central part of the ridge (our study area) was the most affected, receiving between about 1 and 2.5 m of pumiceous fallout, plus some ash emitted during the final phreatomagmatic phase of the eruption. The thickness of the phreatomagmatic ash has never been mapped because of the scarcity of uneroded control sections.

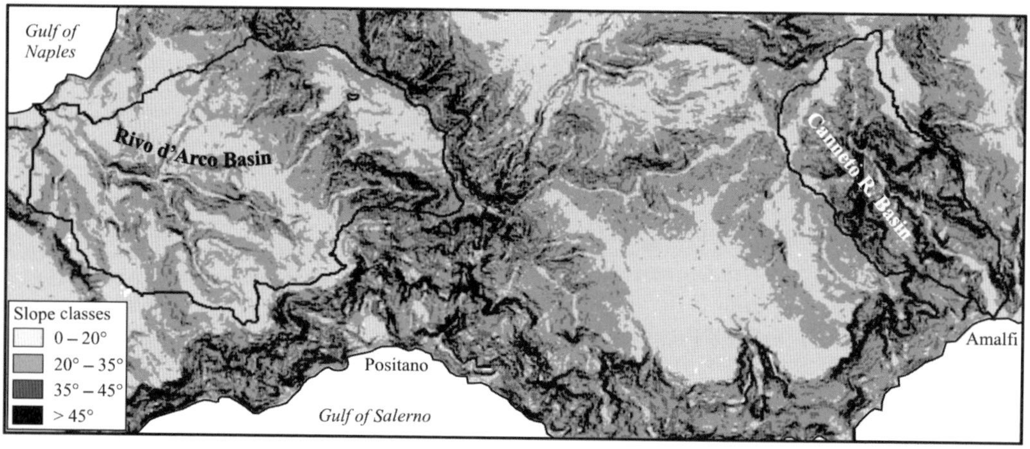

Fig. 2. The study area classified by slope angle. It should be noted that, to avoid a 'salt and pepper' effect, some smoothing was applied to the DEM used before classifying the areas by slope angle. A side effect is that many small cliffs have disappeared. The real extent of the steepest slope classes is given in Table 2.

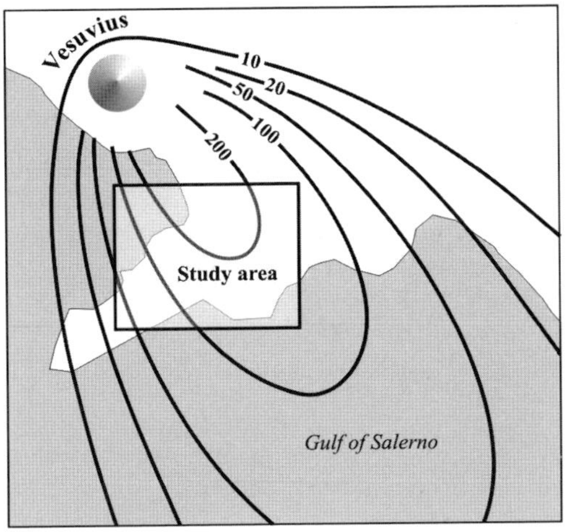

Fig. 3. Isopach map of the AD 79 pumice deposits (redrawn after Sigurdsson *et al.* 1985).

The AD 79 cover currently appears almost intact (with the initial thickness substantially preserved) only on the rare, subhorizontal land units of the study area. In such cases it is composed of a thick basal portion composed of very light, tephritic phonolite pumice fragments (predominantly 1–2.5 cm across, but with some up to 5 cm large) with a much subordinate component of scoriaceous and lithic fragments (normally 1 mm to a few millimetres in diameter, but sometimes up to 3 cm). Superimposed on the pumiceous bed is a fine-textured brownish soil (up to 50 cm thick), which mostly derives from the weathering of the ashy final part of the eruption's fallout (Fig. 4).

This cover has been almost totally removed from the steepest hillslopes of the Lattari Mts., which also show evidence that the volcanoclastic deposits supplied by previous eruptions had already been removed by AD 79. Because of differences in both microclimate (thermal and moisture soil regimes) and vegetation cover, the south-facing slopes appear deprived of the AD 79 cover only where the mean angle reaches at least 30–35°, whereas total removal from north-facing slopes occurs only above 40–45°.

On less inclined hillslopes the AD 79 deposits are variably preserved, depending on the slope angle and length, and on the history of land use during the last two millennia. From both field evidence and scientific reports we know that removal of the AD 79 pyroclastic deposits has continued in modern times by means of rill and gully erosion (noted only on badly managed areas) and by means of sporadic landslides (earthfalls or slumps) that are triggered by periods of exceptional rainfall and typically evolve into wet and rapid debris flows (Guadagno *et al.* 1999; Cinque *et al.* 2000; Esposito *et al.* 2004; Violante *et al.* 2009).

These present-day phenomena give a good picture of how a pyroclastic blanket continues to be removed from the Apennine mountains many centuries after its formation. However, to know what kind of downwasting occurred (and could occur again) soon after an eruption, we have to look at the stratigraphic record; namely, at the sequences representing the earliest phases of reworking of the AD 79 pyroclastic deposits.

The Durece deposits

The word Durece (probably from the Latin verb durescere: to become hard) is the term traditionally used by the local peasants to indicate a typical lithofacies of reworked AD 79 pyroclastic deposits. It refers to a relatively hard (breakable with a pickaxe), cemented mixture of grey ash and pumice. In this paper the terms Durece deposits and Durece unit (informal lithostratigraphic unit) are used in the broader sense of 'the entire succession of deposits (not only its hardened and massive facies) representing the earliest downwasting phases of the AD 79 pyroclastic cover'. Similarly, by Durece event is meant the whole succession of sedimentary phenomena that resulted in the deposition of the Durece unit.

The map in Figure 5 shows the valley reaches where remnants of the Durece unit have been found. As the Durece event was followed by many centuries of dissection, it is probable that other V-shaped valleys had hosted such deposits before being cleaned out. On the same map we also mark, among the quiescent alluvial fans of the northern piedmont, those that probably were reactivated during the event (Durece deposits seen in outcrop and/or boreholes) and others whose reactivation can be suspected because of the presence of residual Durece outcrops slightly upstream (i.e. in the lower reach of the feeding valleys). Dealing with streams flowing directly into the sea, in Figure 5 we have indicated those mouths where the Durece event caused the growth of fan-deltas.

The Durece deposits reach their greatest thickness (from several metres to some tens of metres) in the outcrops located along intramontane valleys. The greatest thicknesses have been measured in V-shaped valley reaches that are narrow, with subvertical flanks and of moderate longitudinal gradient. For example, they are up to 18 m thick in Positano (Site 7 in Fig. 5), about 20 m thick along the Rivo d'Arco (Site 4 in Fig. 5) and up to 40 m thick in the Canneto valley (Site 9 in Fig. 5; see also Fig. 6).

Fig. 4. An example of *in situ* AD 79 fallout deposits well preserved on a flat palaeosurface of the Lattari Mts. (Piani di S. Erasmo; about 990 m a.s.l.). The topmost ash layer appears as the modern dark brown soil into which the tree roots penetrate. The dark material at the base of the white pumices is a reddish palaeosol developed on late Pleistocene pyroclastic deposits.

On the alluvial fans of the northern piedmont, the Durece unit is generally represented by broad blankets only a few metres thick. At Site 1 (Fig. 5), for example, it is represented by about 2 m of debris-flow deposits covering a building of Roman age (the so-called 'Villa Cuomo'). Similar deposits have have also been found in Castellammare di Stabia during some drillings on the end fan of the Rio di Gragnano stream (Fig. 5).

Some of the valleys that descend directly to the sea have a flat-floored final part because of the presence of alluvial fill formed mostly during the post-glacial sea-level rise and the following high-stand. Also on such confined, alluvial coastal

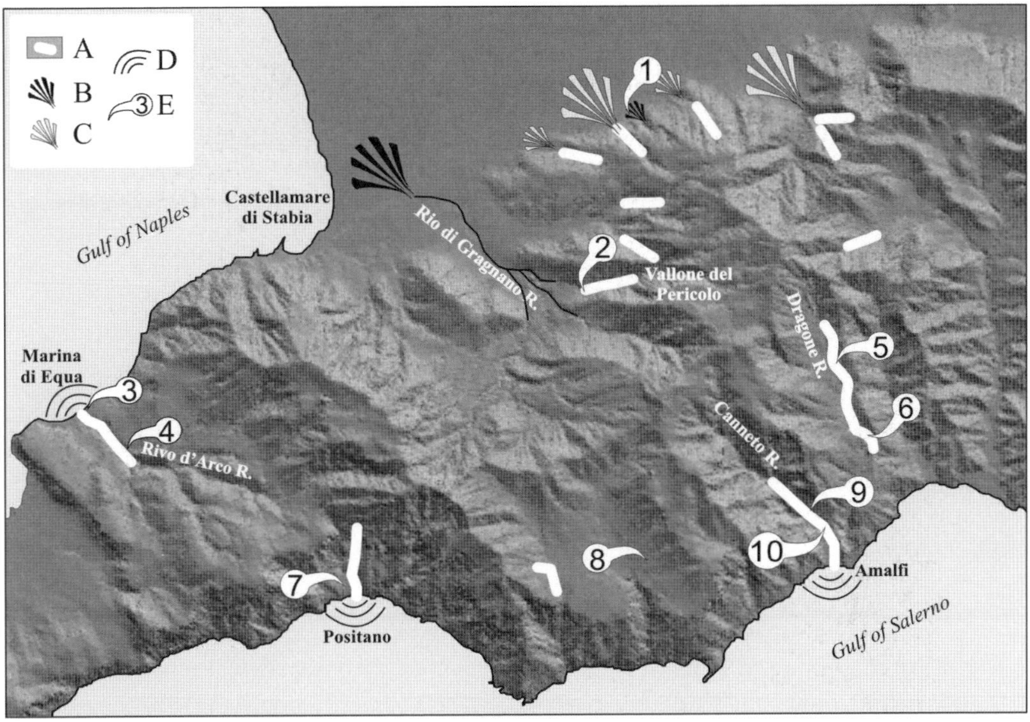

Fig. 5. The distribution of the Durece deposits in the study area. A, Valley reaches containing residual outcrops of Durece sediments; B, alluvial fans with evidence of reactivation during the Durece event; C, alluvial fans that were probably reactivated during the Durece event; D, river mouths where fan-delta progadation occurred; E, outcrops mentioned in the text.

plains, as for the previously noted case of the open alluvial fans of the northern piedmont area, the Durece unit has relatively modest thickness; for example, 5–6 m at Site 3 and about 8 m at Site 10 (Fig. 5), where the flat valley floor is too narrow to be evident in the small-scale map of Figure 5.

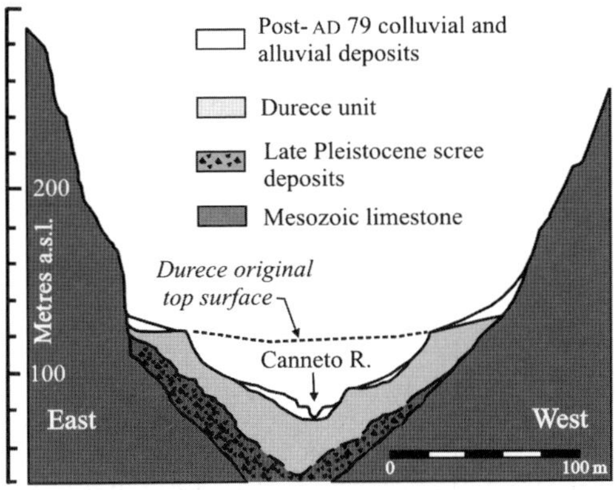

Fig. 6. Cross-section of the lower reach of the Canneto River gorge upstream of Amalfi.

Seismic reflection evidence of the post-AD 79 phase of great fluvial discharge has recently been found off the southern sector of the study area, the Amalfi coast, in the form of coarse-grained deltas at the mouth of the main streams (Sacchi *et al.* 2009).

Facies analysis

Among the investigated sites, the most complete facies sequences can be observed near Amalfi and near Vico Equense, where good exposures occur along the Canneto River and the sea cliff of Marina di Equa, respectively. Based on what these sections show, and on observations made at Sites 2, 4, 5 and 6 (Fig. 5), five facies associations (FAs; Table 1 and Fig. 7) were identified and genetically interpreted.

Facies association 1. The characteristics of the basal portion of the sequence, which locally rests erosively on the AD 79 pumice (Sites 2, 4 and 9 in Fig. 5), are suggestive of deposition from subaerial debris flows (Pierson *et al.* 1996; Segschneider *et al.* 2002). These sediments consist of massive,

Table 1. *Lithofacies of the Durece succession*

Facies	Lithofacies	Sedimentary structures	Interpretation
Gms_p	Matrix-supported massive gravel, with pumice matrix and carbonate angular clasts	None, or weak inverse grading	Carbonate–pumice-rich debris flow
Gms_v	Clast-supported massive gravel with volcanoclastic matrix and subordinate carbonate subangular clasts	Ungraded, erosive-based	Volcanoclastic debris flow
Gms_c	Clast-supported massive gravel with volcanoclastic matrix and carbonate subangular clasts	Ungraded, erosive-based; possible coarse-tail normal grading	Carbonate-rich debris flow
Dd	Massive to crudely bedded, poorly sorted pumiceous gravel; bed thickness is between 40 and 100 cm; minor components are crystals and lithic fragments	Ungraded, slightly erosive- or nonerosive-based; possible coarse-tail inverse grading at the base	Pumiceous debris flow
Ds	Massive, stratified, poorly sorted pumiceous gravel; bed thickness is between 30 and 60 cm; minor components are crystals and lithic fragments	Ungraded; lower contact generally erosive	Pumiceous hyperconcentrated flood flow
Gm	Poorly to moderately sorted stratified gravel with subangular to poorly rounded carbonate and volcanoclastic fragments	Laterally parallel stratification, inverse grading	Coarse-grained hyperconcentrated flood flows
Sh	Moderately sorted coarse sand and pebbly sandstones with poorly rounded clasts of volcanoclastic origin	Parallel stratification; gradational contacts; possible inverse or inverse to normal grading	Fine-grained hyperconcentrated flood flows
Ss	Fine to coarse sand of volcanoclastic origin, with scarce carbonate fragments	Shallow scours	Rapid scour fills by high-concentration flood flows
PGm	Moderately to well-sorted stratified gravel with subrounded to well-rounded pumiceous clasts	Parallel strata, laterally discontinuous; shallow scours	Waning stage flood and scour fills produced during overland flow
Fl	Greenish grey to grey pelitic beds	Diffuse horizontal and/or undulating lamination	Waning stage flood

poorly sorted, ungraded pebble-size pumice conglomerates, forming multi-storey sequences up to many metres thick (Fig. 7a; Dd facies).

Massive and ungraded pebble to cobble conglomerates made of limestone clasts supported by a pumice-rich matrix also occur as an interbedded facies (Gms_p). Their top surface sometimes shows scours up to 0.8 m deep and up to 3.5 m wide, which are filled by clast-supported, moderately to well-sorted, centimetres-thick sheet conglomerates consisting of rounded pumice fragments, of pebble size (PGm facies).

Facies association 2. Many debris-flow units of FA 1 grade upward and downstream into another association, which differs from the first in having well-defined and widely spaced low-angle cross-stratifications. In particular, this facies association, which locally alternates with FA 1, is characterized by non-erosive bases and massive, ungraded pumice pebble conglomerates (Dd facies), which alternate with beds of pumice pebbles and granules, showing scoured lower contacts (Ds facies; Fig. 7a and b). We interpret each Dd–Ds couplet as the result of the deposition of a debris flow and its tail of hyperconcentrated flow, which was more diluted and therefore able to extend further than the frontal debris flow when this eventually stopped (Sohn *et al.* 1999).

Facies association 3. FA 1 and FA 2 pass upward and downstream into a third facies association made of granule to fine pebble conglomerates and stratified sandstones. It consists of volcanoclastic debris whose sandy fraction is made of mafic crystals, whereas the pebbly fraction consists of lava, tuff and subordinate limestone and pumice fragments.

Fig. 7. Details of facies associations of the Durece succession. (**a**) Alternation of facies associations 1 and 2; they consist of stacked sets of Dd facies and Dd–Ds facies, respectively, with subordinate Gms and PGm facies (Site 9 in Fig. 5). (**b**) Laterally extensive, massive, poorly sorted pebble-size pumice conglomerates of facies association 1. In the upper part are alternating Dd and Ds facies, characterizing the facies association 2 (between Sites 9 and 10). (**c**) Coarsening upward sequence whose lower part is made up of facies association 3 (Sh, Ss and PGm facies alternating). It passes upward into facies association 5, characterized by Gm–Sh couplets (Site 3). (**d**) Facies association 5: Gm–Sh couplets, which alternate because of current evolution of a composite sediment gravity flow (Site 3). (**e**) Facies association 5 showing alternation of Sh, PGm and Fl facies as a result of less powerful sandy hyperconcentrated flood flows (Sh facies) followed by waning stage deposits (Site 3).

At Marina di Equa (Site 3 in Fig. 5) FA 3 constitutes the axial portion of the Rivo d'Arco coastal alluvial fan. Facies analysis allows two subfacies association (FA 3a and FA 3b) to be distinguished; they consist of alternating fine and coarse beds that locally form sequences up to 0.5 m thick representative of high- and low-energy flows, respectively.

FA 3a consists of tabular sheets up to 20 cm thick and laterally persistent for metres. Usually, the basal bed consists of a tabular sheet of coarse sand or sands with fine pebbles (Sh facies in Fig. 7c–e); sheet sandstones show gradational interstratal contacts and commonly occurring isolated larger clasts (>3 cm) that have a preferred orientation with the *a*-axis normal to the palaeoflow direction. Sometimes the beds show inverse or inverse-to-normal grading with a sharp bottom, locally scoured. These characteristics, together with the lack of trough cross-bedding, suggest rapid deposition by sandy hyperconcentrated flood flows (Smith 1987, 1988) or shallow fluctuating sheet flows (Maizels 1993). Locally, this facies contains lenses of pebble- to cobble-size conglomerates, which are massive, poorly sorted, ungraded and characterized by erosive bottom contacts (Gms_v facies; Fig. 7c). The Sh facies passes locally upward into scour-and-fill sandy beds (Ss facies; Fig. 7c) with low-angle, onlapping laminae showing a lateral (transverse to palaeoflow) decrease in inclination; these beds consist of coarse- to fine-grained sandstone. This facies is interpreted as resulting from deposition by high-energy, shallow sheet flows (Smith 1988).

Low-energy, tractive flows (FA 3b) were recorded by moderately to well-sorted, low-angle and/or horizontally stratified sandstone pumice conglomerates (Fig. 7c), organized in tabular sheets and/or shallow scours consisting of alternating fine and coarse beds (PGm facies). Subordinate millimetre-thick greenish grey to grey pelitic beds also occur (Fl facies).

Facies association 4. This facies association (Fig. 7d) is particularly well exposed near the NE end of Marina di Equa sea cliff (Site 3 in Fig. 5), where it is up to 5 m thick and contains several tens of sedimentation units, each connected with a single flood event. It is formed of alternating sandy to fine pebble sheets (Sh facies), pumice-rich sheet beds (PGm facies) and thin greenish grey to grey pelitic beds (Fl facies). The PGm facies is made of moderately to well-sorted, alternating coarse- and fine-grained laminae with a clast-supported texture.

This facies association is related to lateral sectors of the coastal alluvial fan that was growing at Marina di Equa during the Durece event. It was fed by many flood flows of moderate energy, whereas the coarser FA3 formed in the fan axial zone.

Facies association 5. At Site 3 (Fig. 5), in the portion of sea cliff that is closer to the Rivo d'Arco mouth, FA 3 abruptly passes upward into FA 5, which is mostly characterized by well-stratified fine and coarse conglomerates organized in tabular sheets up to 30 cm thick (Fig. 7e).

The conglomerates consist of subangular pebble- to cobble-size limestone clasts. They are poorly sorted and have a clast-supported texture. Their matrix consists of volcanoclastic fragments (coarse sand and granules). Inverse grading is well developed (Gm facies), but locally grading is absent (Gms_c facies). Many elongated clasts have the *a*-axes oriented normal to palaeoflow directions. Beds of this facies have sharp, slightly erosive lower contacts and occur in laterally extensive but discontinuous units up to 20 cm thick. Sometimes the bottom contact of the massive facies appears erosive, being characterized by scours up to 30 cm deep.

This lithofacies alternates with tabular sheets characterized by coarse sand to fine pebbles (Sh facies), in which isolated larger clasts may occur. These beds, which are massive or inversely graded, pass locally upward to PGm facies.

The massive and inversely graded lithofacies suggest that depositional events were of very high energy. They can be ascribed to gravelly hyperconcentrated flood flows (Gm facies) alternating with rare clast-rich debris flows (Gms_c facies; Shultz 1984). The observed alternations of coarse (Gms_v and Gm) and fine facies (Sh and PGm) are tentatively related to deposition by composite sediment gravity flows (Sohn *et al.* 1999).

Facies distribution and depositional processes

The nature and distribution of the studied volcaniclastic sediments suggest that, soon after the AD 79 eruption of Vesuvius, the pyroclastic fall deposits arriving at the Lattari Mts. entered a period of fast remobilization that was accomplished mostly (more than 90%) by debris and hyperconcentrated flows. As proven, for example, by outcrop and well-log data from some alluvial fans of the northern piedmont, these phenomena extended to the distal fluvial system. However, the strongest aggradation affected the lower reaches of the intramontane V-shaped valleys, where it was favoured by relatively low longitudinal gradients and laterally limited accommodation area. A good example is given by the outcrops in the Canneto valley (between Sites 9 and 10 in Fig. 5), which show vertically stacked pumice-rich deposits (FA 1 and FA 2) up to 40 m thick. The sequence is indicative of a series of sedimentary events, each producing aggradation of some decimetres to a few metres of the narrow valley floor with a frontal debris-flow body and a more diluted tail. Periods of dissection and reworking intervened between the sedimentary events, but they were very short, as both the low relief of scours and the modest thickness of the PGm lenses indicate.

The presence of some carbonate clasts in the facies associations 1 and 2 is an indication that the gravity flows were also able to entrain coarse detrital material occurring locally on the hillslopes and along the stream beds (karstogenic fragments sometimes developed under a cover of Late Pleistocene pyroclastic soil and local accumulations of fluvial pebbles and boulders). As witnessed by the nature of some residual scree deposits, in the study area the process of slope replacement has been acting mainly during the cold stages of the Quaternary, when the limestone outcrops were subject to gelifraction. In the present climatic conditions (as during previous interglacials) the dominant processes are karstic solution and fluvio-karstic dissection, plus sporadic rockfalls from the steepest rock cliffs.

Other Durece deposits of both the Canneto and the Rivo d'Arco valleys are suggestive of flood facies (hyperconcentrated flow of FA 3 and FA 5) that grade vertically and laterally into waning-stage water-flow deposits (FA 4). Deposits of this second group overlap those of the FA 1–FA 2 group in all the Durece outcrops that are visible along the lower valley reaches and on the alluvial fans. On the other hand, the Durece outcrops that survive along mid-valley reaches are entirely composed of gravity-flow deposits. The sole exception to this rule was found in the upper Dragone valley (Site 5 in Fig. 5), where the Durece event is recorded only by hyperconcentrated-flow deposits that are very poor in pumiceous clasts. This feature is due to the circumstance that this is a mature, hanging valley reach whose side slopes are gentler than in other incisions of the study area. The pumiceous clasts bypassed this sector as suspended load and accumulated further downstream (near the town of Ravello) as a succession dominated by PGm facies.

The above-described relationships lead us to the conclusion that the first part of the Durece event was largely dominated by debris flows that ended between the mid-portion of valleys and the mid-part of fans (or in the shore area, where the mountains plunge into the sea). Subsequently, deposition by hyperconcentrated flood flows and stream flows became the dominant mechanism of aggradation in the lower valley reaches and the alluvial fans. As regards the distal fan areas, our opinion is that deposition by stream and flood flows occurred also in the first part of Durece event, when these areas were reached by the diluted tails of debris flows that extended beyond the denser parts of these flows.

Chronological constraints

As the Durece deposits consist completely of pyroclastic deposits emitted during the AD 79 eruption there is no doubt that they post-date that volcanic event. In favour of a deposition that started soon after the eruption there are two lines of evidence: (1) the Durece basal layers have often been observed in direct contact over *in situ* fallout deposits of AD 79 (e.g. Sites 1, 7 and 10 in Fig. 5); (2) Roman ruins of the first century are buried by the Durece deposits. Such ruins include four cases: a 'ninfeo' recently discovered in Amalfi (Site 10 in Fig. 5, and Fig. 8), the oldest part of the so-called 'Villa del Pezzolo' at Marina di Equa (Site 3 in Fig. 5), the so-called 'Villa Cuomo' at S. Antonio Abate

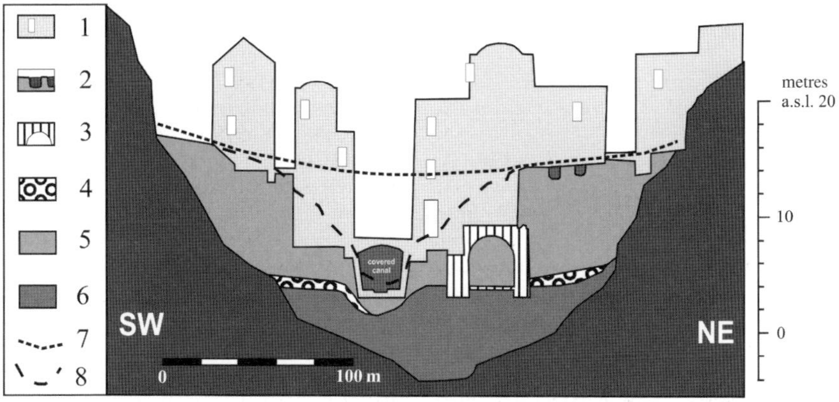

Fig. 8. Geological cross-section of the Canneto valley at Amalfi. 1, Medieval to modern buildings; 2, graves of the third to fourth century AD; 3, buried villa of the first century AD; 4, *in situ* AD 79 fallout deposits; 5, Durece formation; 6, Late Pleistocene and Holocene alluvial deposits; 7, the Durece formation's original top surface; 8, valley profile before urbanization.

(Site 1 in Fig. 5) and the villa of Positano (Site 7 in Fig. 5). In the last case, the columns of the peristyle appear to have been pulled down by the weight of huge debris-flow deposits (Maiuri 1955) whose petrological and sedimentary features perfectly match those of the Durece facies Dd and Ds.

Dealing with the problem of dating the end of the Durece event, a vague old age is suggested by the fact that all the valley fills it generated are now deeply re-dissected (normally down to the bedrock) and sometimes largely removed. However, better age constraints are offered once again by the local archaeology. A first example is from Amalfi, where the Durece top surface (Fig. 8) revealed graves of the third century during foundation work carried out in the 1960s. A second case is that of a test pit we excavated for stratigraphical purposes in Agerola (near Case Coccia; Site 8 in Fig. 5). It revealed pottery fragments of the third to fourth century in a buried soil covering a Durece debris-flow deposit (Dd facies) under which *in situ* AD 79 pumice was found. A third age-constraining case is that of Marina di Equa (Site 3 in Fig. 5), where the already mentioned 'Villa del Pezzolo' was rebuilt during the second Century over the Durece deposits (see also the section on 'The geomorphological consequences on coastal areas', and Cinque *et al.* 2000).

The data from Amalfi and Agerola indicate only that the Durece event had already terminated sometime before the third to fourth century. Therefore, we assign more importance to the constraints coming from Marina di Equa, where the rebuilding of the Roman villa proves that the post-AD 79 aggradation phase lasted no more than 120 years. For comparison, we recall that measurements made in volcanic areas throughout the world after recent eruptions (e.g. Major *et al.* 2000; Hayes *et al.* 2002, and references therein) suggest that, although declining, sediment yields remain elevated for at least a few decades.

Tentative reconstruction of the post-eruption morphodynamic scenario

The thickness and sedimentological features of the studied Durece deposits clearly demonstrate that the central part of the Lattari Mts. was affected by a period of dramatic geomorphological instability after it had been covered by the fallout products of the AD 79 Vesuvius eruption. The consequent deposition in valleys of reworked pyroclastic deposits was an exceptional event that interrupted a long-lasting period of downcutting (which had probably been uninterrupted since the end of the Late Glacial) and was followed by a period of under-charged water flow and erosion (which still continues today). In fact, the Durece valley fills are normally found in direct contact with the rocky valley bottoms and always appear deeply re-dissected. The exceptional case of Durece deposits lying on older Holocene alluvial beds is restricted to some alluvial fans of the northern piedmont area and to the small, nested alluvial plain that characterizes the final reach of those valleys that directly descend to the sea.

To infer the nature of the phenomena that fed the Durece depositional events it is useful to observe that great volumes of Durece deposits are found along those valleys that have large proportions of hillslope elements steeper than 35° (e.g. close to 50% in the Positano valley and about 45% in the Canneto basin). The fact that the Rivo d'Arco basin (with only 17% of its area steeper than 35°; Table 2) also emitted a great volume of Durece deposits can be explained by the high basin relief (1440 m), the very steep longitudinal profiles of the right-hand tributary valleys and the presence of steep valley-side slopes along the main water course.

The fact that the steepest slopes were the main source areas for the Durece deposits appears clear by comparing the Rivo d'Arco and the Canneto basins: although the former had a greater volume of AD 79 pyroclastic deposits than the latter (Va_{tot} in Table 2), it recorded less volcanoclastic sedimentation (Vd). The explanation can be found in the fact that the volume of pyroclastic deposits lying on hillslope elements steeper than 35° (partial values of Va in Table 2) was greater in the Canneto basin.

On this basis and by also taking into account common knowledge about the way a fallout cover forms and degrades, we tentatively reconstruct the events in each catchment of the study area starting with the eruption, as follows.

(1) During the fall of pumice fragments, a carpet 1–2.5 m thick (Fig. 3) formed on all landforms with a slope less than about 40° (the angle of rest of the AD 79 pumice). Fragments falling on steeper slopes immediately rolled down and re-accumulated downslope as much thicker taluses and cones. Because of the general lack of valley-side terraces, most of the re-accumulation bodies grew at the bottom of valleys (especially in the steeper head parts of catchments), producing valley bottom aggradation and ephemeral dams.

(2) When the rain of fine ash occurred (the final phreatomagmatic pulse of the AD 79 eruption), a silty blanket a few decimetres thick formed on slopes up to 55–60°, including the above-mentioned taluses and cones of reworked pumices.

(3) With the obvious absence of grass cover, intense rains of the first years after the eruption (including those that probably occurred during the eruption itself, triggered by the huge Plinian

Table 2. *Basic morphometric parameters of Rivo d'Arco and Canneto basins with estimates of volumes of* AD *79 fallout deposits*

Parameter	Rivo d'Arco basin					Canneto basin				
Basin area (km^2)			19.87					7.95		
Basin relief (m)			1444					1206		
Basin length (km)			8.5					6.2		
Average thickness of received fallout (m)			1.2					1.7		
Slope classes (degrees)	>45	45–40	40–35	35–30	<30	>45	45–40	40–35	35–30	<30
% of basin area	4.0	5.1	6.9	13.4	70.6	18.6	12.7	13.3	18.4	37.0
Extent (km^2)	0.8	1.0	1.4	2.7	14.0	1.5	1.0	1.0	1.5	3.0
Va, volume of fallout (10^6 m^3)	0.96	1.20	1.68	3.24	16.80	2.55	1.70	1.70	2.55	5.10
			Va_{tot} = 24.0					Va_{tot} = 13.6		
Vd_v (10^6 m^3)			1.7 ± 0.5					4.0 ± 0.7		
Vd_f (10^6 m^3)			1.7 ± 0.3					1.0 ± 0.4		
Vd_{tot} (10^6 m^3)			3.4 ± 0.8					5.0 ± 1.1		

V_a, volume of AD 79 fallout deposits originally stored on slopes of various inclination; Vd_v, volume of Durece deposits originally accumulated in valleys; Vd_f, volume of Durece deposits originally accumulated in fan-deltas. The fallout thickness data were calculated from the isopach map of Figure 3. Calculations of Vd_v and Vd_f values were based on control cross-sections combining both stratimetric and geomorphological constraints. The given error margins depend mainly on doubts regarding the pre-Durece valley topography and nearshore bathymetry, as well as the original radius of the deltaic bodies.

cloud) caused severe rilling and gullying in the pyroclastic cover. Gully propagation and deepening were probably accompanied by frequent sliding of gully heads and flanks. As a modern (although isolated) analogue of such rapidly evolving gullies, we cite the one that was cut on a 30° slope near Agerola during the heavy shower of 25 September 2004. Here, in a couple of hours, a deep gully some 80 m long was cut across one of the rare surviving fragments of the AD 79 pumice taluses. The erosion was promoted by pre-existing severe damage to the vegetated soil cover caused by the passage of mules dragging tree logs. The eroded volume was some 500 m^3 and the road passing at the gully base was covered by a debris-flow dominated fan. Gulling was presumably favoured initially by the low permeability of the top ash layer (producing high runoff rates) and then by return-flow waters emerging from the base of the pumiceous blanket. During periods of prolonged rainfall, shallow landslides of another kind (independent of gullying and less localized) probably affected the areas where the slope was near 40° and the precarious stability of the pumice carpet was broken by the increase of shear stress caused by the absorption of rainwater and consequent increase of weight.

Our approximate calculation of the original volume of Durece deposits that accumulated in some valleys and on their fan-deltas (Vd_{tot} in Table 2) suggests that hillslopes above 40° entirely lost the AD 79 cover and less steep hillslopes also were noticeably affected by gullying and possibly some slides. In this regard it is important to note that the rough figures given for Vd_{tot} in Table 2 are underestimates because we did not include the distal subaqueous parts of the fan-deltas.

(4) As years of downwasting went by, on increasing areas of the steep hillslopes the hard carbonate bedrock was re-exposed, and the residual patches of the pyroclastic cover developed inceptisols and became increasingly protected by new vegetation. Rates of denudation progressively decreased (especially because of landslides becoming both less frequent and smaller), and the rate of aggradation of lower valley reaches changed from high to low as debris-flow facies were replaced by flood-flow ones (facies associations 3, 4 and 5). It is possible that during this stage the distal alluvial deposition (lower valley reaches and end fans or fan-deltas) was fed also by some dissection of the debris-flow deposits that had previously formed in the mid-valley reaches.

(5) After many decades of removal, most of the steep portions of each catchment had lost most of their AD 79 fallout cover (apart from small areas trapped in karstic pockets and furrows) whereas other slope portions had their residual pyroclastic

cover well protected by trees rooted to the fractured and karstified bedrocks and by the associated undergrowth. This change caused a major decline in the average soil erosion rate, so as to produce the dominance of undercharged flood flows and to establish the final dissection of both valley fills and fan additions created during the previous Durece event.

The geomorphological consequences on coastal areas

At the mouth of valleys descending directly into the sea, the rapid post-eruption downwasting of the AD 79 pyroclastic deposits caused the growth of fan-deltas that extended the local coastline outward (see Fig. 5; Sacchi *et al.* 2009). The Durece geomorphological crisis had a relatively short duration, and it was followed by the still continuing period of modest fluvial clastic discharge. After sediment yields dropped, the coastal bodies were dismantled by sea waves, while being also longitudinally dissected. The best constrained case of this kind is that of the Marina di Equa coastal plain (Site 3 in Fig. 5). Here the sea cliff that has formed upon the destruction of the ancient fan-delta is 6–8 m high. By extrapolating seaward the slope of the terrace resting above and by taking into account a −1.3 m sea level for the first century (Lambeck *et al.* 2004), we calculate that the shore of the Durece fan-delta was about 500 m from the present coastline at the time of maximum progradation.

At the NE end of the Marina di Equa bay there are Roman ruins that help decipher and date the sequence of geomorphological events that occurred there since the first century (Fig. 9). These ruins show three phases of building. On the basis of the materials and techniques used, the first building phase can be fixed between the first century BC and the middle of the first century AD. This first villa was erected on the Cretaceous limestones flanking the bay and it had a staircase descending to the beach. By excavating small test pits we discovered that the beach predating the AD 79 eruption dipped westward near the ruins, suggesting that the ancient coastline was more embayed than today (Fig. 10a).

As a result of the AD 79 eruption, the first villa was severely damaged and partly buried. In particular, the lower part of the staircase to the beach disappeared under a sequence of deposits that starts with the fallout and surge deposits of that eruption and continues with debris-flow and alluvial facies of the Durece event, reaching a total thickness of about 6 m (see Fig. 9).

The second phase of building occurred when the coastal plain had already been aggraded by the Durece deposits and the resumption of landscape stability encouraged the reconstruction of the seaside villa, probably by the same owner family. Some hundreds of metres to the south, along the Rivo d'Arco stream bed, some remnants of a robust protection wall of Roman age suggest that the entrenchment of the stream in the Durece deposits had already occurred when the place was reoccupied and the villa rebuilt (Cinque *et al.* 2000). Probably part of the fan-delta had already been dismantled by the waves, but the distance of the sea cliff from the new villa must have been great enough to not worry its inhabitants (Fig. 10b).

The masonry works carried out in the second phase included the restoration of some of the pre-existing rooms on the calcareous slope and the extension of the compound, with new rooms constructed on the adjacent valley floor terrace (Fig. 9). Judging from the building materials and the techniques used, the second construction phase can be ascribed to the first half of the second century.

The third building phase consists of some restoration and adaptation to the rooms resting on the terrace and the construction of a new way down to the beach by an inclined tunnel that started from those rooms and emerged from the sea cliff (Fig. 9). Moreover, the building work included the construction of a room (a coastal 'ninfeo' or perhaps simply a store) near sea level, in the space between the old staircase (by then re-exhumed by coastal erosion) and the tunnel to the beach. This building activity coincided with the time when the prominent portion of the Durece fan-delta had already been destroyed by the sea and the resulting sea cliff had retreated to almost its present position (Fig. 10c). Judging from the architectural forms and the materials used, the third phase of building and the correlative geomorphological stage (Fig. 10c) can be ascribed to the third century (Cinque *et al.* 2000). The Marina di Equa geo-archaeological site also reveals ruins of Medieval houses (4 in Fig. 9) whose present state reflects an additional, minor pulse of sea-cliff retreat that occurred in modern times.

On the basis of the data reported above, it can be reconstructed that, at the mouth of the Rivo d'Arco stream, the Durece event was able to cause about 500 m of coastal progradation in no more than 120 years. Progradations of similar timing and magnitude probably also occurred in the bays of Positano and Amalfi, where the Durece unit, instead of tapering to zero toward the coast, faces the modern beaches with a thickness of 18 and 8 m, respectively. These post-Durece sea cliffs (which are both masked by modern buildings) probably have the same significance as that of Marina di Equa. However, south of the Lattari Mts. there are offshore gradients much higher than to the north,

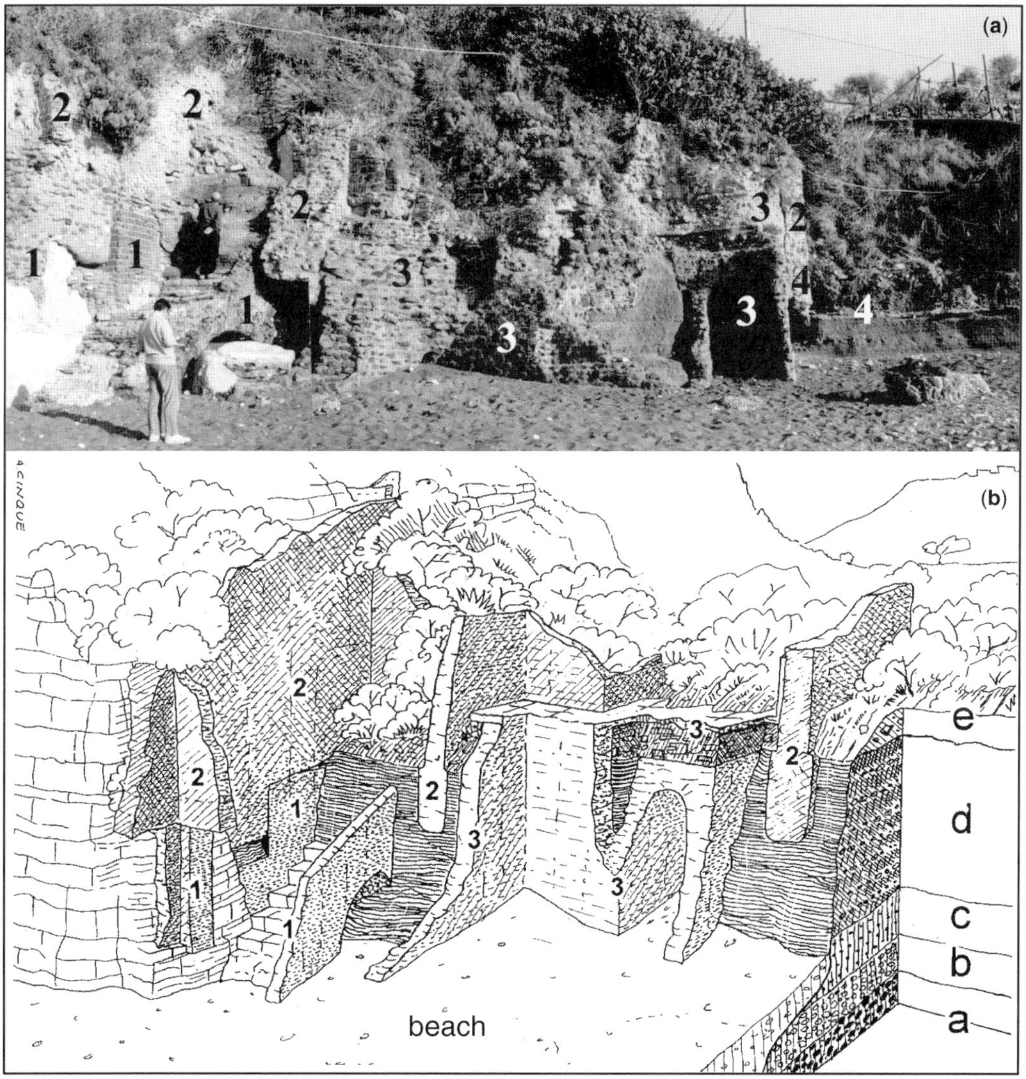

Fig. 9. The Roman ruins of Marina di Equa (**a**, photograph; **b**, sketch), emphasizing their relationships with the local Late Holocene succession. 1, Remains of the first phase of building (first century BC to middle of first century AD); 2, remains of the second phase of building (second century AD); 3, remains of the third phase of building (second century AD); 4, remains of Medieval buildings; a, beach deposits of the first century; b, ashy surge deposits of AD 79; c, Durece debris-flow deposits; d, Durece alluvial deposits; e, post-Roman colluvial cover.

so fan-delta progradation (per unit volume) was certainly less marked at Positano and Amalfi than at Marina di Equa.

Conclusions

The huge Plinian eruption of Vesuvius in AD 79 had severe, although indirect, consequences on the slope, fluvial and coastal dynamics of the Lattari Mts. ridge. Because of the rugged topography of these mountains, the thick cover of loose pyroclastic materials deposited by the eruption experienced a period of accelerated erosion and mass wasting soon afterwards. The lower reaches of valleys were consequently aggraded by volcanoclastic debris-flow and alluvial deposits that reached several metres to some tens of metres in thickness (depending on the local topography) within a span of 20–120 years (see the section on 'Chronological constraints').

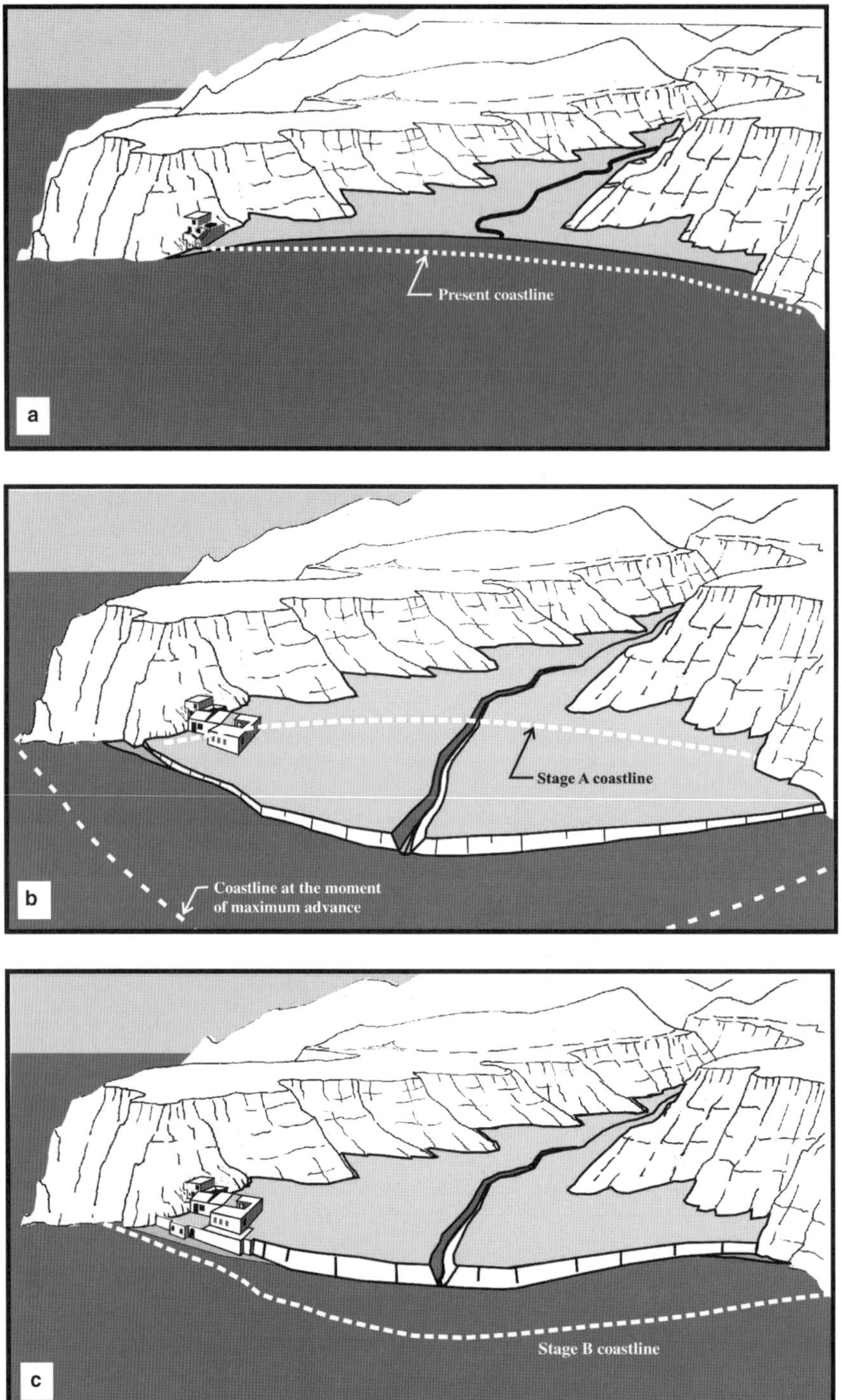

Fig. 10. Schematic representation of the geomorphological change caused at Marina di Equa by the accumulation and subsequent erosion of the Durece deposits. (**a**) The situation in the first century (before AD 79 eruption); (**b**) the situation when the second building phase occurred (second century); (**c**) the situation during the third century.

Fig. 11. The intense urbanization that now characterizes the alluvial coastal plain of the Canneto River (Amalfi). The lower river course flows through a subway tunnel. (Note also the steep topography of the catchment.)

Averaged at the basin scale and over the whole duration of the Durece event, the rates of annual yield were (1.1–10) $\times$ 10^3 m^3 km^{-2} in the Rivo d'Arco basin and (4–38) $\times$ 10^3 $m^3 km^{-2}$ in the Canneto basin. For each basin, the lower value comes from combining the minimum value of Vd (see Table 2) and the maximum duration of the Durece event (120 years), and the higher value comes from the maximum value of Vd combined with the minimum duration (20 years) of the event. As already stressed, these rates are considerable underestimates because our Vd values do not consider the eroded pyroclastic deposits that were carried off the fan-delta fronts: this probably occurred for an high percentage of the very light pumices and the fine-grained ashes of the AD 79 fallout deposits.

At the mouth of V-shaped valleys descending directly to the sea, the Durece deposition caused rapid floor aggradation and the growth of fan-deltas that pushed the coastline out hundreds of metres. Fortunately, these sectors of the study area were little populated at that time (the main settlements were on higher terraces), but the seaside villas erected there by rich Roman families were destroyed and buried.

Nowadays all the small, nested coastal plains of the southern side of the ridge (the Amalfi coast) are occupied by very densely constructed towns that began to be built in the early Middle Ages, when the local morphodynamics were much less of a problem and the good business made by the local merchants across the Mediterranean Sea made the Byzantine Duchy of Amalfi very rich and influential (Fig. 11).

If another Plinian eruption from Vesuvius were to again cover the Lattari Mountains with decimetres of pyroclastic deposits, the Durece scenario would be repeated and, owing to the much increased population density, enormously greater damage could occur.

This historical case shows that volcanic-related hazards occur not only in areas lying on a volcano or near it. When evaluating the hazards of coastal sites located within reach of potential pyroclastic fallout, we must consider not only the immediate and direct damage from pyroclastic fall, but also the damage from the geomorphological response of hillslope and fluvial systems. As we have seen, these indirect effects may be very significant and may induce rapid geomorphological change along the coast.

We are very grateful to the two anonymous referees, whose comments and suggestions improved the quality of this paper.

References

BLONG, R. 2000. Volcanic hazard and risk managment. *In*: SIGURDSSON, H., HOUGHTON, B. F., MCNUTT, S. R., RYMER, H. & STIX, J. (eds) *Encyclopedia of Volcanoes*. Academic Press, New York, 1215–1227.

BRANCACCIO, L., CINQUE, A., ROMANO, P., ROSSKOPF, C., RUSSO, F., SANTANGELO, N. & SANTO, A. 1991. Geomorphology and neotectonic evolution of a sector of the Tyrrhenian flank of the Southern Apennines (Region of Naples, Italy). *Zeitschrift für Geomorphologie, Neue Folge, Supplementband*, **82**, 47–58.

BUESCH, D. C. 1991. Changes in depositional environments resulting from emplacement of a large-volume ignimbrite. *In*: FISHER, R. V. & SMITH, G. A. (eds) *Sedimentation in Volcanic Settings*. Society of Economic Paleontologists and Mineralogists, Special Publications, **45**, 139–153.

CINQUE, A., ROBUSTELLI, G., SCARCIGLIA, F. & TERRIBILE, F. 2000. The dramatic cluster of pyroclastic debris flows occurred on 5–6 May 1998 on the Sarno mountains (Vesuvius regions, southern Italy): a geomorphological perspective. *In*: BROMHEAD, E., DIXON, N. & IBSEN, M. C. (eds) *Landslides in Research, Theory and Practice, Vol. I*, Thomas Telford, London, 273–278.

ESPOSITO, E., PORFIDO, S., VIOLANTE, C., BISCARINI, C., ALAIA, F. & ESPOSITO, G. 2004. Water events and historical flood recurrences in the Vietri sul Mare coastal area (Costiera Amalfitana, southern Italy). *In*: RODDA, G. & UBERTINI, L. (eds) *The Basis of Civilization—Water Science?* International Association of Hydrological Sciences Publications, **286**, 1–12.

GUADAGNO, F. M., CELICO, P., ESPOSITO, L., PERRIELLO ZAMPELLI, S., PISCOPO, V. & SCARASCIA MUGNOZZA, G. 1999. The debris flows of 5–6 May 1998 in Campania, Southern Italy. *Landslide News*, **12**, 5–7.

HAYES, S. K., MONTGOMERY, D. R. & NEWHALL, C. G. 2002. Fluvial sediment transport and deposition following the 1991 eruption of Mount Pinatubo. *Geomorphology*, **45**, 211–224.

LAMBECK, K., ANZIDEI, M., ANTONIOLI, F., BENINI, A. & ESPOSITO, A. 2004. Sea level in Roman time in the central Mediterranean Sea and indications for recent change. *Earth and Planetary Science Letters*, **224**, 563–575.

MACK, G. H., MCINTOSH, W. C., LEEDER, M. R. & MONGER, H. C. 1996. Plio-Pleistocene pumice floods in the ancestral Rio Grande, southern Rio Grande rift, USA. *Sedimentary Geology*, **103**, 1–8.

MAIURI, A. 1955. Le vicende dei monumenti antichi della costa amalfitana e sorrentina alla luce delle recenti alluvioni. *Rendiconti della Accademia di Archeologia, Lettere e Belle Arti in Napoli*, **29**, 87–98.

MAIZELS, J. 1993. Lithofacies variations within sandur deposits: the role of runoff regime, flow dynamics and sediment supply characteristics. *Sedimentary Geology*, **85**, 299–325.

MAJOR, J. J., JANDA, R. J. & DAAG, A. S. 1996. Watershed disturbance and lahars on the east side of Mount Pinatubo during the mid-June 1991 eruptions. *In*: NEWHALL, C. G. & PUNONGBAYAN, R. (eds) *Fire and Mud, Eruptions and Lahars of Mount Pinatubo, Philippines*. Washington University Press, Seattle, 895–919.

MAJOR, J. J., PIERSON, T. C., DINEHART, R. L. & COSTA, J. E. 2000. Sediment yield following severe volcanic disturbance. A two-decade perspective from Mount St. Helens. *Geology*, **28**, 819–822.

MANVILLE, V. 2001. Sedimentology and history of Lake Reporoa: an ephemeral supra-ignimbrite lake, Taupo volcanic zone, New Zealand. *In*: WHITE, J. D. L. & RIGGS, N. R. (eds) *Volcanogenic Sedimentation in Lacustrine Settings*. International Association of Sedimentologists, Special Publications, **30**, 109–140.

MANVILLE, V., NEWTON, E. H. & WHITE, J. D. L. 2005. Fluvial responses to volcanism: resedimentation of the Taupo ignimbrite eruption in the Rangitaiki River catchment, North Island, New Zealand. *Geomorphology*, **65**, 49–70.

MONTGOMERY, D. R., PANFIL, M. S. & HAYES, S. K. 1999. Channel-bed mobility response to extreme sediment loading at Mount Pinatubo. *Geology*, **27**, 271–274.

NAKAYAMA, K. & YOSHIKAWA, S. 1997. Depositional processes of primary to reworked volcaniclastics on an alluvial plain; an example from the Lower Pliocene Ohta tephra bed of the Tokai Group, central Japan. *Sedimentary Geology*, **107**, 211–229.

PIERSON, T. C., DAAG, A. S., REYES, P. J. D., REGALADO, M. T. M., SOLIDUM, R. U. & TUBIANOSA, B. S. 1996. Flow and deposition of posteruption hot lahars on the east side of Mount Pinatubo, July–October 1991. *In*: NEWHALL, C. G. & PUNONGBAYAN, R. S. (eds) *Fire and Mud, Eruptions and Lahars of Mount Pinatubo, Philippines*. Washington University Press, Seattle, 921–950.

RIGGS, N. R., HURLBERT, J. C., SCHROEDER, T. J. & WARD, S. A. 1997. The interaction of volcanism and sedimentation in the proximal areas of a Mid-Tertiary volcanic dome field, Central Arizona, USA. *Journal of Sedimentary Research*, **67**, 142–153.

SACCHI, M., MOLISSO, F. *ET AL*. 2009. Insights into flood-dominated, fan deltas: very high-resolution seismic examples off the Amalfi cliffed coasts, eastern Tyrrhenian Sea. *In*: VIOLANTE, C. (ed.) *Geohazard in Rocky Coastal Areas*. Geological Society, London, Special Publications, **322**, 33–71.

SANTACROCE, R. (ed.) 1987. *Somma–Vesuvius*. Consiglio Nazionale delle Ricerche, Pubblicazlone, **114**.

SCOTT, K. M. 1988. *Origins, behavior, and sedimentology of lahar and lahar-runout flows in the Toutle–Cowlitz River system*. US Geological Survey, Professional Papers, **1447-A**.

SEGSCHNEIDER, B., LANDIS, C. A., MANVILLE, V., WHITE, J. D. L. & WILSON, C. J. N. 2002. Environmental response to a large, explosive rhyolite eruption: sedimentology of post-1.8 ka pumice-rich Taupo volcaniclastics in the Hawke's Bay region, New Zealand. *Sedimentary Geology*, **150**, 275–299.

SHULTZ, A. W. 1984. Subaerial debris-flow deposition in the upper Paleozoic Cutler formation, western Colorado. *Journal of Sedimentary Petrology*, **54**, 759–772.

SIGURDSSON, H., CAREY, S., CORNELL, W. & PESCATORE, T. S. 1985. The Eruption of Vesuvius in AD 79. *National Geographic Research*, **1**, 332–387.

SMITH, G. A. 1987. The influence of explosive volcanism on fluvial sedimentation: the Deschutes Formation (Neogene) in central Oregon. *Journal of Sedimentary Petrology*, **57**, 613–629.

SMITH, G. A. 1988. Sedimentology of proximal to distal volcanoclastics dispersed across an active foldbelt: Ellensburg Formation (late Miocene), central Washington. *Sedimentology*, **35**, 953–977.

SMITH, G. A. 1991*a*. Facies sequences and geometries in continental volcaniclastic sediments. *In*: FISHER, R. V. & SMITH, G. A. (eds) *Sedimentation in Volcanic Settings*. Society of Economic Paleontologists and Mineralogists, Special Publications, **45**, 109–121.

SMITH, R. C. M. 1991*b*. Landscape response to a major ignimbrite eruption, Taupo Volcanic Center, New Zealand. *In*: FISHER, R. V. & SMITH, G. A. (eds) *Sedimentation in volcanic Settings*. Society of Economic Paleontologists and Mineralogists, Special Publications, **45**, 123–137.

SMITH, G. A. & LOWE, D. R. 1991. Lahars: volcano-hydrologic events and deposition in the debris flow–hyperconcentrated flow continuum. *In*: FISHER, R. V. & SMITH, G. A. (eds) *Sedimentation in Volcanic Settings*. Society of Economic Paleontologists and Mineralogists, Special Publications, **45**, 60–70.

SOHN, Y. K., RHEE, C. W. & KIM, B. C. 1999. Debris flow and hyperconcentrated flood-flow deposits in an alluvial fan, northwestern part of the Cretaceous Yongdong Basin, central Korea. *Journal of Geology*, **107**, 111–132.

TACITUS, C. 2001. *The Histories*. (Translated by W. H. Fyfe.) Oxford University Press, Oxford.

VIOLANTE, C., BISCARINI, C., ESPOSITO, E., MOLISSO, F., PORFIDO, S. & SACCHI, M. 2009. The consequences of hydrologic events on steep coastal watersheds: the Costa d'Amalfi, eastern Tyrrhenian Sea. *In*: LIEBSHER, H. J. *ET AL.* (eds) *The Role of Hydrology in Water Resource Management. Capri Italy, 2008*. IAHS Publication, **327**, 102–113.

WALTON, A. W. 1986. Effect of Oligocene volcanism on sedimentation in the Trans-Pecos volcanic field of Texas. *Geological Society of America Bulletin*, **97**, 1192–1207.

The use of documentary sources for reconstructing flood chronologies on the Amalfi rocky coast (southern Italy)

S. PORFIDO[1]*, E. ESPOSITO[1], F. ALAIA[2], F. MOLISSO[1] & M. SACCHI[1]

[1]*Institute for Coastal Marine Environment (IAMC) – CNR, Calata Porta di Massa 80133 Naples, Italy*

[2]*Campobasso State Archive, Via degli Orefici 43, 86100 Campobasso, Italy*

**Corresponding author (e-mail: sabina.porfido@iamc.cnr.it)*

Abstract: Documentary source materials are essential for retrospective reconstruction of flood events occurring in past centuries. This paper presents methods of research and archiving of historical data from the 16th century to the present. The quality and completeness of the various original sources were evaluated and carefully analysed in their historical context, to avoid serious mistakes. Systematic investigation of about 3000 documents, mainly found in national State Archives and libraries, allows us to identify and localize at least 106 flood events occurring along the Amalfi coast (southern Italy) for five centuries between the years 1500 and 2000. The collected data provide useful details on flood dynamics, size of flooded areas, flood duration, damage level, number of victims and induced geological effects. When available in sufficient quantity, the flood data allow determination of very useful parameters such as the severity class, to identify large floods and their recurrence interval.

Worldwide flood events are one of the most common types of natural disasters, accounting for at least one-third of all economic losses. Despite being the most widely available tool for flood risk assessment, a reliable flood record requires long data series. However, the modern instrumental period covers only short-run periods and requires additional information on extreme flooding from historical sources. Hence, long-term flood chronology can be reconstructed from selected data contained in documentary sources. The use of historical data in estimating large flood events has increased in recent years (Brázdil *et al.* 1999; Robson & Reed 1999; SPHERE European Project 2000–2003; Glade *et al.* 2001; Thorndycraft *et al.* 2002) both for risk analysis and climatic variability.

The Italian peninsula is characterized by the largest number of geological–hydraulic instabilities in Europe, whether landslides or floods. During the last 100 years alone, 32 162 landslides and 29 233 floods have taken place in Italy, with a total of 7525 victims. In particular, along the coasts, where most of the people live, 4309 landslides and 6501 floods were recorded, which resulted in 2000 casualties. Campania (Fig. 1) is one of the regions that has experienced the largest number of events: 3152 landslides have affected 1855 localities, and 1640 floods have affected 843 localities (Guzzetti & Tonelli 2004; AVI Project 1989–1996; and the continuing IFFI Project (Italian Landslide Inventory); http://www.sinanet.it; SICI PROJECT: http://sici.irpi.cnr.it/index.htm).

With the aim of improving the hydrogeological hazard and risk assessment along rocky coastal areas, a systematic study of the major known events that have affected the Amalfi coast (Fig. 1) was initiated. This stretch of coast has a prominent role in the region of Campania not only in economic terms as an up-market tourist destination and an area of high-quality agricultural land, but also as a heritage site of outstanding importance (included in the UNESCO World Heritage list since 1997). It is a steep rocky coast trending NE–SW, separating the Bay of Naples to the north from the Bay of Salerno to the south, mainly composed of Mesozoic limestone that has been tectonically uplifted since the early Pleistocene (Fig. 1). The sedimentary cover includes Miocene siliciclastic deposits and Quaternary volcaniclastic and alluvial deposits, which form a discontinuous mantle overlying the carbonate bedrock and are prone to detachment during hydrological events (Esposito *et al.* 2004*a*, *b*). These volcaniclastic and alluvial deposits have an extremely variable thickness (from a few centimetres to some metres), and relate to the Quaternary explosive activity of the Somma–Vesuvius volcanic complex (Santacroce 1983; Sigurdsson *et al.* 1985; Cinque 1986; De Vivo & Scandone 2003; Casciello *et al.* 2004).

From: VIOLANTE, C. (ed.) *Geohazard in Rocky Coastal Areas*. The Geological Society, London, Special Publications, **322**, 173–187. DOI: 10.1144/SP322.8

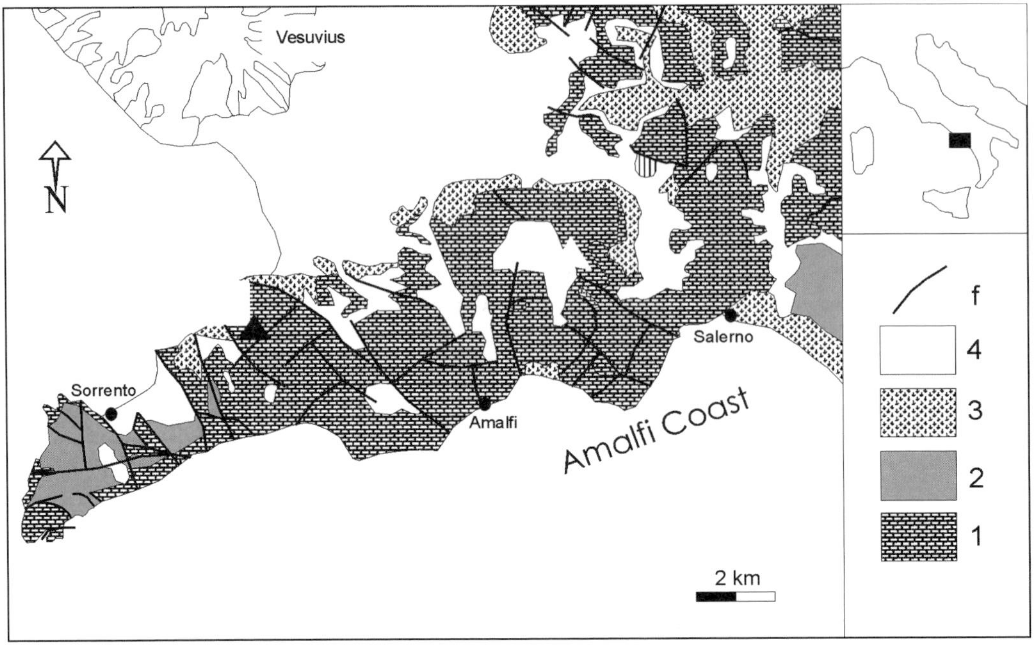

Fig. 1. Location of the study area and geological outline: 1, carbonate platform dolomites and limestones with discontinuous cover of pyroclastic and epiclastic deposits (Late Triassic–Cretaceous); 2, terrigenous deposits (Late Tertiary); 3, continental deposits (Quaternary); 4, pyroclastic deposits and lavas (Quaternary); f, fault.

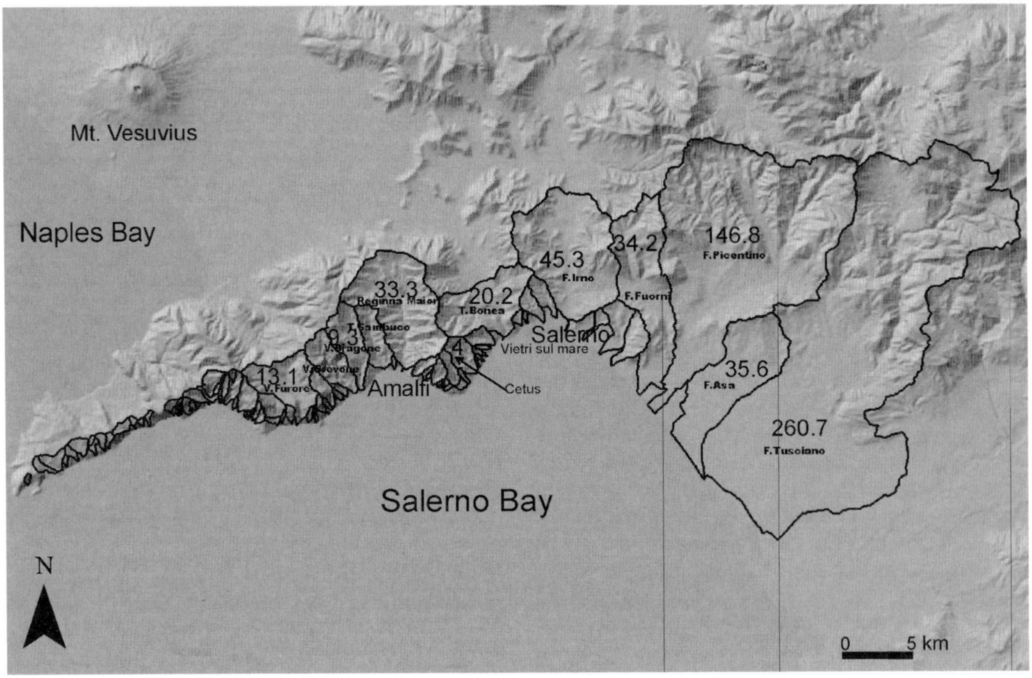

Fig. 2. Fluvial basins of the Amalfi coast and their areal extent (km^2).

From the hydrogeological viewpoint, the whole of the Amalfi coast is characterized by small catchments never exceeding 34 km^2 (Fig. 2), with short and steep stream paths normally characterized by low flow and sediment transport. This can change dramatically during heavy rainfall events, which trigger widespread failures of the sedimentary cover and consequent devastating floods (Esposito *et al.* 2004*a*, *b*; Violante 2009; Violante *et al.* 2009). In the last century the Amalfi area was affected by numerous severe floods in conjunction with exceptional rainfall that caused major damage in terms of lost lives and economic cost.

In this study historical floods since the 16th century were analysed in detail to provide a sounder basis for reliable risk assessment and consequent safer land planning on flood-prone areas.

Methods

Historical documentary sources are an important source of information for reconstructing exceptional flood events occurring prior to the instrumental era. Historical analysis also provides an opportunity to extend the time scale window for flood risk studies. Experience in historical data collection shows that not all documentary sources can provide useful information for flood characterization, but it is necessary to define a selection criterion so as to obtain the best information rather than the best dataset quality. A scientific systematic approach to identify the main groups of sources indicates the easiest approach in documentary research (Ambraseys & Finkel 1993; Vogt 1993; Boschi *et al.* 1995; Benito 2000; Bayliss & Reed 2001; Benito *et al.* 2004). The main groups of sources are generally found in local documentary material, administrative records, special studies, scientific reports, diaries, books and newspapers. They are available in record offices, libraries, public and private documentary collections, and parish archives throughout Europe.

For the purpose of this research we used a rigorous method of source investigation, which has been widely tested and consolidated during recent decades and is suitable not only for hydrological phenomena but also for other natural catastrophes. The method may be summarized as follows: (1) analysis of historical sources, including the completeness and reliability of the document within the historical context; (2) source classification, based on published and unpublished papers, contemporary sources (or otherwise), official reports and general public information (press reports), taking into account the intrinsic quality of the document (such as type, author, date, circumstances of writing, etc.); and (3) evaluation of historical flood data.

This procedure is somewhat complex, but the basic principles have been well described by Bell & Ogilvie (1978), Ingram *et al.* (1978), Bayliss & Reed (2001), Agasse (2002), Barriendos *et al.* (2003) and Telelis (2005).

Documentary data gathering can provide not only qualitative information but also information on flood dynamics, height of the water at precise locations, extent of the flooded area and damage levels. The critical analysis of documentary sources can thus be used to build a scale of flood event magnitude.

As regards the Italian documentary heritage, the archive sources are certainly among the most fertile for first-hand information, especially regarding the most ancient documented events, and are useful up to the 20th century (Boschi *et al.* 1995; Guidoboni & Ferrari 1995). These sources include technical administration reports, financial and legal acts, expert investigations, projects, memoirs, chronicles, other heterogeneous public, private and ecclesiastical documents, contemporary maps and iconographies. They provide two levels of information: (1) general, giving information on the type, sometimes the magnitude, and locality of the event; and (2) detailed (from technical reports, projects), providing the precise location of the event, the size of the flooded area, and the type of flood-induced damage. Occasionally drawings or photographs of the flood-induced effects are also occasionally available. Chiefly, sources contemporary with or chronologically close to the event are considered and, subordinately, second-hand documents. Often the multiplicity of sources allows one to cross-check the data. It is essential to consider that these data, covering a wide time period, are influenced by the evolving social and cultural environments; therefore, although they may represent the best possible dataset, the description of the event may still be incomplete.

Another important type of source is the bibliographic one. This is generally divided into two main groups: (1) direct source, including text written by eyewitnesses and specific studies, such as scientific literature and contemporary newspapers; and (2) indirect source, including texts written after the event by local authors, scientific literature, etc.

New sources of information have been identified by analysing pictures, particularly for the 20th century, postcards, prints, drawings and art reproductions, which are often able to effectively capture the impact of disasters. Finally, *ad hoc* legislation has not been overlooked, as it sometimes also provides very useful information on damage distribution and extent of destruction affecting local communities. In particular, the main laws enacted to assist the local population hit by flooding have been reviewed from the unification of Italy (1861) to 1955, including their technical and financial provisions.

Data collection and evaluation

The study area is of special interest because of both its geological and geomorphological features as well as the tourist industry. The Amalfi coast includes more than 20 towns dating back to medieval times, which are of major architectural value as well as spectacular morphology, being considered worldwide an outstanding example of a Mediterranean landscape. These towns generally developed along the main streams dissecting the coastal slopes, and the inhabitants channelled and covered these streams to use them as sewer systems. This resulted in an unstable equilibrium that becomes critical with consequent disruption and damage during heavy rainfall.

For a reconstruction of the time–space distribution of extreme floods and/or of intense rainfall events that have affected the province of Salerno, a thorough search of the documentary sources from the 16th century was carried out at the State Archives in Naples and Salerno, and at the historical archives of several comuni in the province of Salerno, with over 3000 documents being consulted. In particular, interesting information of an administrative nature was gathered, allowing reconstruction of the damage caused by floods or extensive landslides.

This particularly applies to the documents conserved in the State Archives coded as *Protocolli Notarili*, *Intendenza*, *Prefettura* and *Genio Civile* (Fig. 3a and b). The *Protocolli Notarili* (1500–1798) contain a broad collection of local news believed to be of possible relevance to notary activity, such as earthquakes, floods, famines, and uncommon climatic conditions, such as prolonged droughts or heavy snowfall.

For the first half of the 19th century, the main archival sources are documents of the *Intendenza* (1806–1860). These include wide-ranging correspondence between the Intendente (a sort of High Sheriff) of Salerno province and the mayors of the villages affected by flooding, reporting the need for first aid and appraising damage for planning restoration and reconstruction works.

Data on the hydrological context of the province and news on important rainfall episodes have also been found in the archives entitled *Prefettura* (1861). The Prefettura is the principal administration bureau, representing the Government at the province level.

Finally, in the *Genio Civile* collection (office for engineering works) there are documents from the directorate of bridges, roads, waters and forestry for the years 1859–1929, reporting technical interventions following heavy rainstorms.

Among the oldest sources, the local chronicles are particularly interesting (Carraturo 1784; Greco 1787; Camera 1881), as they chronologically list the most important local historical occurrences, such as Vesuvius eruptions, the worst famines and pestilences, exceptional climatic periods, sea storms, floods and earthquakes.

There are copious very informative bibliographic sources for the 20th century, consisting of technical papers associated with descriptions given by eyewitnesses as well as technical–scientific reports (Lazzari 1954; Penta *et al.* 1954; Cancellara 1955; Frosini 1955; Amarotta 1994; Fumanti *et al.* 2002; Rischia *et al.* 2003; Esposito *et al.* 2003*a*, *b*, 2004a, *b*, and references therein), which summarize a wealth of useful data to precisely reconstruct the dynamics of the most devastating events. Newspapers and photographs (Fig. 3c and d) are a fundamental source, particularly for the 20th century. They give detailed accounts and comments prepared by journalists sent to disaster areas immediately after the events and in the following weeks and months. They vividly illustrate the dramas suffered by the local inhabitants, and give accounts of the living conditions of the homeless, and of the technical and administrative interventions by the local and central governments.

Reconstruction of the chronology of Amalfi coast historical floods is based on many different types of sources. To evaluate the quality and completeness of information, particular attention was paid to the original sources, according to the historical context. The weight of each source was based on the reliability of the writer as well as how contemporary the document was to the event. In particular, three kinds of source were distinguished (Fig. 4), as follows.

Type 1 (*highest quality of reference*). This indicates documentary or bibliographic sources, chronologically contemporary with the event, written by a local or regional administrator, lawyer, historian, parish priest, journalist, scientist, academician or technician.

Type 2 (*medium–high quality of reference*). This indicates documentary or bibliographic sources, subsequent to the event (from 5 to 50 years), written by a local historian, parish priest, journalist, or scientist or technician.

Type 3 (*medium quality of reference*). This indicates bibliographic sources, subsequent to the event (over 50 years), written essentially by local literary figures and journalists.

Although this research cannot be considered complete, the data concerning the 16th–18th centuries are mainly represented by archival sources as well as contemporary bibliographic documents. The special nature of these documents, which directly refer to the authority producing the document, allows the material to be classified as the

Fig. 3. Archival and photographic sources: (**a**) *Protocolli Notarili*, 25 January 1736 flooding and destruction along the Bonea stream in Molina di Vietri; (**b**) *Prefettura, I Serie*, 7 October 1899 flooding and government subsidy to the Vietri sul Mare homeless; (**c**) 1954 flood event destruction in Marina di Vietri, at Bonea stream mouth (Foto Pariso 1954, Coll. EPT, Salerno); (**d**) mud deposit along the Salerno coast (Foto Pariso 1954, Coll. EPT, Salerno). Reproduced with permission of the Archivio di Stato di Salerno (a, b) and Ente Provinciale del Turismo di Salerno (c, d).

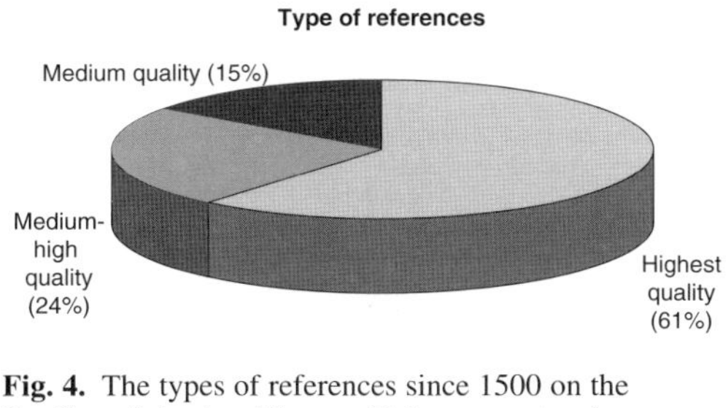

Fig. 4. The types of references since 1500 on the flooding of the Amalfi coast (Salerno province).

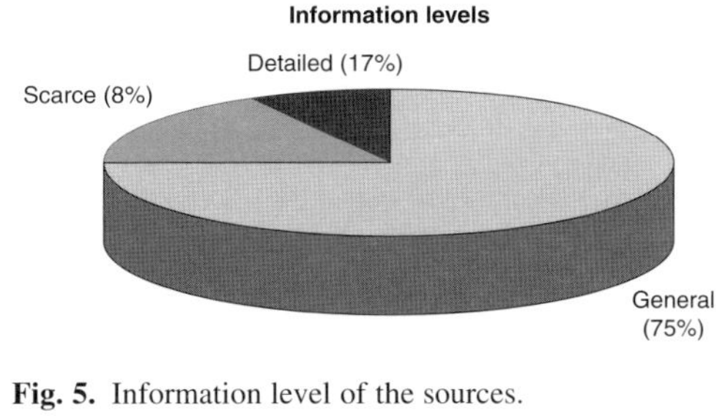

Fig. 5. Information level of the sources.

Table 1. *Major floods occurring since 1500 along the Amalfi coast (Salerno province) identified on the basis of the available sources*

No.	Date	TR	IL	Location	Stream/river	TBA (km^2)	SC
1	30 09 1581	1, 3	a	Salerno, Cava T., Vietri M., Castiglione G., Giffoni C., Giffoni V. P.	Irno, Cavaiola, Bonea, Picentino, Fuorni, and ephemeral streams	336	III
2	31 08 1588	1, 3	b	Atrani	Dragone	9	II
3	20 12 1683	3	b	Maiori	Regina Major	33	I
4	15 10 1696	3	b	Minori	Regina Minor	5, 5	I
5	09 11 1735	1	b	Cetara, Cava T., Salerno, Vietri M.	Cetus, Bonea, Cavaiola, Irno and ephemeral streams	162	I
6	25 01 1736	1	b	Vietri M.	Bonea	20	I
7	26 09 1736	1	b	Vietri M.	Bonea	20	I
8	11 1738	1	b	Vietri M.	Bonea	20	I
9	03 11 1750	1	b	Vietri M., Salerno	Bonea, Irno	65	II
10	10 10 1751	1, 2	b	Amalfi	Canneto	6	I
11	01 09 1753	1, 2	c	Amalfi	Canneto	6	I
12	23 01 1757	1	a	Vietri M.	Bonea	20	I
13	09 10 1757	1, 2	a	Amalfi	Canneto	6	I
14	25 05 1762	2	a	Cetara	Cetus	4	I
15	19 01 1764	1, 3	b	Naples and Salerno province	Sarno, Dragone, Regina Major, Regina Minor	264	II
16	11 1770	1	b	Salerno	Irno, and ephemeral streams	50	I
17	11 11 1773	1, 2, 3	a	Salerno, Coperchia, Cava T., Vietri M., Tramonti, Cetara, Nocera, Mercato S. S.	Irno, Cavaiola, Fuorni, Solofrana, Bonea, Cetus, Regina Major and ephemeral streams	361	III
18	02 1780	1	b	Atrani	Dragone	9	I
19	25 12 1796	1	a	Cava T.,Vietri M., Salerno	Cavaiola, Bonea, Irno and ephemeral streams	158	I
20	12 11 1817	1	a	Cava T., Vietri M., Salerno	Cavaiola, Bonea, Irno and ephemeral streams	158	I
21	12 1822	1	b	Salerno,Vietri M.	Irno, Bonea and ephemeral streams	71	I
22	24 01 1823	1	a	Amalfi, Maiori, Cetara, Cava T., Nocera, Vietri M., Salerno, Bracigliano	Canneto, Regina Major, Cetus, Bonea, Cavaiola, Solofrana, Irno and ephemeral streams	336	II
23	13 09 1834	1	b	Cetara	Cetus	4	I
24	18 07 1835	1	b	Conca marina, Salerno, Cava T.	Irno, Bonea, Cavaiola and ephemeral streams	158	I
25	27 09 1837	1	b	Salerno, Vietri M.	Irno, Bonea and ephemeral streams	71	I
26	01 01 1841	1	c	Salerno province		–	I
27	26 10 1843	3	b	Cetara, Maiori, Vietri M., Salerno	Cetus, Regina Major, Bonea, Irno and ephemeral streams	108	I
28	18 03 1845	3	b	Maiori, Vietri M.	Regina Major, Bonea	53	I
29	05 01 1853	1	b	Vietri M.	Bonea	20	I
30	11 10 1866	1	b	Vietri M.	Bonea	20	I
31	08 1866	3	a	Tramonti	Regina Major	33	I
32	11 11 1866	1	a	Vietri M.	Bonea	20	I
33	16 03 1867	1	b	Vietri M., Salerno	Bonea, Irno and ephemeral streams	70	I
34	12 11 1868	1	b	Salerno	Irno and ephemeral streams	50	I
35	01 04 1875	1	b	Conca M, Salerno	Irno and ephemeral streams	56	I
36	12 1875	3	c	Salerno province		–	I
37	01 02 1878	1	a	Conca M.	Ephemeral streams	4	I
38	11 1881	3	c	Salerno province			I
39	15 09 1882	1	b	Salerno	Irno and ephemeral streams	56	I
40	05 02 1885	1	b	Amalfi	Canneto	6	I
41	1891	3	b	Tramonti	Regina Major	33	I

(*Continued*)

Table 1. *Continued*

No.	Date	TR	IL	Location	Stream/river	TBA (km²)	SC
42	1896	3	b	Conca M., Castiglione G., Baronissi, Bracigliano, Salerno	Picentino, Fuorni, Irno and ephemeral streams	232	I
43	1898	3	c	Salerno province		–	I
44	1899	3	b	Conca M.	Ephemeral streams	4	I
45	07 10 1899	1, 2, 3	a	Castiglione G., Giffoni C., Giffoni V. P., Montecorvino R., Montecorvino P. Vietri M., Cava T., Salerno, Caposele, Calabritto, Quaglietta, Pontecagnano, Battipaglia	Sarno, Picentino, Fuorni, Tusciano, Asa, Cavaiola, Bonea, Irno, Sele and ephemeral streams	900	III
46	02 1903	1	b	Vietri M.	Bonea	20	I
47	07 10 1904	2	b	Ravello	Dragone	9	I
48	23 06 1905	3	c	Salerno province		–	I
49	01 09 1905	3	c	Salerno province		–	I
50	11 12 1908	3	c	Salerno province		–	I
51	24 10 1910	1, 2, 3	A	Ravello, Tramonti, Furore, Amalfi, Scala, Cetara, Maiori, Minori, Vietri M., Salerno, Ischia	Cetus, Canneto, Regina Major, Regina Minor, Bonea, Irno and ephemeral streams	130	II
52	02 01 1911	1	a	Cetara, Vietri M., Salerno	Cetus, Bonea, Irno and ephemeral streams	79	I
53	21 09 1912	1	c	Salerno province		–	I
54	03 01 1915	2	a	Minori	Regina Minor	5	I
55	06 11 1916	1	b	Vietri M.	Bonea	20	I
56	13 11 1921	1	a	Furore, Salerno	Ephemeral streams, Irno	62	I
57	26 03 1924	1, 2, 3	a	Positano, Agerola, Vettica M., Praiano, Amalfi, Atrani, Furore, Minori, Maiori, Vietri M., Salerno province	Dragone, Regina Minor, Canneto, Furore, Regina Major, Bonea, Irno and ephemeral streams	127	II
58	21 09 1929	2	B	Montecorvino R., Giffoni, Vietri sul M.	Asa, Picentino, Bonea and ephemeral streams	202	I
59	01 03 1935	2, 3	b	Conca M., Minori, Tramonti, Ravello, Cava T.	Dragone, Regina Minor, Regina Major, Cavaiola and ephemeral streams	141	I
60	18 11 1935	2	b	Salerno, Vietri M.	Bonea, Irno and ephemeral streams	70	I
61	14 09 1939	1, 2	b	Conca M., Amalfi, Maiori, Salerno, Pontecagnano	Canneto, Regina Major, Irno, Picentino and ephemeral streams	231	I
62	1940	3	c	Salerno province		–	I
63	18 06 1944	2	b	Minori	Regina Minor	5	I
64	09 12 1946	2	b	Minori	Regina Minor	5	I
65	02 03 1947	2	b	Minori	Regina Minor	5	I
66	25 10 1947	2	b	Minori	Regina Minor	5	I
67	23 05 1948	2	b	Minori	Regina Minor	5	I
68	05 09 1948	2	b	Minori	Regina Minor	5	I
69	28 10 1948	2	b	Minori	Regina Minor	5	I
70	01 10 1949	1, 2	b	Praiano, Maiori, Vietri M., Salerno, Giffoni	Regina Major, Bonea, Irno, Picentino	245	I
71	21 01 1951	2	b	Minori	Regina Minor	5	I
72	09 11 1951	1, 2	b	Montecorvino R., Giffoni	Asa, Picentino	183	I
73	11 09 1953	2, 3	b	Agerola, Ravello, Salerno	Dragone, Irno and ephemeral streams	65	I
74	25 10 1954	1, 2	a	Positano, Vettica M., Praiano, Amalfi, Atrani, Minori, Tramonti, Maiori, Vietri M., Cava T., Nocera, Salerno	Regina Minor, Regina Major, Canneto, Dragone, Cavaiola, Bonea, Irno, Picentino and ephemeral streams	550	III

(*Continued*)

Table 1. *Continued*

No.	Date	TR	IL	Location	Stream/river	TBA (km^2)	SC
75	11 09 1955	2	b	Tramonti, Agerola, Pellezzano	Regina Major, Irno and ephemeral streams	85	I
76	22 10 1957	2	b	Tramonti, Minori, Cava T.	Regina Major, Regina Minor, Cavaiola	126	I
77	03 1960	2	c	Salerno province		–	I
78	16 02 1963	2	b	Tramonti, Cava T., Pellezzano	Regina Major, Cavaiola, Irno and ephemeral streams	170	I
79	25 09 1963	2	b	Cetara, Minori, Cava T., Pellezzano	Regina Minor, Cavaiola and ephemeral streams	120	I
80	08 10 1963	1, 2	b	Amalfi, Cetara, Cava T., Salerno	Canneto, Cetus, Cavaiola, Irno and ephemeral streams	142	I
81	16 12 1963	1, 2	b	Tramonti, Pellezzano	Regina Major, Irno	78	I
82	26 10 1966	1, 2	a	Giffoni, Salerno, Cava T., Baronissi,	Fuorni, Picentino, Cavaiola, Irno and ephemeral streams	313	II
83	09 01 1968	2	c	Salerno province		–	I
84	19 12 1968	1, 2	b	Amalfi, Tramonti	Canneto, Regina Major and ephemeral streams	50	I
85	15 03 1969	2	b	Agerola, Cava T.	Furore, Cavaiola and ephemeral streams	110	I
86	02 10 1970	1, 2	b	Salerno, Pellezzano, Baronissi, Giffoni	Irno, Picentino and ephemeral streams	200	I
87	25 12 1970	1, 2	b	Amalfi, Minori, Baronissi, Pellezzano	Canneto, Regina Minor, Irno and ephemeral streams	60	I
88	15 10 1971	1, 2	b	Tramonti, Cava T.	Regina Major, Cavaiola	120	I
89	23 11 1971	1, 2	b	Amalfi, Minori	Canneto, Regina Minor	10	I
90	06 03 1972	1, 2	b	Tramonti, Cava T.	Regina Major, Cavaiola	120	I
91	21 10 1972	1, 2	b	Tramonti, Cava T.	Regina Major, Cavaiola	120	I
92	21 11 1972	1, 2	b	Cava T., Baronissi, Pellezzano	Cavaiola, Irno	132	I
93	02 01 1973	1, 2	b	Amalfi, Tramonti, Minori, Maiori, Cava T.	Regina Minor, Regina Major, Cavaiola	125	I
94	28 06 1976	2	b	Salerno			I
95	09 04 1978	2	c	Salerno province			I
96	12 10 1980	1, 2	b	Tramonti, Minori, Maiori, Cava T.	Regina Major, Regina Minor, Cavaiola	125	I
97	15 11 1980	2	b	Cava T.	Cavaiola	87	I
98	17 11 1985	1, 2	b	Tramonti, Maiori, Cava T., Salerno	Regina Major, Cavaiola, Irno	170	I
99	13 03 1986	1, 2	b	Cava T., Pellezzano, Pontecagnano	Cavaiola, Irno	120	I
100	24 11 1986	1, 2	b	Tramonti, Cava T.	Regina Major, Cavaiola	120	I
101	16 10 1987	1, 2	b	Baronissi, Pellezzano	Irno	45	I
102	10 11 1987	1, 2	b	Positano, Ravello, Tramonti, Minori, Cava T.	Regina Major, Cavaiola, Cavatola and ephemeral streams	130	I
103	15 09 1988	1, 2	b	Tramonti, Pellezzano, Baronissi, Salerno	Regina Major, Irno and ephemeral streams	90	I
104	25 09 1992	1, 2	b	Tramonti, Cava T., Salerno	Regina Major, Cavaiola, Irno and ephemeral streams	170	I
105	04 10 1992	1, 2	b	Cava T., Baronissi, Salerno	Cavaiola, Irno and ephemeral streams	90	I
106	20 09 1996	1, 2	b	Tramonti, Cava T., Salerno, Giffoni	Regina Major, Cavaiola, Irno, Picentino	300	I

TR, type of reference (1, highest quality of reference; 2, medium–high quality of reference; 3, medium quality of reference); IL, information level (a, detailed level; b; general level; c, scarce level); SC, severity class (I, small flood; II, intermediate flood; III, catastrophic or large flood); TBA, total area of hydrological basins affected.

Table 2. *The most intense flooding events occurring along the Amalfi coast*

Date	FD	Locations	SC	Geological effect	Deaths	Comments and contemporary descriptions
30 09 1581	–	Salerno, Cava T., Vietri M., Castiglione G., Giffoni C., Giffoni V. P.	3	Extensive inundation, landslides, shoreline progradation in Salerno	300	**Extensive damage; property destroyed in several localities**; 'there was a very great, unusual flood in the area of Castiglione di Giffoni in several localities and in the village of Ogliara belonging to the city of Salerno due to rainfall and the high mountains . . . which caused . . . many deaths and much destruction to houses and property from their very foundations, together with bridges, and uprooted trees and carried off many goods and firewood down to the sea, and drowned animals. . . . almost 300 people died' (Salerno State Archive 1581–1584)
31 08 1588	–	Atrani	2	Extensive inundation, landslides, shoreline progradation in Atrani	2–3	**Severe damage to property**; 'at the end of the past month of August 1588 . . . much lava* having fallen . . . it destroyed the Seggio building . . . the force of the lava loosened trees, wood, earth and rocks and . . . it filled the harbour and pushed the sea back seven rods (14 m) thereby enlarging the harbour' (Cronaca Amalphitana, Ignoti auctoris 1588, cited by Camera 1881)
11 11 1773	3	Salerno, Coperchia, Cava T., Vietri M., Mercato S. S., Cetara, Nocera, Tramonti	3	Extensive inundation, landslides, shoreline progradation	400–450	**Extensive damage; some villages completely destroyed in Cava de'Tirreni**; 'in Salerno a large stretch of the fields near the bridge was flooded . . . many bodies were taken away and the sea was covered with innumerable pieces of wood, tossed by the river, like . . . in Marina di Vietri, when the paper-mills were washed away, and a paper mill, . . . all kinds of stuff . . . covered the waters, and mixing with them, formed very long thick ridges of the same material. There were 450 deaths in Cava: . . . 16 in Coverchia . . . the damage was considerable . . . in Nocera the fields were flooded' (Greco 1787)
7 10 1899	7–10	Castiglione G., Giffoni C., Giffoni V. P., Cava de' T., Montecorvino R., Salerno, Montecorvino P., Vietri M., Caposele, Battipaglia, Quaglietta, Pontecagnano Calabritto	3	Extensive inundation, landslides	86	**Extensive damage; Some villages completely destroyed in Castiglione and Giffoni**; 'Along the Cava–Vietri road, many landslides fell' (*L'Irno* 1899)

(*Continued*)

Table 2. *Continued*

Date	FD	Locations	SC	Geological effect	Deaths	Comments and contemporary descriptions
						'Vietri: destruction of most of the factories along the course of the Bonea . . . Salerno: the quarter of Fieravecchia severely damaged where several buildings collapsed' (Salerno State Archive 1899–1900)
24 10 1910	1–2	Ravello, Tramonti, Furore, Amalfi, Scala, Cetara, Maiori, Minori, Vietri M., Salerno, Ischia	2	Extensive inundation, landslides	200	**Severe damage; Casale village (Cetara) was completely destroyed**; 'the night before 24 October 1910 violent . . . cloudburst was unleashed on the Amalfi coast, especially at Cetara, Maiori, Amalfi and Minori . . . the disaster was serious: flooding, landslides, collapse, deaths (200)' (Minori Municipal Archive 16.11.1910) 'We walk along Corso Regina amid an enormous heap of pebbles, tree trunks and roots, mixed with earth and mud that reach a height of 1 to 3 metres. Half-way along the avenue, which is built entirely and along its whole length on the River Regina . . . we note the subsidence of the cover on the river for a length of about 60 metres' (*Corriere d'Italia* 1910)
26 03 1924	1–2	Positano, Agerola, Vettica Min., Praiano, Amalfi, Atrani, Furore, Minori, Maiori, Vietri M., Salerno	2	Extensive inundation, In Vettica (Amalfi) a wide landslide destroyed 25 buildings	60	**Severe damage to public and private property**; 'given the very severe floods lasting throughout the day and all last night . . . works are needed to repair and re-lay the municipal roads, destroyed and damaged' (Minori Municipal Archive, 26.3.1924–11.4.1924)
25 10 1954	1	Cava T., Salerno Praiano, Amalfi, Atrani, Positano, Tramonti, Vietri M., Vettica M., Minori, Maiori, Nocera	3	Extensive inundation, landslides, shoreline progradation	318	**Extensive damage; property destroyed in several localities**; 'A disaster of inestimable proportions, perhaps without precedent . . . has befallen Campania in the past 48 hours, in particular the Province of Salerno . . . The disaster was caused by the violent rains that fell yesterday and the day before in the Salerno area, causing the collapse of houses, rivers to overflow and landslides, almost the complete interruption of telegraph and telephone communications, of roads, of many railway lines, bridges collapsing, flooding . . . deaths' (*L'Unità* 27.10.1954) 'Solid transport discharge into the sea has altered the coastline, creating a broad delta in the area of Marina di Vietri' (Penta *et al.* 1954)

*The Italian term 'lava' in this context refers to a mixture of rainwater and sediment, forming liquid mud.
FD, Flood duration.

highest quality of references, representing 63% of the total sources (total number 200). The medium–high quality of references consists of bibliographic sources amounting to 11% of the total, and the remaining 26% includes bibliographic sources written over 50 years after the events. The data for the 19th century also come from archival sources and contemporary national and local newspapers, and there is a total of about 1500 sources. Of these, 61% are classified as the highest quality of references, 4% medium–high quality, and 35% as medium quality.

As regards the 20th century most of the documents (about 1300) have been classified as medium–high quality of reference, coming from both archival and bibliographic sources as well as local and national newspapers (from 1899 to 2000), amounting to 55% of the total, whereas the highest quality of reference consists of a huge repository of documents (32% of the total) from archival sources to scientific and technical reports. Only 8.5% of the documents have been classified as the third type of source.

Once the information was extracted, the content was analysed taking into account the level of detail of the information (Fig. 5), and grouped into three categories, as follows.

Type a, Detailed. This category (from technical reports, projects, etc.) gives the precise location of the event, the extent of the flooded area, and the type of flood-induced damage. Occasionally drawings or photographs of the flood-induced geological effects are also available.

Type b, General. This category gives information on the event type, sometimes the size, and the locality of the event.

Type c, Scarce. This category gives very poor information regarding the occurrence of flooding in a wide area; sometimes the date of the flood is the only information available.

Historical flood chronology and severity classes

The systematic search for historical sources has led to the identification and classification of 106 floods, which affected the whole province of Salerno, and specifically the Amalfi coast (Table 1). Such pooling of information made it possible to determine some flood characteristics, such as duration of the event, river location, damaged localities, deaths, induced geological effects (Table 2), and seasonal and monthly distribution (Fig. 6a and b).

Figure 6a shows that autumn (51%) is the season with the highest number of floods, followed by winter (28%), and October (21 cases) is the month with the highest number, followed by November (19 cases) (Fig. 6b).

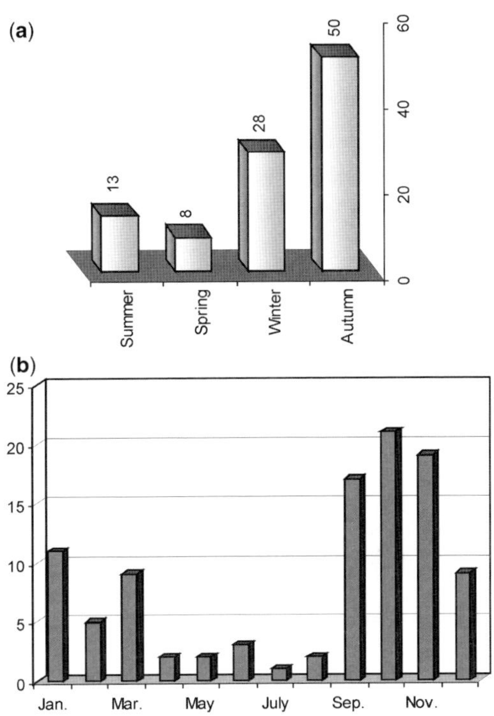

Fig. 6. Flood distribution along the Amalfi coast (Salerno province): (**a**) seasonal; (**b**) monthly.

As can be seen from Figure 7, the areas most frequently hit by floods in the period in question are Salerno (35 floods: 25 small floods, six intermediate floods and four catastrophic floods) and Vietri sul Mare (29 floods: 21 small floods, four intermediate floods and four catastrophic floods).

The distribution of the 106 floods against years is represented in Figure 8. The absolute cumulative frequency curve for all the data shows that the sample is significant for the events that have occurred since 1735. Prior to this date the number of floods decreases considerably, which suggests that further research is required for the period from the Middle Ages to the end of the 17th century. This curve also indicates some interesting correlations with the explosive activity of neighbouring Vesuvius: many severe or catastrophic floods occurred in the years of eruptions or immediately following. Interestingly, during the period from 1707 to 1872, a period of the most intense explosive activity of Vesuvius, about 27 floods occurred. The magnitude of the analysed floods was estimated on the basis of the recurrence interval, size of the flooded area, and damage caused to both buildings and infrastructure, giving three classes of distribution, as follows.

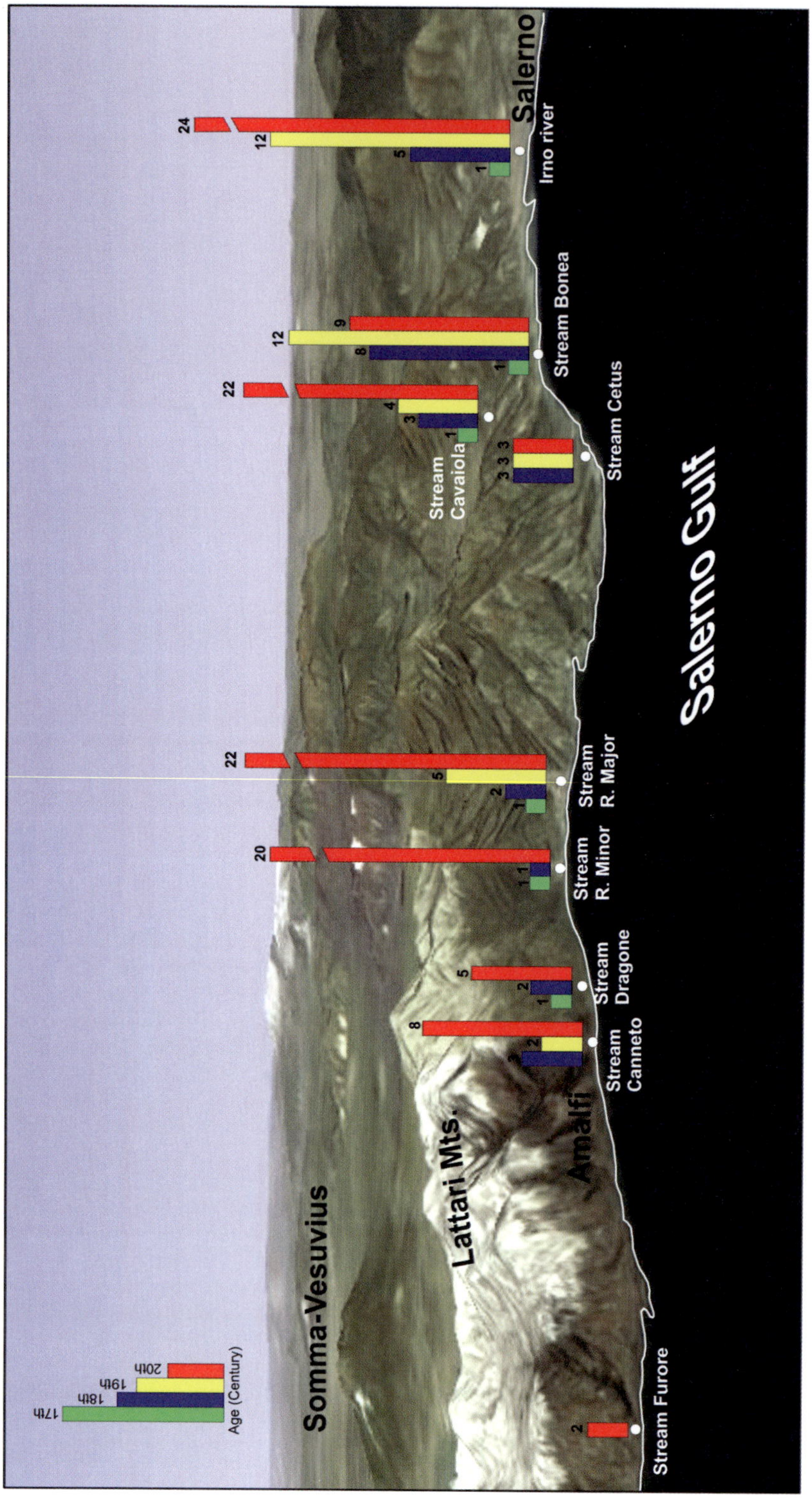

Fig. 7. Areas most frequently affected by floods in the past 500 years.

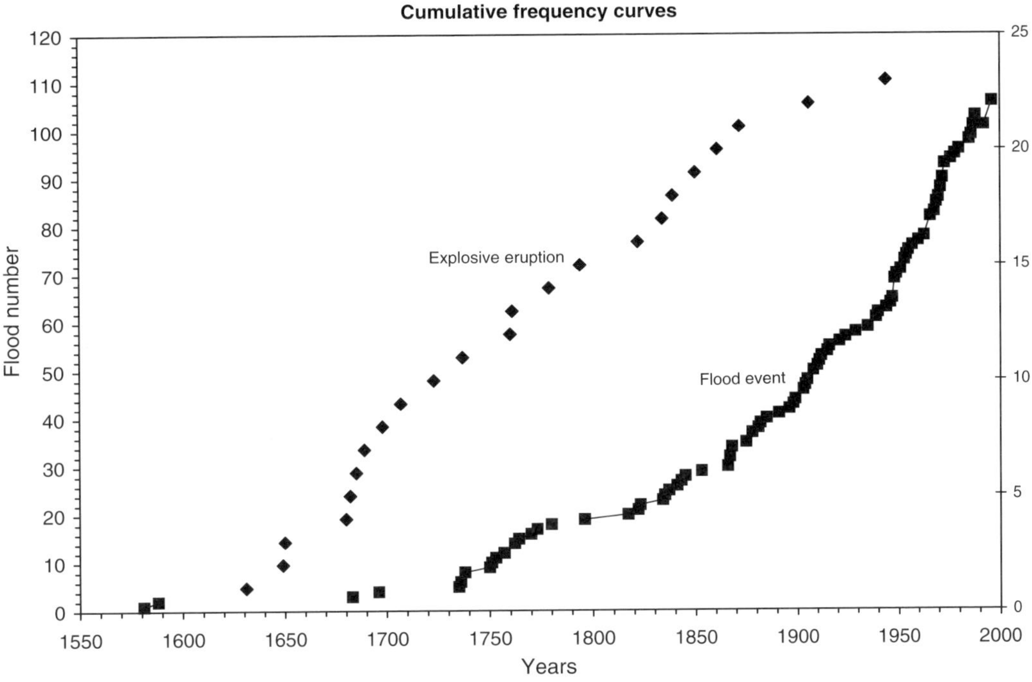

Fig. 8. Absolute cumulative frequency curves: ◆, Vesuvius explosive activity; ■, all the flood events reported in Table 1.

Class I (small flood). There is a restricted area of flooding, minor damage to buildings located adjacent to the river and no serious damage to the population. Overflows depend on the river bed obstruction and on the state of the embankment. Recurrence interval is <20 years.

Class II (intermediate flood). There is a large area of flooding, severe damage to and partial destruction of buildings located adjacent to or along the river. Infrastructure is destroyed along several hundred metres. Other damage is caused by the overflow with its heavy sediment transport. Bankful discharge is exceeded in several places. Recurrence interval is <100 years.

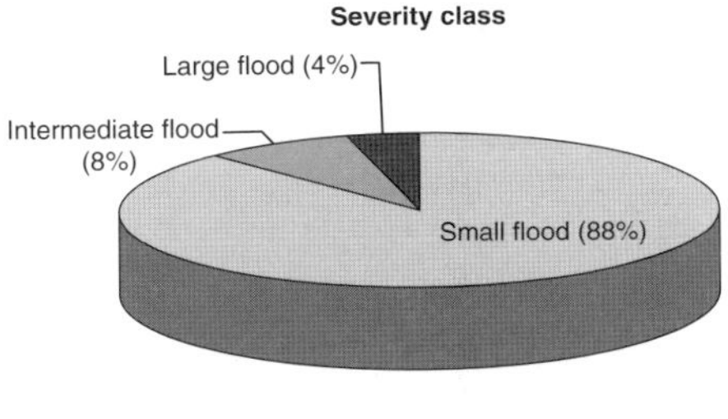

Fig. 9. Severity class of floods occurring in the past 500 years (total number of events 106).

Class III (catastrophic or large flood). There is a large area of flooding, severe damage or complete destruction of infrastructure close to the river, and sections of roadways may be swept away. Overflowing affects zones away from the river bed. Large morphological changes with river bed transformation are also possible. Recurrence interval is >100 years (Fig. 9).

In all, four events were estimated in severity class III, seven in severity class II and the rest in severity class I, which chiefly hit the coastal areas in autumn and winter, with peak occurrences in October. The most severe events occurred in 1581, 1588, 1773, 1899, 1910, 1924 and 1954, producing severe damage to buildings and many victims as well as morphological changes (see Table 2).

Conclusions

In countries such as Italy with a considerable heritage endowment, historical research may be very useful for a large number of scientific activities. Not least of these is natural hazard assessment. In particular, from a perspective of land use planning, detailed knowledge of historical floods is useful to assess hydrogeological hazard, with the possible reduction of hydrogeological risk, damage mitigation and reduction of human fatalities.

In this paper our historical research was chiefly focused on acquiring information on a selected rocky coastal area chosen because of both its geomorphology and its local climatic features. To study historical floods we collected all the information concerning the period between the 16th and the 20th centuries by analysing both published and unpublished sources. The great variety of historical sources made it necessary to formulate an *ad hoc* scientific procedure that would take account not only of the completeness and reliability of documents relating to the period, but that would above all take into consideration the document's intrinsic quality, namely the type, author, period and motivation of the author. Using this approach we examined about 3000 documents, most of which were classified as the highest quality of reference. Analysis of the data in question allowed us to achieve a chronological reconstruction of 106 floods. In this task, the level of information was decisive: we were able to carry out space–time identification, to estimate the affected area and the type of damage to public and private structures, and the geological effects induced. On the basis of the size of the areas affected by flooding, the type of effects induced on the urban and physical environment, and the recurrence intervals, we estimated the magnitude of the events. This measure is fundamental for hydrogeological risk assessment and for estimating the return times of extreme events in the area.

In conclusion, although this study is not exhaustive (indeed, further research is required for period from the Middle Ages to the 16th century), it sets out to make a methodological contribution to data search and analysis. Such information may then be used for initial assessment and subsequent reduction of hydrogeological risk in all coastal areas with ephemeral high-gradient fluvial basins.

Dedicated to the everlasting memory of my sister, Angelina, and my father.

Archival sources

Archivio comunale di Minori (1910–1954).

Archivio comunale di Vietri sul Mare (1899–1954).

Archivio di Stato di Napoli, *Regia Camera della Sommaria* (1573–1661).

Archivio di Stato di Salerno, *Protocolli Notarili* (1500–1798).

Archivio di Stato di Salerno, *Regia Udienza dei Principato Citra* (1620–1827).

Archivio di Stato di Salerno, *Atti Demaniaili* (1806–1961).

Archivio di Stato di Salerno, *Intendenza* (1806–1860).

Archivio di Stato di Salerno, *Prefettura* (1860–1932).

Archivio di Stato di Salerno, *Genio Civile* (1859–1929).

References

AGASSE, E. 2003. Flooding from the 17th to 20th centuries in Normandy (western France) methodology and use of historical data. *In*: THORNDYCRAFT, V. R., BENITO, G., BARRIENDOS, M. & LLASAT, M. C. (eds) *Proceedings of the PHEFRA (Palaeofloods, Historical Data and Climatic Variability), International Workshop, Barcelona*, 99–105.

AMAROTTA, A. R. 1994. *L'alluvione nel Salernitano. Un'esperienza disattesa.* Arti Grafiche Boccia Fuorni, Salerno.

AMBRASEYS, N. N. & FINKEL, C. 1993. Material for the investigation of seismicity of eastern Mediterranean region during the period 1690–1710. *In*: STUCCHI, M. (ed.) *Historical investigation of European earthquakes.* CNR, **1**, 173–194.

AVI Project 1989–1996. World Wide Web Address: http://avi.gndci.pg.cnr.it/welcome_en.htm

BARRIENDOS, M., COEUR, D., LANG, M., LLASAT, M. C., NAULET, R., LEMAITRE, F. & BARRERA, A. 2003. Stationarity analysis of historical flood series in France and Spain (14th–20th centuries). *Natural Hazards and Earth System Sciences*, **3**, 583–592.

BAYLISS, A. C. & REED, D. W. 2001. The use of historical data in flood frequency estimation. Report to MAFF, CEH. World Wide Web Address: www.nwl.ac.uk/feh/historical_floods_report.pdf

BELL, W. T. & OGILVIE, A. E. 1978. Weather compilations as a source of data for the reconstruction of European climate during the Medieval period. *Climatic Change*, **1**, 331–348.

BENITO, G. 2000. Palaeoflood studies in Spain: state of the art and future perspective. *The fragile territory. Research and application on hydrogeological disarray in the world.* Consiglio Nazionale dei Geologi, Rome, Italy, 103–111.

BENITO, G., LANG, M. ET AL. 2004. Use of systematic, palaeoflood and historical data for the improvement of flood risk estimation. Review of scientific methods. *Natural Hazards*, **31**, 623–643.

BOSCHI, E., FERRARI, G., GASPERINI, P., GUIDOBONI, E., SMRIGLIO, G. & VALENSISE, G. 1995. *Catalogo dei forti terremoti in Italia dal 461 a.C. al 1980.* ING–SGA, Bologna, Italy.

BRÁZDIL, R., GLASER, R. ET AL. 1999. Flood events of selected rivers of Europe in the sixteenth century. *Climatic Change*, **43**, 239–285.

CAMERA, M. 1881. *Memorie storico diplomatiche dell'antica città e ducato di Amalfi.* (Reprinted Centro Cultura e Storia Amalfitana, Amalfi, 1999.)

CANCELLARA, E. 1955. Sulla genesi del disastro alluvionale nel Salernitano del 25–26 ottobre 1954. *Annali della Facoltà di Scienze Agrarie, Università di Napoli, in Portici*, 3–22.

CARRATURO, A. 1784. Ricerche storico-topografiche della città e territorio della Cava. *In*: DI MAURO (ed.) *Cava dei Tirreni*, 1976, **III**, 45.

CASCIELLO, E., CESARANO, M. & PAPPONE, G. 2004. Contesto geologico regionale delle aree emerse.

In: ESPOSITO, E., PORFIDO, S. & VIOLANTE, C. (eds) *Il nubifragio dell'ottobre 1954 a Vietri sul Mare–Costa di Amalfi Salerno*. Pubblicazione Gruppo Nazionale per la Difesa dalle Catastrofi Idrogeologiche, **2870**, 33–44.

CINQUE, A. 1986. *Guida alle escursioni geomorfologiche. Penisola sorrentina, Capri, Piana del Sele e Monti Picentini*. Gruppo Nazionale Geografia Fisica e Geomorfologia, **33**, 1–119.

DE VIVO, B. & SCANDONE, R. (eds) 2003. The ignimbrites of the Campanian Plain. *Mineralogy and Petrology*, **79**, 1–2.

ESPOSITO, E., PORFIDO, S. & VIOLANTE, C. 2003*a*. Reconstruction and recurrence of flood-induced geological effects: the Vietri sul Mare case history (Amalfi coast, Southern Italy). *In*: PICARELLI, L. (ed.) *Fast Slope Movements, Prediction and Prevention for Risk Mitigation*. Associazione Geotecnica, **1**, 169–172.

ESPOSITO, E., PORFIDO, S., VIOLANTE, C. & ALAIA, F. 2003*b*. Disaster induced by historical floods in a selected coastal area (Southern Italy). *In*: THORNDYCRAFT, V. R., BENITO, G., BARRIENDOS, M. & LLASAT, M. C. (eds) *Proceedings of the PHEFRA (Palaeofloods, Historical Data and Climatic Variability), International Workshop, Barcelona*. CSIC-Centro de Ciencias Medioambientales Serrano, Madrid, Spain, 143–148.

ESPOSITO, E., PORFIDO, S., VIOLANTE, C., BISCARINI, C., ALAIA, F. & ESPOSITO, G. 2004*a*. Water events and historical flood recurrences in the Vietri sul Mare coastal area (Costiera Amalfitana, southern Italy). *In*: RODDA, J. C. & UBERTINI, L. (eds) *The Basis of Civilization—Water Science?* Proceedings of the UNESCO/IAHS/IWHA Symposium. IAHS Publications, **286**, 95–106.

ESPOSITO, E., PORFIDO, S. & VIOLANTE, C. (eds) 2004*b*. *Il nubifragio dell'ottobre 1954 a Vietri sul Mare–Costa di Amalfi, Salerno*. Pubblicazione Gruppo Nazionale per la Difesa dalle Catastrofi Idrogeologiche, **2870**.

FROSINI, P. 1955. Il nubrifagio di Salerno del 25–26 Ottobre 1954. *Giornale Genio Civile*, **3–4**, 179–188.

FUMANTI, F., RISCHIA, I., SERVA, L., TRANFAGLIA, G., TRIGILA, A. & VIOLANTE, C. 2002. Effetti sul territorio dell'evento meteorico del 7–8 ottobre 1899 nel Salernitano. *Atti dei Convegni Lincei—Il dissesto idrogeologico: inventario e prospettive*, **181**, 395–410.

GLADE, T., ALBINI, P. & FRANCÈS, F. (eds) 2001. The use of historical data in natural hazard assessments. *In*: *Advances in Natural and Technological Hazards Research*. Kluwer, Dordrecht, **17**, 35–54.

GRECO, M. 1787. *Cronaca di Salerno*. (Reprinted Palladio, Salerno, 1985.)

GUIDOBONI, E. & FERRARI, G. 1995. Historical cities and earthquakes: Florence during the last nine centuries and evaluations of seismic hazard. *Annali di Geofisica*, **38**, 617–647.

GUZZETTI, F. & TONELLI, G. 2004. SICI: an information system on historical landslides and floods in Italy. *Natural Hazards and Earth System Sciences*, **4**, 213–232.

INGRAM, M. J., UNDERHILL, D. J. & WIGLEY, T. M. L. 1978. Historical climatology. *Nature*, **276**, 329–334.

LAZZARI, A. 1954. Aspetti geologici dei fenomeni verificatisi nel Salernitano in conseguenza del nubrifagio del 25–26 Ottobre 1954. *Bollettino della Società dei Naturalisti*, **63**, 131–142.

PENTA, F., LUPINO, R., CAMOZZA, F. & ESU, F. 1954. Effetti dell'alluvione del 26 ottobre 1954 nel Salernitano. *Rivista Italiana di Geotecnica*, **6**, 245–257.

RISCHIA, I., FUMANTI, F. *ET AL*. 2003. *Analisi e modellazione dei fenomeni di debris flow nel bacino di Cetara (Sa)*. FIST–Geoitalia, Bellaria, 532–534.

ROBSON, A. J & REED, D. W. 1999. *Statistical procedures for flood frequency estimation*. Institute of Hydrology, Wallingford, Flood Estimation Handbook, **3**.

SANTACROCE, R. 1983. A general model for the behavior of the Somma–Vesuvius volcanic complex. *Journal of Volcanology and Geothermal Research*, **17**, 237–248.

SIGURDSSON, H., CAREY, S., CORNELL, W. & PESCATORE, T. S. 1985. The Eruption of Vesuvius in A.D. 79. *National Geographic Research*, **1**, 332–387.

SPHERE EUROPEAN PROJECT 2000–2003. World Wide Web Address: http://www.ccma.csic.es/dpts/suelos/hidro/sphere/index.html

TELELIS, I. G. 2005. Historical–climatological information from the time of the Byzantine Empire (4th–15th centuries AD). *History of Meteorology*, **2**, 41–50.

THORNDYCRAFT, V. T., BENITO, G., BARRIENDOS, M. & LLASAT, M. C. (eds) 2003. *Proceedings of the PHEFRA (Palaeofloods, Historical Data and Climatic Variability) International Workshop, Barcelona*. CSIC-Centro de Ciencias Medioambientales Serrano, Madrid, Spain.

VIOLANTE, C. 2009. Rocky coast: geological constraints for hazard assessment. *In*: VIOLANTE, C. (ed.) *Geohazard in Rocky Coastal Areas*. Geological Society, London, Special Publications, **322**, 1–31.

VIOLANTE, C., BISCARINI, C., ESPOSITO, E., MOLISSO, F., PORFIDO, S. & SACCHI, M. 2009. The consequences of hydrologic events on steep coastal watersheds: the Costa d'Amalfi, eastern Tyrrhenian Sea. *In*: LIEBSHER, H. J. *ET AL*. (eds) *The Role of Hydrology in Water Resource Management. Capri Italy, 2008*. IAHS Publication, **327**, 102–113.

VOGT, J. 1993. Historical seismology. Some notes on sources for seismologists. *In*: STUCCHI, M. (ed.) *Historical Investigation of European Earthquakes*. CNR, **1**, 15–24.

Application of a method to assess coastal hazard: the cliffs of the Sorrento Peninsula and Capri (southern Italy)

T. DE PIPPO[1]*, C. DONADIO[1], M. PENNETTA[1], F. TERLIZZI[1] & A. VALENTE[2]

[1]*Dipartimento di Scienze della Terra, Università degli Studi 'Federico II' di Napoli, Largo S. Marcellino 10, 80138 Napoli, Italy*

[2]*Dipartimento di Studi Geologici e Ambientali, Università del Sannio, Via dei Mulini, 59a, 82100 Benevento, Italy*

**Corresponding author (e-mail: depippo@unina.it)*

Abstract: A systematic method to quantify, rank and map the distribution of hazards is applied to the coastal cliffs of the Sorrento Peninsula and Capri (Campania, southern Italy). For such cliffs, which have previously been characterized in terms of types and processes, and therefore compartmentalized, the predisposition to a particular hazard (or indicator), based on its nature, magnitude and recurrence, is evaluated by assigning a code: the higher the predisposition, the higher the code for each compartment. Moreover, hazards can influence one another, and the number of such interactions indicates the seriousness of each hazard, to which a weighting is assigned. By comparing each code in a specific compartment using an interaction matrix, which takes the weighting into consideration, we have calculated a resultant, which is the overall hazard for the compartment. This resultant can also be expressed cartographically. In this application six primary hazards (parameters) are considered: cliff retreat, riverine flooding, storms, landslides, seismicity and volcanism, and man-made structures. The last is the most hazardous parameter, which is weighted highly, owing to its extensive influence on the other hazards. In contrast, riverine flooding and seismicity and volcanism are the least interactive.

Coastal areas are generally dynamic environments because continental and marine processes converge along them to produce a landscape that is subject to rapid changes. Such changes could be attributed to single catastrophic events, as well as to continual events and processes, which contribute to the modelling of the coastal landscape. The evolution of this landscape, in which waves, tides and marine currents interact, may modify the intensity of one of these processes by increasing or reducing the effects of another, in time and space. Moreover, some recent coastal changes are the result of human activity, which is able to heighten the effect of coastal processes.

In coastal areas, which vary greatly in topography, climate and vegetation as well as land use (tourism, industry, agriculture, etc.), the rate of coastline retreat has assumed significant proportions. This rate can represent a high degree of coastal hazard, especially in populated areas near the coast, because of frequent losses of money and property, and even of human life. In this case, knowledge of coastal hazards as well as their distribution becomes a basic tool for supporting planning and management decisions.

This paper proposes an application of a semi-quantitative method for assessing coastal hazard along a stretch of the Campania coast in southern Italy, where steep and rocky coasts are prevalent. This stretch, comprising a peninsula and its geologically related island, is known for its holiday resorts, such as Sorrento, Positano, Amalfi and Capri.

Coastal setting

The investigated coastal area is on the eastern side of the Tyrrhenian Sea in the region of Campania (Fig. 1) and has a very complex topography, which reflects the neotectonic activity affecting this portion of the Apennine chain during the Quaternary, when extensive subsidence and uplift resulted in a horst and graben structure (Brancaccio *et al.* 1991). Along this coast, the horst is represented by a narrow mountainous transverse ridge, oriented nearly east–west, with prevalent steep and rocky coastal cliffs, as observed along the Sorrento Peninsula and on Capri, whereas the graben consist of broad alluvial plains with sandy beaches, as seen on the Campania and Sele plains, respectively to the north and south of the Sorrento Peninsula (Fig. 1). In both plains the filling of the graben is mainly due to the contribution of alluvial sediments of the Volturno and Sele rivers, as well as to marine transgression. However, only on the

From: VIOLANTE, C. (ed.) *Geohazard in Rocky Coastal Areas*. The Geological Society, London, Special Publications, **322**, 189–204. DOI: 10.1144/SP322.9

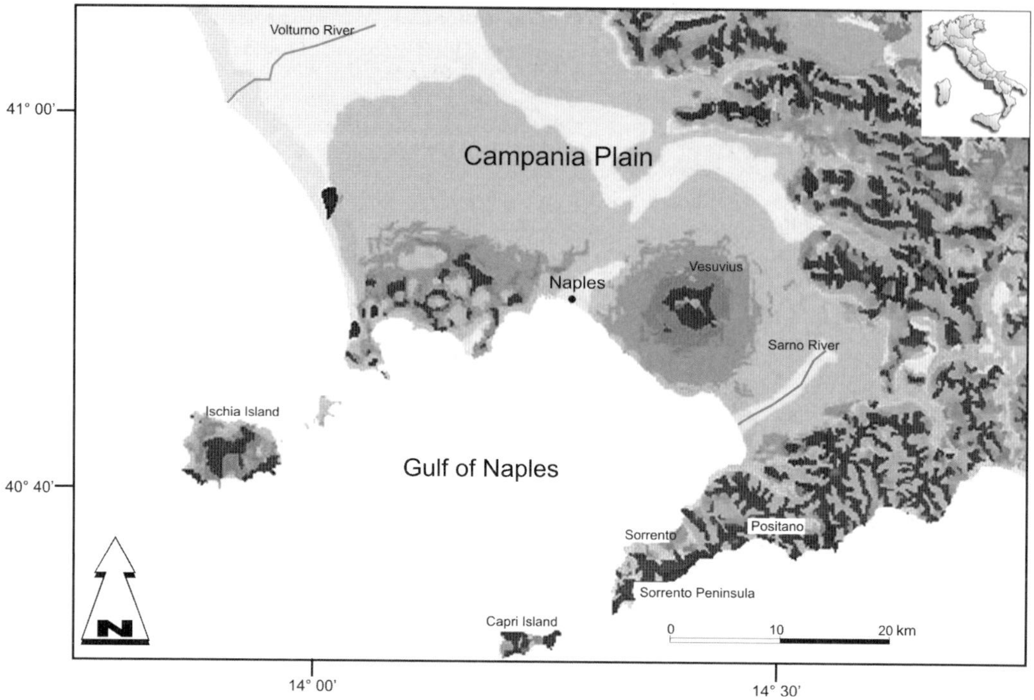

Fig. 1. Location of the study area. The Sele Plain and the Sele River are south of the area shown in the figure.

Campania Plain is there a major contribution in terms of thickness and distribution of deposits deriving from the volcanic centres of the Phlegrean Fields and Somma–Vesuvius, whose activity began in the Early–Middle Pleistocene (De Vivo *et al.* 2001). These products, in particular tuff and lava, reach the sea and characterize the rocks exposed in cliffs as well as part of the submarine floor in the Bay of Naples. Pyroclastic deposits are also found on hillslopes and in valleys bordering the Campania Plain as well as in the Sorrento Peninsula and on Capri.

Urbanization of the coastal area started in Classical times and has since expanded markedly, with the coast being used for several purposes (thermal bath complexes, fishing, shipping, tourism, etc.); thus some of the structures and infrastructure sited on the coast were protected from wave action. In time, these protected stretches have been extended because of the critical rate of coastal erosion and the greater occupation of the coastal zone for residential and industrial purposes. In most cases, the urbanized coastal stretches have lost the character of the natural environment as a result of engineering structures being built, at sea or on the coast, to protect the waterfront ('technocoast'; De Pippo *et al.* 2008).

The rainfall and temperature recorded for the coast of Campania are typical for a Mediterranean climate, with a dry season between June and September and a wet season from October to May. For this coast the westernmost areas experience lower rainfall (Capri and the tip of the Sorrento Peninsula: <1000 mm a^{-1}) compared with the southeasternmost area (southern side of the Sorrento Peninsula: >1200 mm a^{-1}). The latter condition is due to a barrier effect to air masses from the south, especially in autumn, caused by the presence of high-altitude relief close to the coast (increasing eastward to 1440 m above sea level (a.s.l)).

Records from the Capri weather station (267 m a.s.l.) show a significant percentage of northerly winds, although the frequency of southerly winds, especially from the SE, increases in spring and in autumn, as already noted for rainfall. Probably related to the widest southern fetches, above all from the SW (>500 km), are the prevalent storm waves coming from the south; northwesterly storm waves are much less common.

Wave height generally ranges between 0.9 and 2.2 m, although heights of up to 4.7 m may be reached, especially in winter. Lower values occur in sheltered stretches, such as some areas in the Bay of Naples. Along the Campania coast the littoral drift is generally from NW to SE (Cocco *et al.* 1989, 1992).

Sea-cliff features

The morphological features and evolution of the rocky cliffs along the Sorrento Peninsula and Capri island have been defined through field observations, study of aerial photographs and satellite images, as well as single beam bathymetric surveys down to −20 or −30 m depth. Along this coastline, some rocky cliffs are characterized by a shore platform at their foot, whereas others continue vertically below sea level. On top of the cliff, a predominantly convex or almost uniform gradient (at times with crags) and occasionally a concave slope develops. There are also cases where the cliff represents the edge of a nearly horizontal or slightly inclined surface, such as Pleistocene marine terraces (Brancaccio *et al.* 1991).

The presence of the coastal platform (*sensu* Sunamura 1992) in most of the analysed morphologies indicates persistent erosion processes along the cliff, which have also occurred during sea-level stands other than the present. The physiography of the coastal platform is related both to the intensity of the erosional processes and to the time span in which they have acted, as well as to the structure and durability of the outcropping rocks (hard or soft). More specifically, cliffs with a platform up to 200 m wide, and occasionally up to 500 m, have been observed, with a bottom slope between 3% and 10%. Other cliffs with a platform, or rather a ramp, which extends for less than 100 m and with a slope exceeding 10% have been recognized. They are classified as sloping shore platforms (type A of Sunamura 1992), but our observations allow two categories to be discriminated within this type: A1, large coastal platforms with low gradient; A2, narrow coastal platforms with a significant gradient.

On the northern side of the Sorrento Peninsula, which is less subject to intense sea storms, coasts develop mainly with the type A1 profile, showing a large platform and reduced gradient (Fig. 2). The height of the cliffs varies greatly in relation to the lithology: between 25 and 70 m for carbonate formations; between 120 and 200 m for terrigenous (sandstone and pelite) units overlain by carbonate rocks; between 25 and 60 m for arenaceous–silty layers; less than 25 m for coastal slopes formed by pseudo-coherent talus deposits; up to 50 m for tuffs (Fig. 3) (De Pippo *et al.* 2007).

This variability is frequently associated with a geotechnical characterization of soft consistency rocks. Indeed, along the northern side of the Sorrento Peninsula the mechanical resistance tested on some rock masses, such as highly fractured limestones, proved mediocre (Budetta *et al.* 1991).

The profile classified as A2 develops frequently in the case of very high cliffs, essentially in carbonate rocks. The degradation materials usually accumulate at the cliff toe in the area of connection with the coastal platform, which is inclined at a few degrees. In some cases, the platform develops down to −7 or −8 m depth, with a 10° tilt. This situation is common along the Sorrento Peninsula, especially to the west of the village of Sorrento (Fig. 4). In some cases, a cliff can be located behind a fault offshore, as suggested by relict forms still anchored to the substratum (stacks to the south of Capri and Li Galli islets off Positano, on the southern side of the Sorrento Peninsula).

The initial lack of homogeneity in physico-mechanical characteristics of some lithotypes along the study cliffs is linked to marked tectonic movements and to the effects of intense degradation phenomena that are active along the slopes above them. Tectonics has acted on the massive rock formations leading to rock breakdown into blocks or, in some cases, to such a dense network of fractures that the rock resembles a cataclasite. On these rocks degradation phenomena, and thus the erosive action of the waves, are facilitated. Dissolution of carbonates with the development of microforms (rock pools and lapiés) and macroforms (caves) are among the most common weathering phenomena observed along the investigated coast. Moreover, detrital cover commonly develops along sea-cliffs, with characteristics suggesting that it is not due to present-day morphological conditions and degradation processes, but is inherited from the frost-wedging conditions existing during the latest Pleistocene glaciations.

Another cliff type has heights varying from 10 m to just over 100 m and is frequently characterized by slopes with medium gradients. At the foot of these coastal slopes, terraced surfaces are found, which may form exposed surfaces during low tide. On these surfaces, strips of well-cemented calcarenites with abundant mollusc fragments of a beach environment, probably of Tyrrhenian age (Brancaccio *et al.* 1991), are found. The surfaces are shaped by processes of chemical and physical weathering of the rocks, which are favoured by the frequent drying to which they are subjected. This morphotype can be observed along the northern sector of the Sorrento Peninsula and Capri, mainly on carbonate formations and secondarily on terrigenous deposits (De Pippo *et al.* 2007). The erosive action of waves has strongly shaped the coastal platform and therefore this type must be classified as a different form.

The extension of the emerged coastal profile in an underwater environment to −30 m, without significant erosive forms, highlights the role of tectonic movements that mainly occurred during the Pleistocene. The cliffs that develop along faults are usually steep, so the erosive action of waves vanishes or drastically decreases as a result of the well-known

Fig. 2. Carbonate cliff with a pocket beach developed at its base. Bathymetric data reveal a wide and gently sloping shore platform offshore (Vico Equense village; northern side of the Sorrento Peninsula).

phenomena of reflection. These cliffs belong to the plunging cliff type of Sunamura (1992) and occur mainly in the carbonate formations at the tip of the Sorrento Peninsula (Punta Campanella) and in some segments of its southern side (Fig. 5; Brancaccio 1968; De Pippo *et al.* 1998, 2007), as well as in the west of Capri (Barattolo *et al.* 1992; De Pippo *et al.* 2007). These cliffs can be classified as hard rock cliffs. In particular, along their exposed surfaces palaeo sea-notches at different heights are found; the genesis of these is probably connected with recent sea-level highstands (Brancaccio *et al.* 1978; Pirazzoli 1993; Riccio *et al.* 2001), suggesting stability of the cliff and the persistence of the wave

Fig. 3. Cliff composed of volcanic tuff, about 50 m high. In this area a submerged shore platform reaches about 2 km width (S. Agnello on the northern side of the Sorrento Peninsula). (Note the large structures built on the top of the cliff.)

action. The formation of sea-notches on carbonate cliffs is mainly due to the dissolution of carbonate rocks and to organism grazing and perforation activity (biokarst). The role of chemical processes is also emphasized by the widespread presence of marine caves, especially along the southern cliffs of the Sorrento Peninsula, at sea level or slightly above or below it (De Pippo *et al.* 1998).

Fig. 4. Cliff composed of carbonate deposits, with a narrow and steep shore platform (Punta Gradelle on the northern side of the Sorrento Peninsula). (Note the cliff composed of tuff on the lower right.)

Fig. 5. Example of a plunging cliff. The photograph also shows truncated incisions (arrows indicate some of them). Seno di Ieranto near the western end of the Sorrento Peninsula.

Structural cliffs showing a continuity along the stream incisions orthogonal to the coast have great importance in the coastal landscape of Capri and the southern Sorrento Peninsula, thus defining a characteristic ria morphology. Such incisions partly extend underwater and often correspond to tectonic alignments (De Pippo *et al.* 2007). In other cases, they may be related to the degree of fracture of outcropping rocks or to varying resistance of rock formations along the slope. The latter case is observed, for example, along the southern side of the Sorrento Peninsula, where a structural slope is composed of Cretaceous limestone on Miocene arenaceous–pelitic deposits.

In conclusion, the morphology of the coast of the Sorrento Peninsula and Capri represents a dynamic result of past and long-term evolution and can be considered rather homogeneous, being mostly steep and rocky. Moreover, the stretch around Sorrento and Castellammare di Stabia in the Sorrento Peninsula, as well as the localities known as Marina Grande and Marina Piccola on Capri, are somewhat modified by anthropogenic action expressed by several manmade structures and by use of the waterfront, which have greatly interfered with the natural environment. The evolution of such cliffed coasts, which have lost most of their natural features, is less predictable than that of natural cliffs.

Method

In recent decades many attempts have been made to seek the elements that could be essential in assessing coastal hazard. At first, the erosion rate of a beach or along a cliff was measured to give the degree of hazard (e.g. in Italy, by Caputo *et al.* 1991). Later, other researchers recognized in the features of a coast (i.e. geological setting, coastal slope) as well as in the occurrence of some natural events (coastal erosion, wave damage from storm, rockfalls from the cliff) the potential hazard for an area (Berger 1997; Berger & Iams 1996; Bush *et al.* 1999). Recently, we proposed a method that followed the latter trend: the primary hazards were identified and ranked for a specific coastal area using an interaction matrix (De Pippo *et al.* 2008).

Here this method is applied to rocky coastal areas, which are characterized in terms of physical and anthropogenic features as well as morphological processes. An arrangement of these features as well as a high magnitude and a short recurrence interval of such processes could indicate the tendency to become a hazard. Therefore, a detailed knowledge of the coast is important to choose the most significant attribute to consider in ranking the matrix parameters. In other coastal studies (Bush *et al.* 2001; Richmond *et al.* 2001) the importance of the overall hazard has been given by a simple sum of the ranks, whereas in this study we opted to use a matrix in which primary hazards interact with each other in a given coastal area (Hudson 1992). This matrix, which has been tested in other environmental studies (e.g. Simeoni *et al.* 1999*a*, *b*), takes account of the impact of each parameter on the others.

In this application six parameters are considered: cliff retreat, riverine flooding, storm waves, landslides, seismicity and volcanism, and manmade structures; these are the most common hazards recognized along this coast. These parameters represent the basic components of the matrix, as they are plotted in the boxes of its diagonal. The other boxes in the matrix are filled by following a clockwise scheme of interaction. In particular, each element of the row that crosses one of the parameters on the mean diagonal shows the influence of this parameter on the system, thus indicating the cause of the phenomena, whereas each element of the column that crosses the same parameter shows the influence of the system on this parameter, thus focusing on the effect of the phenomena (Fig. 6). For instance, in the interaction between two diagonal terms such as cliff retreat and storm

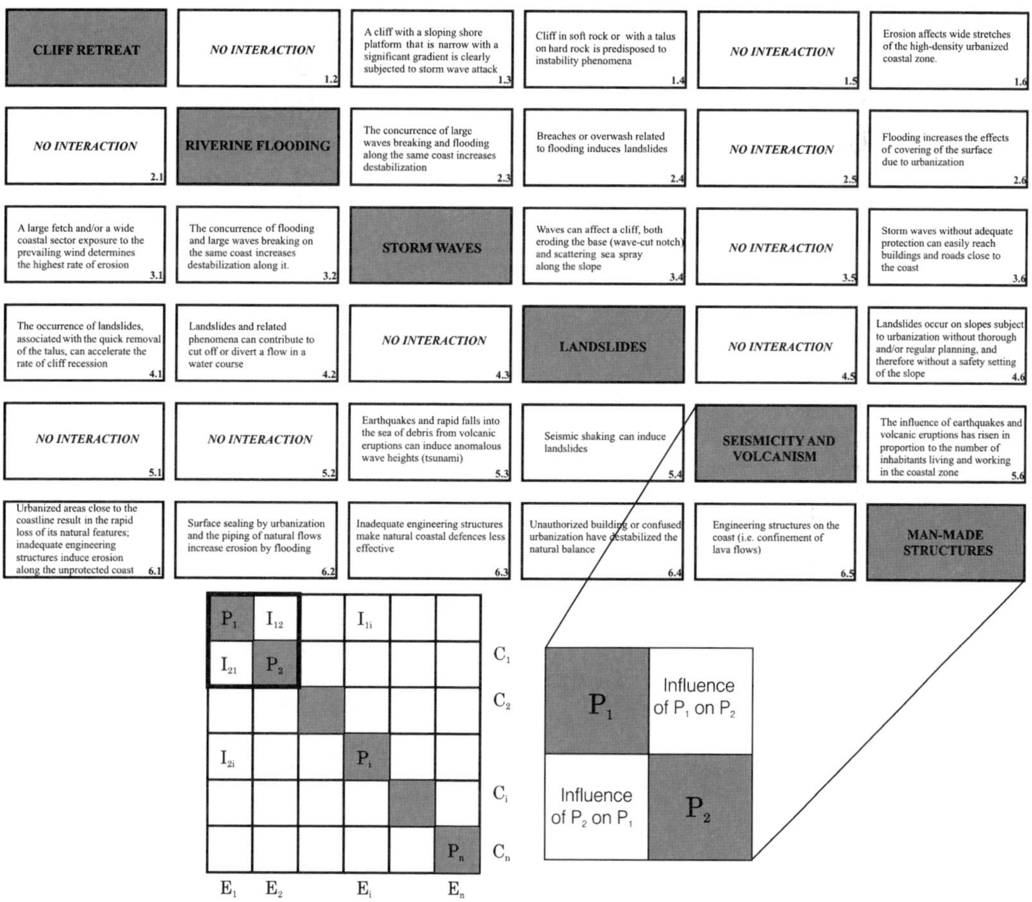

Fig. 6. Descriptive matrix showing the interactions of hazards. There are six major diagonal terms: cliff retreat, riverine flooding, storm waves, landslides, seismicity and volcanism, and manmade structures. This matrix indicates the influence of the morphological parameters on the system (the cause of the phenomena) or the influence of the system on each of the parameters (the effect of the phenomena) (lower-right scheme). The cause–effect diagram for *N* parameters (lower-left scheme) indicates how the interaction matrix is filled: the box I12 represents the influence of P1 on P2 (cause); conversely, the box I21 represents the influence of P2 on P1 (effect). This mechanism is repeated for each parameter recognized, shown as the diagonal terms in the interaction matrix.

waves, a cliff with a sloping shore platform, which is narrow with a significant gradient (cause), is clearly subject to storm wave attack, whereas a large fetch exposed to the prevailing wind, and therefore to the highest waves, results in the highest rate of erosion (effect).

Both causes and effects contribute at the same rate to coastal hazard assessment. However, in analysing the number of matches for the chosen parameter the most interactive parameters (which were subordinate in the study by Hudson 1992) are represented by the manmade structures and the landslides, whereas the least interactive are riverine flooding and seismicity and volcanism (which were dominant in the study by Hudson 1992). The most interactive factors account for the influence that the system has on them. In contrast, the least interactive, which are strictly related to the physical nature of the area, occur without affecting any changes from the surrounding context.

The importance of each parameter describing the cause–effect interaction is expressed through a semiquantitative score from zero to four (*Ip*), from nil to critical (Table 1). This score is attributed according to the features of a compartment of the coastal stretch examined, which differ from those of the adjacent ones. Such features (forms and processes) determine the score on this stretch, which reflects their predisposition to increase the hazard. Quantification of specific forms (indicators) could be fixed with a range of values or assigned a precise value for a particular stretch or left at the discretion of the operator. For instance, a slope with a low gradient (<20°) free of erosional features has the minimum rank, a medium gradient (20–45°) with minor gullies or rills, and a steep gradient (45–70°) with several discontinuities are given respective scores of two and three, and lastly, an oversteepened to steep slope (>70°) with slump scars has a score of four. Several indicators with high values give a final maximum score, but indicators of different signs could lower the final score. For instance, the presence of a toe protection of the cliff (beach or talus) lowers the score, notwithstanding other indicators able to increase erosion along the subaerial slope.

The score is also assigned on the basis of previously defined indicators of physical changes and/or events occurring in a suitable time range on a particular coastal stretch. Obviously, the probability of an event being repeated in a very short time range, of the order of a couple of years (high storm waves or landslides induced by heavy rainfall), makes the score significantly worse, whereas those events with a recurrence that is indefinable (volcanic eruption, seismicity, tsunami) or too long for the planning timetable (>10 years) weigh negligibly on the score. Lower recurrence times from 5 to 10 years and from 2 to 5 years respectively worsen the weight by scores of one and two.

For instance, a cliff exposed to a large fetch has a higher score than one exposed to a limited fetch, but this score could also change according to the presence or absence in the submerged profile of a sloping platform. Furthermore, if the recurrence of the storm waves is less than 5 years, the score of this stretch fixed at two (i.e. moderate fetch, cliff with a reduced and high-degree sloping shore platform, and so on) can be raised to four.

To avoid the excessive weighting of a single parameter, such as a locally induced retreat or a single exceptional event, this method considers the sum of the score both for the row and column. Such a sum, equal to the intensity of interaction as positive (cause) and negative (effects), does not yet correspond to the overall hazard, as the highest weight of parameters must be considered. We therefore established for the different geomorphological units the percentage of influence of each parameter (*Xp*) on the overall hazard (on a decreasing scale from a maximum of six to a minimum of zero). For instance, landslides are assigned a weight of five, as the number of causes and effects is lower than that of man-made structures; cliff retreat and storm waves are assigned a weight of four and three, respectively. The lowest weights are for riverine flooding and seismicity and volcanism (Table 2). The last factor has a reduced importance on the coast in question, although proximity to Vesuvius, an active volcano, or to the seismogenetic area of the Apennines, indicates some influence, which is why we decided to assign it the minimum weight. Although this consideration could emphasize the other weights and slightly increase the degree of overall hazard, we deem it none the less appropriate.

To define the overall hazard (*Ht*), the score (*Ip*), already defined for each coastal stretch and for each parameter, will be multiplied by the coefficient (*Xp*) related to the importance of one parameter over another, so that the weighted sum of the product of each parameter for this coefficient of hazard on the investigated coastal stretch is

$$Ht = \sum_{n=1-6} Ip_n Xp_n.$$

The degree of this hazard has been expressed cartographically from Low to Extreme (Stauble 2003): below the minimum value, that is absent or negligible, no map symbol is used (Table 3). This analysis could also be developed through a geographic information system (GIS), where each feature class has a relative database to which a weight could be attributed.

Table 1. *Rank of incidence of the most important indicators*

Indicators	Rank of incidence (*Ip*)			
	1	2	3	4
Erosion rate	Stable or low (near 0)	Moderate (<1 m a^{-1})	High (<5 m a^{-1})	Severe (>5 m a^{-1})
Nearshore and offshore topography	Natural protection (bars, reefs)	Coastal shelf of moderate width with discontinuous offshore bars	Wide shelf; shallow open water with steep slope without bars	Very narrow shelf; high wave energy; rapid deepening of nearshore; closeness to canyon heads offshore
Fetch	Limited (<50 n.m.)	Moderate (<150 n.m.)	Large (<300 n.m.)	Very large (>300 n.m.)
Geomorphological features	Upland (>6 m); very distant from a river mouth	Floodplain or low-elevation terraces (>3 m); within sight of an inlet or river mouth	Wide coastal plain (<3 m); near an inlet or river mouth	Wide coastal plain (<3 m); adjacent to an inlet or river mouth or close to a lagoon
Cliff topography	Low slope angle (<20°); free of erosional features	Medium slope (20–45°) angle; minor gullies or rill	Steep slope angle (45–70°); irregular and stepped; possible slump scars	Oversteepened to steep slope angle (>70°); stepped and gullied; slump scars
Mechanical condition of the rocks exposed on the cliff	No fractured or weathered 'hard' rocks	'Hard' rocks with a limited number of fractures or a little weathering with a seaward dip: no weathered 'soft' rocks	'Soft' rocks (clays, unconsolidated sands, volcanoclastic rocks) or moderately fractured or weathered 'hard' rocks	Saturated soft rocks (clays, unconsolidated sands, volcanoclastic rocks) or deeply fractured or weathered 'hard' rocks
Vegetation on the cliff	Mature, dense and undisturbed	Mature, undisturbed to minor rotation	Poor and ephemeral; toppled or rotated	Bare; burned or artificially removed
Toe protection of the cliff (beach or talus)	Wide beach or old large talus, well vegetated	Mainly wide beach or vegetated talus	Narrow beach, or eroded or fresh talus	Absent to narrow
Anthropogenic covered surface	<25%	>25%	>50%	>75%
Engineering structures	No structures	Few structures and fronted by beach	Structures grouped along the shoreline	Numerous structures on coast or offshore

n.m., nautical miles/1852 m.

Table 2. *Coding of hazard parameters for the steep and rocky shores*

Parameter number	Hazard parameters	Causes	Effects	Causes + effects	*Xp*
1	Cliff retreat	17	10	27	4
2	Riverine flooding	7	1	8	1
3	Storm	9	10	19	3
4	Landslide	8	24	32	5
5	Seismicity and volcanism	12	1	13	2
6	Manmade structures	19	26	45	6
Total		72	72	144	

Modified from De Pippo *et al.* (2008).

Table 3. *Overall hazard degree according to the weighted sum*

Overall hazard degree (*Ht*)	Weighted sum
Absent or negligible (N)	<21
Low (L)	>21
Medium (M)	>27
High (H)	>33
Extreme (E)	>39

Source: De Pippo *et al.* (2008).

Application

Six coastal hazards have been recognized in the study area, as follows.

(1) Cliff retreat resulting from erosion produced by wave action and weathering. One of the most important factors in cliff evolution is the lithology exposed: a mostly homogeneous hard rock, such as some limestones, will be modelled more slowly than a soft rock, such as tuff or alternating sandstone and shale. This first division produces a lower rank for hard rock (*Ip* = 1 or 2) and a higher rank for soft rock (*Ip* = 3 or 4). However, the presence of elements in hard rock such as fractures and faults, as well as notches and caves, may act as 'catalysts' in the rate of retreat, in which case the rank will be higher. Rock resistance tends to decrease with the increase in weathering phenomena. For instance, on the coastline studied, and especially on the carbonate outcrop, the climatic conditions allow karstic phenomena to develop. During periglacial conditions on the studied coast in the Pleistocene abundant detritus was produced, which now covers the coastal slopes. The great thickness of this talus, often with a significant contribution of pyroclastic elements, pushes the rank up to the maximum (*Ip* = 4).

The last element in cliff retreat is the replacement of rock and vegetation on the slope to construct buildings, roads or tourist facilities. This replacement yields greater dishomogeneity in the rock mass and leads to rainwater infiltration. The assigned rank, in this case, changes according to the type of lithology, the degree of weathering and the extent of the replaced surface.

(2) Riverine flooding occurring in conjunction with intense rainfall. Rapid rise in water level and flash-flooding along steep coastal streams have a series of hazardous implications, as reported for the 1954 catastrophic flood that occurred in the stream Bonea located at the southeastern stretch of the Sorrento Peninsula (the Amalfi coast; Esposito *et al.* 2004*a*, *b*; Violante *et al.* 2009), outside the study area. Indeed, in some steep streams along the northern side of the Sorrento Peninsula (Santo *et al.* 2002), as well as in the central portion of Capri, high-frequency events occur that affect the back area of the coast and could cause new open water to form. This could be a major hazard especially where the area has been urbanized (Rossi & Villani 1994), as in Castellammare di Stabia. Given its recurrence interval (<10 years) and intensity this hazard is ranked as medium close to the stream (*Ip* = 2). However, complete or partial removal of the detritus covering coastal slopes may increase the distance of the shoreface from the cliff, thus reducing the overall hazard.

(3) Storms are well known to those living along the coast, especially when high waves, usually produced by severe winter storms and consequent surges, cause severe property damage and are sometimes fatal.

Regardless of the exposure of the coast, some morphological traits of the shoreface area are important in assigning the rank. For instance, gravel beaches or debris cones lying at the foot of the cliff increase the distance between the shoreface and the cliff, thus averting the energy of the waves from the cliff and lowering the rank. Also, the presence of a sloping shore platform may lower the rank, especially when the platform is wide and has a low gradient (type A1 in this study), whereas a narrow coastal platform with a significant gradient (type A2 in this study) increases the rank, because of the short distance between the shoreline and the depth of breakers.

As regards coastal orientation, exposure of the coast to waves is related mainly to the distance the wave travels after leaving the generating area (the fetch) and secondarily to wave refraction. As mentioned above, the longest fetch and the direction of more frequent storms (40%) is from the SW. Hence swell strongly affects the southern side of the Sorrento Peninsula (*Ip* = 2–3). However, on this side of the peninsula the rank may differ depending on whether the cliff has a sloping shore platform or a plunging one. In contrast, on the northern side of the Sorrento Peninsula the fetches are shorter and affect only the westernmost area directly, especially Capri. Nevertheless, some local storms can come from the north in the Bay of Naples and generate waves that reach the coastal area of the port of Sorrento in approximately the same form as they are generated. In this case high waves strongly attack the cliffs, which are frequently of a sloping shore platform type (*Ip* = 2). In the Bay of Naples the phenomenon of refraction is also common, as a result of both the presence of several islands (i.e. Capri to the south, and Ischia and Procida to the north) and submarine topographic features. Indeed, volcanism in the Phlegrean Fields and on the Somma–Vesuvius edifice has created a very irregular sea floor with terraces and canyons (D'Argenio *et al.* 2004; Violante 2009), which may cause waves to converge or diverge.

(4) Landslides affecting the coastal slopes. The main parameters controlling the probability of rockfalls and mudflows relate to steepness and mechanical conditions of the exposed rocks (Hutchinson 1983; Sunamura 1983). The cliff profile is fundamental in attributing a rank of hazard, mainly on the basis of steepness and secondarily on height. The cliffs in soft rock are clearly less high and less steep, but the propensity to fail is high (*Ip* = 3–4), especially where there is undermining at the foot of the cliff, which is frequent on a narrow sloping shore platform. Steep and plunging cliff frequently occurs in hard rock (Fig. 7), where height and steepness generally reach the highest value, but the rank is low unless a significant intersection of fractures occurs. As noted above, discontinuities in the rock mass, such as overlapping of hard rock on soft rock and vice versa, interlayering of thick pelitic strata, and columnar structures, can be important predisposing factors leading to collapse of large blocks and debris (Budetta *et al.* 2000). In particular, the presence of sea-notches and tension cracks has been found to be responsible for the instability of even hard rock coastal cliffs (Kogure *et al.* 2006), whereas wide wave-cut terraces protect the

Fig. 7. Example of a carbonate cliff affected by slide phenomena (Marina Grande on the northern side of Capri).

cliff from mechanical erosion. The latter feature has controlled, for instance, the development of cliff profiles and hence their evolution generally towards a more stable condition (Sunamura 1983; Trenhaile 2002). Another characteristic that increases the rank is the existence of previous instability phenomena occurring along the cliff. According to recent data several landslides along the coastal cliffs both on the Sorrento Peninsula and on Capri have been reactivated. Indeed, significant damage and avoidable deaths occur both at the top of cliffs and at their base, often during autumn and winter. This may be explained by the greater intensity of the processes that affect the cliffs, such as wave action at the base, and the large amount of rain, which penetrates the weathered rocks at the top or along the cliff (Fiorillo & Wilson 2004). For these reasons the rocky cliffs of the Sorrento Peninsula are given a high score for this hazard, especially on the southern side (Budetta & Santo 1993; Budetta *et al.* 1994). A high score is also assigned to stretches affected by major forest fires, which reduce the vegetated area of the cliff (e.g. maquis), especially when the exposed lithology is unconsolidated sediment, which can easily slide seaward (Russo & Valletta 1995).

(5) The seismicity and volcanism hazard is concentrated in the area close to the volcanic edifice of Somma–Vesuvius, and is thus not so important for the Sorrento Peninsula and Capri. However, the volcanic activity of Vesuvius has strongly modified the topography in the easternmost part of the coast studied, so that a medium to low rank ($Ip = 1–2$) can be assigned. Examples of such modifications include lava flows reaching the sea (e.g. in 1805 on the Vesuvian coastline to the north of the study area) during eruptive phases, as well as pyroclastic products related to explosive phases covering the sea-cliffs and filling stream paths (Cinque & Robustelli 2009). The former change could lead to a rapid retreat of the slope over the cliff, as shown in some cases.

Seismic events related to both volcanism and tectonics (Campania is classified in Italy as a major seismogenetic region) must also be considered, but only in the easternmost area of the Sorrento Peninsula, and with a low degree ($Ip = 1$). For example, the village of Castellammare di Stabia was affected by the seismic waves produced by the earthquake that hit the Campania region on 23 November 1980 and the damage was classified as VIII on the MSC. However, the possibility of landslides and tsunamis being triggered should not be overlooked, as indicated by the National Catalogue of the INGV (Istituto Nazionale di Geofisica e Vulcanologia), which has reported at least two events that produced tsunamis, in 1631 and 1805, in the Bay of Naples. Other events, both ancient and recent, are considered unlikely to be related to such phenomena (Milia *et al.* 2003).

(6) Manmade structures. This primary hazard, which has a varying distribution throughout the study area, consists of buildings, roads, other infrastructure and services in proximity to the coastal area. In some cases, the impact comes from engineering structures close to the coastline designed to defend the coast against erosion and to protect infrastructure and services, to satisfy the growing demands of the tourist industry. For instance, the presence of a road along a coastal slope concentrates the infiltration of rainwater uphill and increases the possibility of a landslide, thereby raising the rank. Also, cliff lying on the lee shore of small harbours, or sea-walls, breakwaters and groynes can be affected by the channelling of energy from waves and currents. Here the score is greater than that for the adjacent cliff protected by construction.

Overall hazard assessment

The results obtained from the comparison of various hazards in each stretch of the studied coast were mapped, making the distribution of every hazard degree (from none (zero) to critical (IV)) easy to observe. In the area where this method was applied, these values were considered so that we could assign a degree (from negligible (N) to extreme (E)) to the overall coastal hazard (Ht).

Let us examine the cliffs of the Sorrento Peninsula in this regard (Fig. 8). This high rocky coast, formed both of limestones (varying from hard to soft rock) and clastic and tuffaceous deposits (soft rock) ($Ip = 1–2$ and 3–4, respectively), is affected by landslides to different degrees: lower ($Ip < 2$) for the former, unfractured rocks and higher ($Ip > 2$) for the latter, especially when they contain significant discontinuities. The hazard is greater, for instance, in the stretch eastward from the village of Sorrento, which has thick detritus loosely cemented onto limestone, or in other cases weathered rock, as a result of the occurrence of remobilization along the steep streams after exceptional rainfall ($Ip = 2$).

Moreover, the presence of a particular cliff type can be decisive for the rate of retreat, as well as for its influence on other parameters: a cliff with a limited, strongly dipping shore platform will have an increased erosion rate in comparison with cliffs with broad, gently dipping shore platforms. Thus the score may be higher for the former and lower for the latter, under the same conditions. In applying this method, it is not possible to omit human impact, here consisting of residential and tourist facilities, which are often located above sheer drops to the sea, as well as the heavy traffic load on the only (winding) road cutting into the steep cliff ($Ip = 2$). Likewise, the effect on the

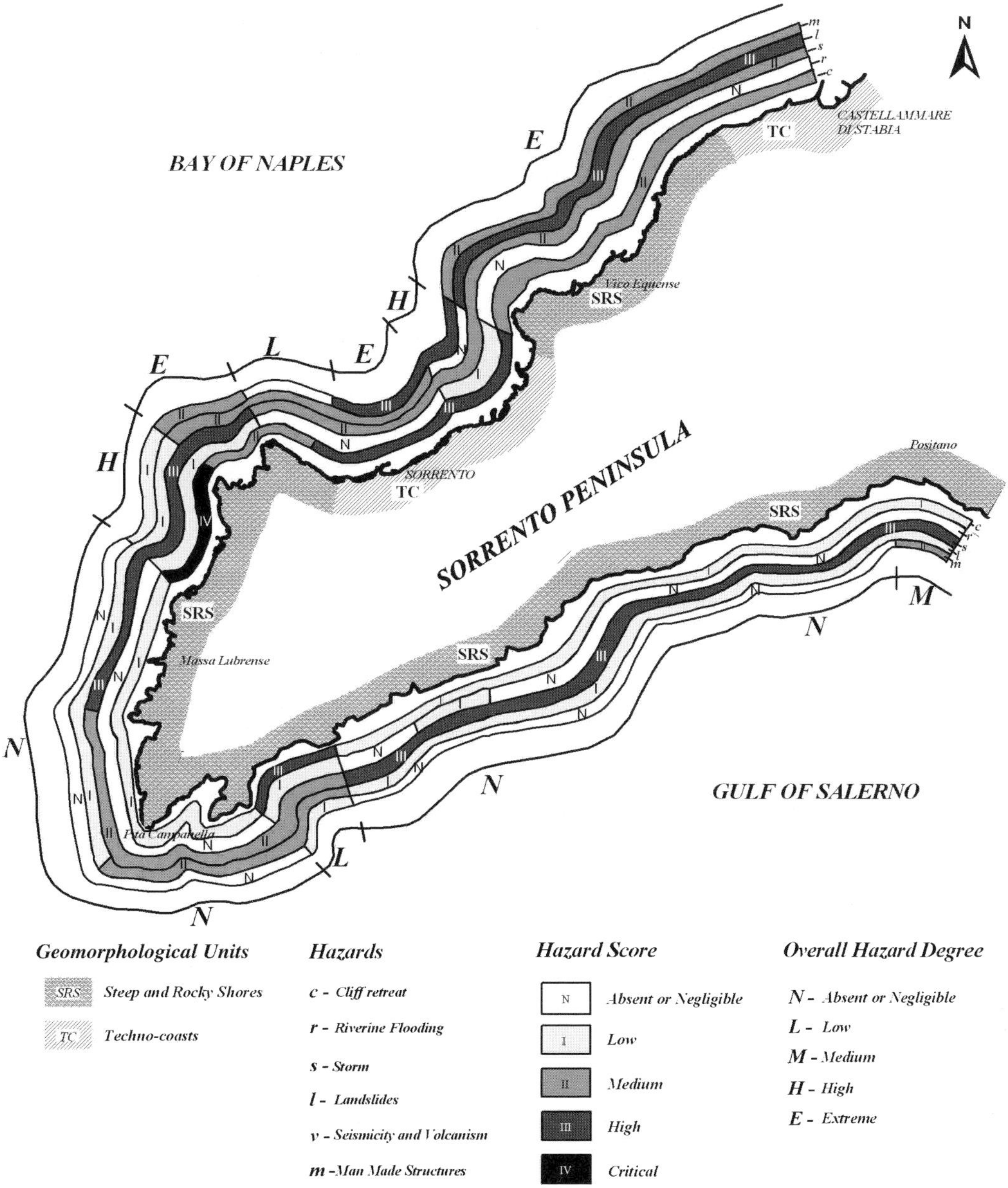

Fig. 8. Map of the Sorrento Peninsula coastline giving hazard scores for each parameter and overall hazard degree (slightly modified from De Pippo *et al.* 2008).

cliffs of strong wave action during winter storms is far from negligible, as a result of wave refraction, diffraction and reflection caused by the alternation of headlands and embayments, and at times by the absence of a shore platform ($Ip = 1$). As mentioned above, the interactive matrix shows that man-made structures and landslides have the highest scores ($Xp = 6$ and 5, respectively), whereas storm waves and riverine flooding are given medium ($Xp = 3$) and close to minimum scores ($Xp = 2$), respectively. Therefore the degree of overall hazard ranges from negligible ($Ht < 20$) on the southern side of the Sorrento Peninsula to extreme ($Ht = 41–42$) along the cliffs adjacent to the town of Vico Equense or in front of Sorrento.

Along the coastline on Capri the type of cliffs as well as the processes developing along the coastline vary depending on the side of the island (Fig. 9). For

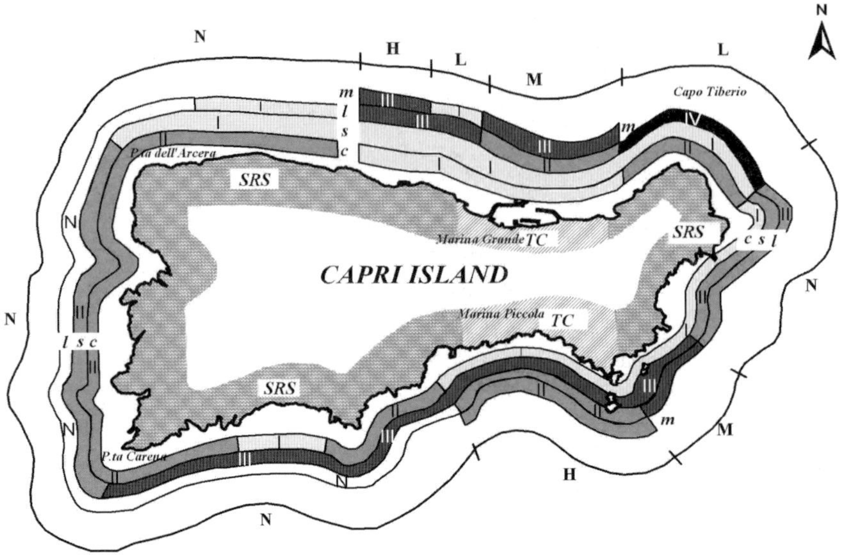

Fig. 9. Map of Capri coastline giving hazard scores for each parameter and overall hazard degree. (See the legend in Fig. 8.)

instance, plunging cliffs, mainly in limestone, are widespread on the western side ($Ip = 2$, as these cliffs are in a highly fractured condition), where retreat is much slower than on the eastern side, where cliffs with a subaerial face composed of soft rock (clastic deposits) occur, albeit with a gently sloping shore platform. Moreover, the superimposition of limestone on terrigenous deposits in turn covered by clastic rocks raised the score in the central portion of Capri, both on the north and south sides ($Ip = 2$). Importantly, a high or critical score may be assigned for landslides on the north-eastern side ($Ip = 3$–4), a high score for storm waves, also related to the widest fetch, is given to the southern side ($Ip = 3$), and a high or medium score for the development of buildings, roads and tourist facilities in the central portions ($Ip = 3$). After allowing for the weight of each hazard, the overall hazard is negligible on the western sides ($Ht < 14$) and extreme or high in the central portion ($40 < Ht < 35$).

Conclusions

The rocky coast of the Sorrento Peninsula and Capri forms very attractive scenery and, locally, a unique cultural landscape, which is one of the major economic factors in the region. However, in recent decades, the whole coast of Campania has experienced increasing human pressure as a result of residential and industrial construction as well as tourist infrastructure, resulting in a hazard for humans, damaged habitats and losses of landscape heritage. To establish best conservation practices and sustainable use of marine and coastal areas, integrated assessment, planning and management of marine and coastal areas are required. This is to prevent, control or mitigate adverse impacts from human activity in the marine and coastal environment, and to contribute to restoring degraded areas. Rocky coastal cliffs have so far aroused little attention relative to sandy beaches and dune systems, despite their regional importance as ecological systems, hazard areas and economic factors. Analysis of coastal evolution is often based exclusively upon the assessment of the rate of local and short-term hydrodynamic–geomorphological cliff modelling processes. In the proposed method we consider the various coastal hazards on a coastal stretch varying greatly in physical features and land use. This method is not limited to defining the rate of coastal retreat so as to assign a relative degree of coastal hazard, but is carried out in the knowledge that such characteristics and processes occur in the coastal zone and there may be interaction between them.

Along the studied coast six hazards are recognized: manmade structures, landslides, cliff retreat, storm waves, riverine flooding, and seismicity and volcanism. This list is given in descending order of influence derived from a dedicated interaction matrix, thus representing the significance of each parameter (its weight) along this kind of coast. Evaluation of each parameter is based on the frequency of particular events as well as the type and intensity of processes, which are inferred from historical, literature and archive data from several sources. The nature of some events, although not scientifically treated, is quantified according to a rating derived from analogous identifiable elements.

Evidently, the weight of each parameter related to hazard is strongly connected to the geology of the area, which includes sedimentary succession and tectonic structures (type and age of tectonic phase), as well as morphological features. This controls the development of current geomorphological processes, such as cliff retreat, and hence the degree of each hazard on a stretch of coast. Moreover, another parameter has to be considered in every analysis of the current landscape: that relating to manmade structures and human activity. This parameter usually has a negative influence on the system and sometimes is non-determinable, hence its weight is the highest.

The results obtained from several applications of the proposed interaction matrix along the Campania and Basilicata coastal zones, in a homogeneous geological context, encourage us to continue with further studies. Indeed, the occurrence of some events on the coast as well as preventive action in a planning project has confirmed the effectiveness of this method. Further applications to other geological contexts and other suggestions leading to the development of the matrix are welcome.

References

BARATTOLO, F., CINQUE, A., D'ALESSANDRO, E., GUIDA, M., ROMANO, P. & RUSSO ERMOLLI, E. 1992. Geomorfologia ed evoluzione tettonica quaternaria dell'Isola di Capri. *Studi Geologici Camerti, volume specide*, **1**, 221–229.

BERGER, A. R. 1997. Assessing rapid environmental change using geoindicators. *Environmental Geology*, **321**, 36–44.

BERGER, A. R. & IAMS, W. J. 1996. *Geoindicators: Assessing Rapid Environmental Change in Earth Systems*. Balkema, Rotterdam.

BRANCACCIO, L. 1968. Genesi e caratteri delle forme costiere nella Penisola Sorrentina. *Bollettino della Società dei Naturalisti in Napoli*, **77**, 247–274.

BRANCACCIO, L., CAPALDI, G., CINQUE, A., PECE, R. & SGROSSO, I. 1978. $^{230}Th-^{238}U$ dating of corals from Tyrrhenian beach in Surrentine Peninsula (southern Italy). *Quaternaria*, **20**, 175–183.

BRANCACCIO, L., CINQUE, A., ROMANO, P., ROSSKOPF, C., RUSSO, F., SANTANGELO, N. & SANTO, A. 1991. Geomorphology and neotectonic evolution of a sector of the Tyrrhenian flank of the Southern Apennines (Region of Naples, Italy). *Zeitschrift für Geomorphologie, Supplementband*, **82**, 47–58.

BUDETTA, P. & SANTO, A. 1993. Morphostructural evolution and related kinematics of rockfalls in Campania (Southern Italy). *Engineeering Geology*, **36**, 197–210.

BUDETTA, P., CALCATERRA, D. & DUCCI, D. 1991. Caratterizzazione geomeccanica degli ammassi rocciosi cartonatici tra Castellammare di Stabia e Vico Equense (Penisola Sorrentina—Campania). *Geologia Tecnica*, **3**, 13–22.

BUDETTA, P., CALCATERRA, D. & SANTO, A. 1994. Engineering-geological zoning of potentially unstable rock slopes in Sorrentine Peninsula (Southern Italy). *In*: OLIVEIRA, R. *ET AL.* (eds) *Proceeding of the 7th International Congress of the IAEG, Lisbon*. Balkema, Rotterdam, 2119–2126.

BUDETTA, P., GALLETTA, G. & SANTO, A. 2000. A methodology for the study of the relation between coastal cliff erosion and the mechanical strength of soils and rock masses. *Engineering Geology*, **56**, 243–256.

BUSH, D. M., NEAL, W. J., YOUNG, R. S. & PILKEY, O. H. 1999. Utilization of geoindicators for rapid assessment of coastal-hazard risk and mitigation. *Ocean and Coastal Management*, **42**, 647–670.

BUSH, D. M., RICHMOND, B. M. & NEAL, W. J. 2001. Coastal-zone hazard maps and recommendations: eastern puerto rico. *Environmental Geosciences*, **8**, 38–60.

CAPUTO, C., D'ALESSANDRO, L., LA MONICA, G. B., LANDINI, B. & LUPIA PALMIERI, E. 1991. Present erosion and dynamics of Italian beaches. *Zeitschrift für Geomorphologie, Supplementband*, **81**, 31–39.

CINQUE, A. & ROBUSTELLI, G. 2009. Alluvial and coastal hazards caused by long-range effects of Plinian eruptions: the case of the Lattari Mts. after the AD 79 eruption of Vesuvius. *In*: VIOLANTE, C. (ed.) *Geohazard in Rocky Coastal Areas*. Geological Society, London, Special Publications, **322**, 155–171.

COCCO, E., DE MAGISTRIS, M. A., DE PIPPO, T., EFAICCHIO, M. T. & VALENTE, A. 1989. Coastal dynamics along the shores of Campania and Lucania (Southern Italy). *In*: MAGOON, O. T. (ed.) *Coastal Zone '89, Vol. 3*. American Society of Civil Engineers, Dallas Miner, 2794–2807.

COCCO, E., DE MAGISTRIS, M. A., EFAICCHIO, M. T. & BOSCAINO, F. 1992. Geoenvironmental features of the Sele River Plain littoral (Gulf of Salerno, Southern Italy). *Bollettino di Oceanologia Teorica ed Applicata*, **X**, 235–246.

D'ARGENIO, B., AIELLO, G. *ET AL.* 2004. Digital Elevation Model of the Bay of Naples and adjacent arcas (Eastern Tyrrhenian Sea). *In*: PASQUARE, G. & VENTURINI, C. (eds) *Mapping Geology in Italy*. APAT – Dipartimento Difesa del Suolo Monography Roma, 21–28.

DE PIPPO, T., DONADIO, C. & SGROSSO, A. 1998. Geomorfologia sottomarina di un settore della Penisola Sorrentina (Campania). *Geografia Fisica e Dinamica Quaternaria*, **21**, 103–110.

DE PIPPO, T., PENNETTA, M., TERLIZZI, F. & VALENTE, A. 2007. Principali tipi di falesia nella penisola Sorrentina e nell'Isola di Capri: caratteri e lineamenti morfoevolutivi. *Bollettino della Società Geologica Italiana*, **126**, 181–189.

DE PIPPO, T., DONADIO, C., PENNETTA, M., PETROSINO, C., TERLIZZI, F. & VALENTE, A. 2008. Coastal hazard assessment and mapping in Northern Campania, Italy. *Geomorphology*, **97**, 451–466.

De Vivo, B., Rolandi, G., Gans, P. B., Calvert, A., Bohrson, W. A., Spera, F. J. & Belkin, H. E. 2001. New constraints on the pyroclastic eruptive history of the Campanian volcanic Plain (Italy). *Mineralogy and Petrography*, **73**, 47–65.

Esposito, E., Porfido, S. & Violante, C. (eds) 2004*a*. *Il Nubifragio dell'Ottobre 1954 a Vietri sul mare, Costa d'Amalfi, Salerno. Scenario ed effetti di una piena fluviale catastrofica in un'area di costa rocciosa*. Pubblicazione Gruppo Nazionale per la Difesa dalle Catastrofi Idrogeologiche, **2870**.

Esposito, E., Porfido, S., Violante, C., Biscarini, C., Alaia, F. & Esposito, G. 2004*b*. Water events and historical flood recurrences in the Vietri sul Mare coastal area (Costiera Amalfitana, southern Italy). *In*: Rodda, G. & Ubertini, L. (eds) *The Basis of Civilization—Water Science?* International Association of Hydrological Sciences Publications, **286**, 1–12.

Fiorillo, F. & Wilson, R. C. 2004. Rainfall induced debris flows in pyroclastic deposits, Campania (Southern Italy). *Engineering Geology*, **75**, 263–289.

Hudson, J. A. 1992. *Rock Engineering System*. Ellis Horwood, Chichester.

Hutchinson, J. N. 1983. A pattern in the incidence of major coastal mudslides. *Earth Surface Processes and Landforms*, **8**, 391–397.

Kogure, T., Hisashi, A., Akira, M., Takashi, H. & Matsukura, Y. 2006. Effect of the development of notches and tension cracks on instability of limestone coastal cliffs in the Ryukyus, Japan. *Geomorphology*, **80**, 236–244.

Milia, A., Torrente, M. M. & Zuppetta, A. 2003. Offshore debris avalanches at Somma–Vesuvius volcano (Italy): implications for hazard evaluation. *Journal of the Geological Society, London*, **160**, 309–317.

Pirazzoli, P. A. 1993. Global sea-level changes and their measurement. *Global and Planetary Change*, **8**, 135–148.

Riccio, A., Riggio, F. & Romano, P. 2001. Sea level fluctuations during oxygen isotope stage 5: new data from fossil shorelines in the Sorrento Peninsula (Southern Italy). *Zeitschrift für geomorphologie Neue Folge*, **45**, 121–137.

Richmond, B. M., Fletcher, C. H., Grossman, E. E. & Gibbs, A. E. 2001. Island at risk: coastal hazard assessment and mapping in the Hawaiian Islands. *Environmental Geosciences*, **8**, 21–37.

Rossi, F. & Villani, P. 1994. *Valutazione delle piene in Campania. Gruppo Nazionale per la Difesa delle Catastrofi Idrogeologiche, Previsione e prevenzione degli eventi idrologici estremi e loro controllo Linea1*. Consiglio Nazionale delle Ricerche, **310**.

Russo, F. & Valletta, M. 1995. Il rischio geologico *sensu lato* al Somma–Vesuvio. *Rendiconti Accademia Scienze Fisiche e Matematiche*, **LXII**, 125–186.

Santo, A., Santangelo, N., Beneduce, A. & Iovine, F. 2002. Pericolosità connessa a processi alluvionali in aree pedemontane: il caso di Castellamare di Stabia in Penisola Sorrentina. *Quaternario*, **15**, 23–37.

Simeoni, U., Calderoni, G., Tessari, U. & Mazzini, E. 1999*a*. A new application of system theory to foredunes intervention strategies. *Journal of Coastal Research*, **15**, 457–470.

Simeoni, U., Mazzini, E., Tessari, U. & Ciavola, P. 1999*b*. Variazione integrata, a fini gestionali, delle dune costiere: metodologia e caso studio. *Studi Costieri*, **1**, 85–102.

Stauble, D. K. 2003. The use of shoreline change mapping in coastal engineering project assessment. *Journal of Coastal Research*, **38**, 178–206.

Sunamura, T. 1983. Processes of sea cliff and platform erosion. *In*: Komar, P. D. (ed.) *CRC Handbook of Coastal Processes and Erosion*. CRC Press, Boca Raton, FL, 233–265.

Sunamura, T. 1992. *Geomorphology of Rocky Coasts*. Wiley, Chichester.

Trenhaile, A. S. 2002. Rock coasts, with particular emphasis on shore platform. *Geomorphology*, **48**, 7–22.

Violante, C. 2009. Rocky coast: geological constraints for hazard assessment. *In*: Violante, C. (ed.) *Geohazard in Rocky Coastal Areas*. Geological Society, London, Special Publications, **322**, 1–31.

Violante, C., Biscarini, C., Esposito, E., Molisso, F., Porfido, S. & Sacchi, M. 2009. The consequences of hydrologic events on steep coastal watersheds: the Costa d'Amalfi, eastern-Tyrrhenian Sea. *In*: Liebsher, H. J. *et al.* (eds) *The Role of Hydrology in Water Resource Management. Capri Italy, 2008*. IAHS Publication, **327**, 102–113.

Index

Figures are indicated in *italic* font, tables in **bold**.